François Arago

Astrophysics and Space Science Library

More information about this series at http://www.springer.com/series/5664

James Lequeux

François Arago

A 19th Century French Humanist and Pioneer in Astrophysics

Springer

James Lequeux
Paris Observatory
Paris, France

Original French title: ***François Arago, un savant généreux***
(c) 2008 by EDP Sciences

ISSN 0067-0057 ISSN 2214-7985 (electronic)
Astrophysics and Space Science Library
ISBN 978-3-319-20722-3 ISBN 978-3-319-20723-0 (eBook)
DOI 10.1007/978-3-319-20723-0

Library of Congress Control Number: 2015944988

Springer Cham Heidelberg New York Dordrecht London

Portrait of Francois Arago provided courtesy of Paris Observatory

Printed on acid-free paper

Springer International Publishing AG Switzerland is part of Springer Science+Business Media (www.springer.com)

Preface

Arago had more ideas by himself than a full generation.

(Léon Foucault 1853)

Many streets and places in France bear the name of Arago, and there are statues of him in Paris, Perpignan, and Estagel, the small city near Perpignan where he was born. The prestigious École Polytechnique has named an amphitheater from him, and one also finds his name written on the front of several high schools and of the Laboratory of Marine Biology of Banyuls, in the south of France. But if one asks: who was François Arago, one obtains in general only embarrassed answers. Some physicists remember him as a collaborator of Fresnel; a few persons interested in history cite his participation in the ephemeral government of France that followed the 1848 Revolution. But no one knows that Arago was basically an astronomer.

Arago has had no luck with History. Historians know Arago the political man, but his importance is too often underestimated,[1] and Arago the scientist is almost forgotten. However, the "Great Arago" was probably the best-known French scientist of his time. His contributions to science are far from negligible, and he played an eminent role in the promotion of science and of its applications. Arago is a rare case of a scientist who was also a statesman: It is exactly that which does harm to his posthumous reputation. There are other French scientists who were politically active during the French Revolution and later, for example, Jean-Sylvain Bailly, an astronomer who was the mayor of Paris, or Nicolas de Condorcet or Joseph Fourier, but none was as far-reaching as Arago in both domains. The only person who can be compared to him in this sense is Benjamin Franklin.

Arago had a prodigious scientific knowledge and a vast intellectual curiosity. His centers of interest were very diverse: he explored astronomy, geophysics, meteorology, and also physics in the emerging domains of thermodynamics, optics,

[1] For details on the political activity of Arago, see Sarda (2000).

electromagnetism, and photography. It was still possible for a single mind of his time to cover the whole science as scientists were generally not very specialized. Ever generous, Arago never used his fame and his dominant position in French science for his own promotion; conversely, when he collaborated with other scientists like Fresnel or Ampère, he managed to make them known and let them have the full glory of their discoveries, even if his own contribution was important. This is another reason for which he has been somewhat forgotten. The physicist Charles Fabry wrote about him[2]:

> *He had a beautiful and generous personality, able to understand and to take interest in everything, with a devouring activity, passionate in his friendships as well as in his antipathies, always ready to defend his friends in all circumstances and to crush his enemies. Remarkable professor, outstanding popularizer, he had an enormous influence on all the audiences.[…] He managed to keep a large influence whatever the political regime, even under those he did not like, and this not by flattery, but because it was considered advisable not to have him as a declared opponent.*

Guillaume Bigourdan summarizes Arago's activities as a scientist in his *History of the Board of Longitudes*[3]:

> *The discoveries of Arago in optics and in electricity could be claimed by the Board [sic!]; however I will only cite them here: they are in particular the electromagnet, which is presently very widespread—the discovery of chromatic polarization which dates from 1811, then of rotational polarization—the magnetism of rotation—finally a considerable number of applications, in particular the determination with Dulong of the elastic force of water vapor, a method to locate reefs at sea in 1835, etc. I will emphasize what concerns meteorology and [terrestrial] magnetism.*

The contemporaries of Arago would speak of him as an astronomer rather than as a physicist. Indeed, he spent his entire scientific career at the Paris Observatory, from 1805 to his death in 1853. Still, the astronomer is even less known than the physicist, and it is significant that Guillaume Bigourdan does not mention his astronomical activities. The excellent biography[4] of Arago by Maurice Daumas does not contain many details on them, and this is regretted by Jean Dhombres in his preface to this book.

Our knowledge of Arago's scientific activity has improved since Daumas, and I naturally wanted to place it within the science of the time, which was in a kind of ebullition. I also found it desirable to describe in some detail his instruments and his experiments, which could pose problems of interpretation to the readers of his publications. Finally, I found it interesting to discuss of Arago's considerable contributions to the technical developments of this period of intense industrial activity. Unfortunately, many people only remember a negative aspect: his inaction concerning the construction of railroads. This book will provide an opportunity to

---

[2] Fabry, C. (1938) in « La vie et l'œuvre scientifique d'Augustin Fresnel », paper inserted in *La vie et l'œuvre de Charles Fabry, Œuvres choisies publiées à l'occasion du jubilé scientifique de M. Charles Fabry*, Paris, Gauthier-Villars.

[3] *Bigourdan (1928–1932).

[4] Daumas (1987).

examine physics in a fascinating period when most of what serves in our daily life was created, with of course the exception of the telephone, the radio, and the electronics: optics, electricity, heat, thermodynamics, and photography.

This book is therefore mainly devoted to the scientific activity of Arago, with a survey of research during his lifetime and somewhat later. The first chapter describes the scientific institutions, which were often created during the French Revolution, and the conditions of research at the time; the second chapter summarizes the life of Arago; the two following chapters describe his work on the nature and velocity of light; the next one examines the beginning of electromagnetism, with Arago's discovery of "magnetism of rotation," which led Faraday to his discovery of induction. Chapters 6–9 are devoted to Arago's astronomical activity, in the broad sense as defined in his time: geodesy, then astronomical instrumentation and astronomy proper. The following chapters discuss his activities related to geophysics and applied physics, and Chap. 11 his actions to promote science. The last chapter summarizes the heritage of this great scientist. These chapters are independent and can be read in any order, but it is recommended to consult first Chaps. 1 and 2, and to read Chap. 7 before Chap. 8.

The works of Arago and many other documents cited in this book are accessible on Gallica, the Internet site of the *Bibliothèque Nationale de France* (http://gallica.bnf.fr); they are indicated by an asterisk.

Paris, France James Lequeux

Acknowledgements

I thank the Archives of the French Academy of Sciences, the library of *École polytechnique*, the Rectorat of the Academy of Paris, the Museum of Suresnes, Prof. Gilbert Amat, Mrs. Danièle Blouet, Dr. Danielle Briot, Dr. Jean-Charles Cuillandre, Prof. Jean-Louis Le Mouël, and Dr. William Tobin for giving me the permission to reproduce documents or photographs. Many other illustrations come from Wikipedia Commons, which I also thank heartily. My work was enormously eased by Gallica, by CNUM, the Internet site of the *Musée des arts et métiers/CNAM,* by the SAO/NASA Astrophysics Data System, and by Wikipedia: I am grateful to these institutions. I also express my gratitude to the librarians of the Paris Observatory for their help and their kindness. Finally, I want to thank Mr. Evan Roth for his efficient polishing of the language of the text.

Contents

Chapter 1
Scientific Life in France at the Time of Arago

Arago presents the details of the photographic process of Daguerre at the public meeting of the Academy of Sciences and of the Academy of Fine Arts on 19 August 1839. Private collection.

J. Lequeux, *François Arago*, Astrophysics and Space Science Library 421,
DOI 10.1007/978-3-319-20723-0_1

The scientific life in France during the first half of the nineteenth century was dominated by several powerful institutions. Some were ancient: the Academy of Sciences was created in 1666, the *École des Ponts et Chaussées* (literally School of Bridges and Roads, in fact a school of civil engineering) in 1747, and the *École des Mines* (School of Mines) in 1783. Still older was the *Collège de France*, a high-level teaching institution created in 1530. Others date from the French Revolution: the *École Polytechnique*, the *École Normale Supérieure* and the *Bureau des longitudes* (Board of Longitudes). One also finds among these new institutions the *Conservatoire National des Arts et Métiers* (National Conservatory of Arts and Crafts), which housed from the beginning a very large collection of scientific items and where scientists gave lectures. Several *Écoles des Arts et Métiers* were to some extent subsidiaries of this Conservatory in various cities. All these institutions are still in existence.

At a lower level, teaching was reformed during the Revolution with the creation of *écoles centrales*, equivalent to high schools, which often replaced earlier schools. Primary education was in principle compulsory (but only in principle, for there was a lack of teachers). However, access to education became possible for many boys coming from the middle or lower social classes; several actually reached important positions, especially in science. This did not last for very long, however, but it is clear that the possibilities of social ascent were very beneficial to the development of science at the beginning of the nineteenth century.

The scientific world of the nineteenth century was a world of men. Very few women could benefit from higher education, and even fewer could become scientists: In France, one can cite only the mathematician Sophie Germain (1776–1831), who made important works on the theory of elasticity. No woman belonged to the institutions I listed above. If a few women played a role in science, this was by helping a member of their family, as was the case for Arago's niece Lucie Laugier, who assisted him at the end of his life.

Let us now examine the scientific institutions, before describing the working conditions of the scientists in the first half of the nineteenth century.

The Great Scientific Institutions

The Academy of Sciences[1]

The Academy of Sciences was created in 1666 by King Louis XIV and his prime minister Colbert. In spite of its royal origin, it crossed the revolutionary period without many problems, disappearing in 1793, and reappearing in 1795 as the *Première classe de l'Institut*. It recovered its initial name in 1816. The Academy then comprised two divisions, one devoted to "Mathematical Sciences," which included

[1] For more information on the Academy of Sciences, see http://www.academie-sciences.fr

Table 1.1 Composition of the Academy of Sciences after 1816

Division	Section	Members	Correspondents
Mathematical Sciences	1. Geometry	6	6
	2. Mechanics	6	6
	3. Astronomy	6	16
	4. Geography and Navigation	6	8
	5. General Physics	6	6
Physical Sciences	6. Chemistry	6	12
	7. Mineralogy	6	8
	8. Botany	6	10
	9. Rural and veterinary economy	6	10
	10. Anatomy and zoology	6	10
	11. Medicine and surgery	6	8
	12. Free section	10	
	Non-residing members	6	
	Foreign associates	8	

mathematics, astronomy, geography, navigation and physics properly speaking, and the other to "Physical Sciences," actually chemistry and natural sciences. These denominations were respectively replaced in 1955 by the more appropriate ones of "Mathematical and Physical Sciences" and of "Chemical and Natural Sciences."

The weekly meetings of the Academy were (and still are) organized by its president, who is replaced every year; but the most important members are the two (life-long) Perpetual Secretaries, one per division. In Arago's time, the Perpetual Secretary for Mathematical Sciences was the astronomer Jean-Baptiste Delambre (1749–1822), replaced after his death by the physicist Joseph Fourier (1768–1830), then in 1830 by Arago himself. For Physical Sciences, the Perpetual Secretary was Georges Cuvier (1769–1832), replaced by Pierre Dulong (1785–1838), who however belonged to the other division; he resigned in 1833 to be replaced by the physiologist Pierre Flourens (1794–1867), who was on good terms with Arago.

At this time, the Mathematical Sciences division was itself divided into five sections, and the Physical Sciences division into six sections (Table 1.1). Each section was formed of six residing members, who were supposed to live in Paris and to attend the meetings regularly. There were also correspondents for each section, who were often quite active, in particular the astronomers who regularly sent the results of their observations to the Academy. A "free" section was composed of ten old academicians, survivors from the pre-revolutionary times. There were also six non-residing members and eight French or foreign associates. (Figure 1.1 gives the list of the residing members from 1795 to the death of Arago in 1853 for sections 1, 2, 3 and 5, which are of interest for us.)

The eight foreign associate members were not attached to any particular section. Their position was essentially honorary, but several associates were very active, in particular Alexander von Humboldt during his stay in Paris. Table 1.2 gives the list of those associates who were mathematicians, astronomers or physicists. All of them are still remembered today, testifying for the excellent judgement of the Academy.

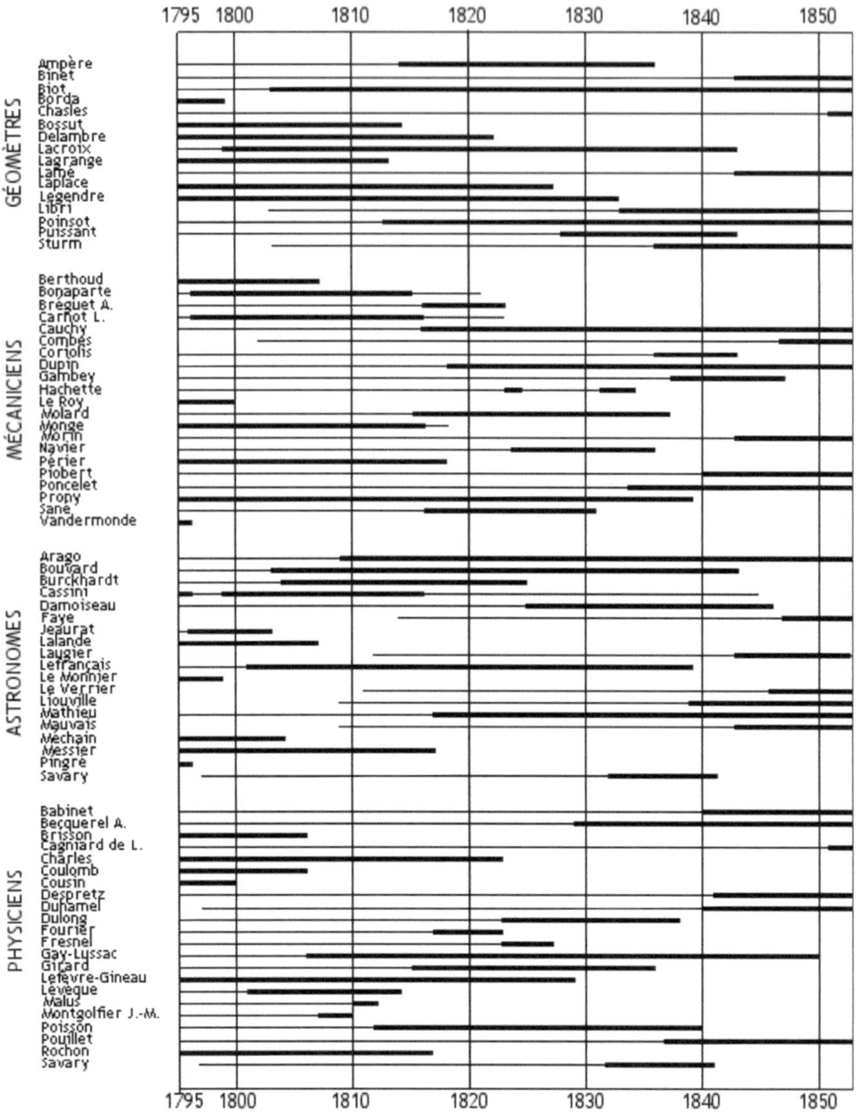

Fig. 1.1 Resident members of sections 1, 2, 3 and 5 of the *Première classe de l'Institut* from 1795 to the death of Arago in 1853. Bonaparte, the future Napoléon, who was elected in 1796, and Jean-Dominique Cassini (Cassini IV) resigned before their death. Several members were dismissed for political reasons: Cassini IV from 1796 to 1799, Lazare Carnot and Monge (For a biography of Monge and details on the persecution he suffered after 1816 due to his political opinions, see *Arago, F. *Œuvres complètes*, t. 2, p. 427–592) in 1816. Hachette was dismissed soon after being elected in 1823, then re-elected in 1831

Table 1.2 Foreign associate members of the Academy of Sciences from 1795 to 1853, excluding those specialized in chemistry and natural sciences

Main speciality	Name	Nationality	Birthdate	Date of election	Death
Mathematics	Gauss	German	1777	1820	1850
	Bessel	German	1784	1840	1846
	Jacobi	German	1804	1846	1851
Astronomy	Maskelyne	English	1732	1802	1811
	Herschel	English	1738	1802	1822
	Piazzi	Italian	1746	1817	1826
	Olbers	German	1758	1829	1840
Physics	Cavendish	English	1731	1803	1810
	Volta	Italian	1745	1803	1827
	Rumford	American	1753	1803	1814
	Watt	English	1736	1814	1819
	Wollaston	English	1766	1823	1828
	Young	English	1773	1827	1829
	Œrsted	Danish	1777	1842	1851
	Faraday	English	1791	1844	1867
	Brewster	Scottish	1781	1849	1868
Various	Humboldt	German	1769	1810	1859

Note the election of several English scientists at times when the relations between France and England were very bad

The communications to the Academy could be read in their entirety or as summaries. The scientific memoirs written by non-members were examined by an ad hoc committee of several members. For those memoirs considered as particularly important, the committee presented a report, which was often a summary written by the author. It was also possible to write to the Academy, and the Perpetual Secretaries read summaries of the letters at the beginning of each meeting. Finally, the Academy accepted *plis cachetés* or *paquets cachetés*, sealed documents containing discoveries or ideas considered by their authors as insufficiently developed to be published; the authors could ask the Academy to disclose their content if some competitor was to present a similar idea.

The Academy meetings were not public until Arago decided in 1835 that they would be open to anyone, especially journalists. Among the latter were Alfred Donné (1801–1878), then Léon Foucault (1819–1868), who wrote reports of the meetings in the *Journal des Débats*, a daily paper very influential among politicians.

When a communication was judged by the Academy to deserve unabridged publication, it was printed in the *Mémoires de l'Académie des sciences* (or, for non-members, in the *Mémoires des savants étrangers*). This could take years; in the meantime, the author was forbidden to publish his results elsewhere. We will see how Arago managed to get around this problem. But above all, in 1835 he created the *Comptes rendus hebdomadaires des séances de l'Académie des sciences* (Weekly Reports of the Meetings of the Academy of Sciences), prepared by the Perpetual Secretaries following each meeting, which allowed a rapid diffusion of scientific results to a wide audience.

One cannot underestimate the importance of the Academy of Sciences in the scientific life of the time. It is clear that most of the active research was done by its members. The list of the members during the first half of the nineteenth century is impressive: only a few of them are forgotten today by historians or scientists or by the general public. It is easy to understand that active Perpetual Secretaries, as was Arago, could influence the scientific life in France.

Some Other Scientific Societies

There were a number of scientific societies in France. Many provincial cities had their own. In Paris, for example, there were the *Société d'encouragement pour l'industrie nationale* (Society for the Promotion of National Industry) and the *Société de géographie* (Geographical Society). The most interesting for us was the *Société philomathique* (from the Greek: philo = to like and mathique = science). It was founded in 1788 by young scientists who publicized and discussed recent discoveries. When the Royal Academy of Sciences was temporarily suppressed in 1793, the academicians met in the premises of this society. There, one could meet Laplace, Cuvier and Lavoisier, who was guillotined the following year. The number of members of this society soon reached about 60, stabilizing in the following years. When the Academy reappeared, the *Société philomathique* became a kind of waiting room for the Academy. Arago, Malus, Fresnel, Gambey and many others were members before their election to the Academy. The Society used to publish minutes of its meetings and a *Bulletin des sciences*.

Less durable but perhaps more active was the *Société d'Arcueil*. Arcueil is a suburb of Paris where two famous scientists, the chemist Claude Berthollet (1748–1822, Fig. 1.2) and the mathematician and physicist Pierre-Simon Laplace (1749–1827, Fig. 1.3) were living next to each other. They opened a door between their domains, and they invited young scientists to meet every 2 weeks, generally on Thursdays or Sundays. An informal society emerged in 1801, and acquired a semi-official status in 1805 under the protection of Emperor Napoléon. The members could use Berthollet's laboratory for their experiments, and above all could present their ideas to the other members of the society. Quite often, a memoir discussed on a Sunday would be presented the following day at the Academy, with changes suggested during the meeting. The *Société d'Arcueil* was discontinued in 1816, partly for political reasons, but not before having published three volumes of scientific papers. The first one appeared in 1807. That year, the Society comprised Laplace, Berthollet and his son Amédée, the chemists Louis Joseph Gay-Lussac (1778–1850) and Louis Jacques Thénard (1777–1857), the Swiss botanist Augustin Pyrame de Candolle (1778–1841), the mineralogist Alphonse Victor Collet-Descotils (1773–1815), as well as Humboldt. Arago entered the Society in 1807 together with Étienne-Louis Malus (1775–1812) and Pierre-Louis Dulong, with whom he would collaborate later.

Fig. 1.2 Claude-Louis Berthollet. Wikimedia Commons

The Board of Longitudes[2]

The Board of Longitudes (*Bureau des longitudes*) was an important creation of the Revolution. It was founded by a law dated 8 messidor An III (in the revolutionary calendar, corresponding to 25 June 1795), with the purpose of "making the Navy and Commerce flourish through the development of Astronomy." It was supposed "to chase Superstition and to create Chronology," and to give (my translation):

a fundamental basis to Geography and Navigation through the solution, still imperfect, of the problem of longitudes.
The tasks of the Board are:

- *to continue the efforts of the previous government to solve this problem;*
- *to improve the maps and hydrography;*
- *to study the terrestrial magnetism, in particular at sea;*
- *to develop clock-making;*
- *to make navigation safer, in order to better protect the life of sailors;*

For this purpose, the Board

---

[2] There is a detailed history of the Board of Longitudes in *Bigourdan (1928–1932). See also Lamy (2007).

Fig. 1.3 Pierre-Simon Laplace, posthumous portrait by Sophie Feytaud (1841). Wikimedia Commons, Luesting

- *shall organize every year a course in Astronomy;*
- *shall improve astronomical tables;*

The Board of Longitudes shall also publish the Connaissance des Temps *[yearly almanacs for astronomers and sailors]... as well as astronomical and meteorological observations....*
[The Board] has among its prerogatives: the National Paris Observatory, the most beautiful monument raised to Astronomy... almost disorganized, the observatory of the previous Military School, the attached housings, and all the astronomical instruments belonging to the Nation.

In 1796, the old observatory of Nicolas-Louis de La Caille (1713–1762) located on the roof of the *Collège des Quatre-Nations* was added to the list.

The Board kept close contact with instrument builders like Noël-Jean Lerebours (1761–1840), who was appointed in 1814 "optician of the Observatory." The foundation decree stipulates that "one will take in the depository of books belonging to the Nation and in the duplicates of the National Library those books considered necessary to complete the library started at the Paris Observatory." This is the origin of the library of the Observatory, which is very rich in old astronomy books.

Also, the Board "will designate the observatories to be created, maintained or developed, amongst which those of the harbours of Brest and Toulon." Amongst the observatories to be maintained were those of Marseille and Toulouse, to which the Board used to send some Parisian instruments after they were replaced by more modern ones. The Board also occasionally sent some instruments to private observatories like that of Honoré Flaugergues (1755–1835) at Viviers, in Ardèche.

The Board planned to construct "a reflecting telescope of 60 ft length and 6 ft diameter," probably following the 1787 visit of Cassini IV to William Herschel at Slough, England, where he could see Herschel's giant telescopes.[3] But the reflecting telescope was never realised. Finally the Board was in charge of "fixing the first meridian." This refers to the new measurement of the Paris meridian from Dunkirk to Barcelona, then in progress thanks to Jean-Baptiste Joseph Delambre (1749–1822) and Pierre-André Méchain (1744–1804). I will discuss this in Chap. 6.

Although the role of the Board of Longitudes was in principle centered on navigation at sea and geodesy, with some incursion on meteorology, it covered in fact the whole of astronomy because it had the observatories under its responsibility.

At the beginning, the Board had an annual budget of 12,000 livres[4] "for maintenance of the instruments." Its personnel consisted of titular members with an annual salary of 8000 livres and of adjunct members paid 4000 livres per year. The equipment budget was attributed on request for construction of buildings and instruments. Figure 1.4 shows the evolution of the personnel of the Board, which was smaller than that of its English counterpart. Most members worked at the Paris Observatory. The Board administrators were: Delambre in 1800, then Méchain en 1801, followed in 1803 by the geographer Buache de la Neuville (1741–1825) and finally by the astronomer Alexis Bouvard (1767–1843) from 1814 to (perhaps) his death. Perhaps, because the title of administrator was less and less in use. On 9 April 1834, Arago was appointed "Director of observations" of the Paris Observatory, a position created for him, which made him the *de facto* director of this institution. He could then have the observations made by "young men who had the title of *élèves-observateurs*."

After Arago was replaced in 1854 by Urbain Jean-Joseph Le Verrier (1811–1877), the prerogatives of the Board were considerably diminished: the Paris Observatory became independent, and the Marseilles observatory subsequently became a "subsidiary" of the one in Paris.

The Grandes Écoles

The *grandes écoles* seem to be a specialty of France. Some existed before the 1789 Revolution, but more were created in 1794, among them the *École Polytechnique* (Polytechnic school) and the *École Normale Supérieure*. Both were to play a major role in high-level teaching and scientific research. They are still in existence and very active.

---

[3] Cassini IV, *Mémoires de l'Académie royale des sciences* for 1788, p. 706–717.

[4] The *livre* was replaced by the *franc* in 1795, with approximately the same value. One can estimate this value from the prices of various products at the beginning of the nineteenth century: a kilogram of bread cost between 0.25 and 0.40 franc, a litre of milk 0.50 franc, a book about 6 francs and the annual subscription to a rather expensive daily journal from 60 to 70 francs. From this, one derives that 1 franc was roughly equivalent to 3.5 euros or 4 US dollars.

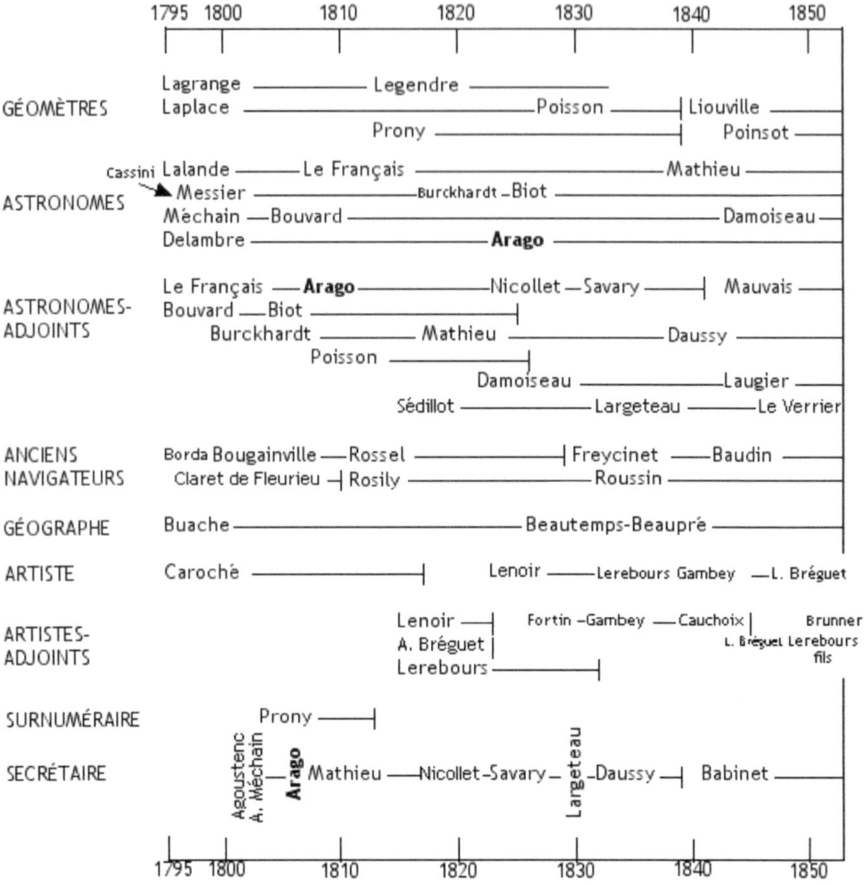

Fig. 1.4 The members of the Board of Longitudes, from its creation to the death of Arago. (From **Connaissance des temps* and **Annuaire du Bureau des longitudes*.) There was another *surnuméraire* not indicated in the figure: Sédillot, "adjunct for the history of Astronomy in Orient," who would later occupy an ordinary position of adjunct astronomer as indicated

The École Polytechnique[5]

After the revolutionary enthusiasm of 1789, France was by 1794 in a very difficult situation. It lacked engineers and scientists, many of them having fled the country or, like Lavoisier and Condorcet, died during the Terror. Under the instigation of several prominent scientists who adopted to the republican ideas, in particular Gaspard Monge (Fig. 1.5) and the chemist Antoine François de Fourcroy (1755–1809), the *Comité de Salut public* (Committee of public safety) created a *Commission*

[5] For the history of the school, in English, see https://www.polytechnique.edu/en/history

Fig. 1.5 Gaspard Monge (1746–1819), lithography by Françoise Séraphin Delpech. Wikimedia Commons, Kelson

des Travaux publics (Commission of public works), which recommended the founding of a new school for the training of civil and military engineers. It was created on 7 vendémaire an III (28 September 1794) as the *École centrale des travaux publics.* Teaching started in December. One year later the school took the name of *École polytechnique*.

The mission of the school was clearly defined: to give to the students a good basic knowledge of mathematics, physics and chemistry, and to prepare them to enter more specialized public schools like the *École d'application de l'artillerie et du génie* (School of Applied Artillery and Military Engineering), the *École des ponts et chaussées* (School of bridges and roads), and the newly created *École des mines* (School of mining). The entrance examination consisted only of an oral test in mathematics. The examiners toured France to find candidates, and to talk to them directly. The students received travel expenses to go to Paris and an stipend of 900 francs. They were lodged outside the school in houses of *pères sensibles*, people known to have "good" political opinions. All this allowed real social advancement as many candidates belonged to relatively lower classes.

Studies lasted for 3 years. There were about 400 students at a time. The teaching was by the best scientists (although the teaching of physics was not very good initially). Almost all the young scientists of the time went through the École polytechnique. Its excellence explains why Bonaparte asked its main promoter, Monge, to lead the scientific expedition in Egypt in 1797–1798; Berthollet, who was teaching chemistry, was also of the party together with no less than 42 students who had just received their diplomas or were still at school. However, the school rapidly became a hotbed of political agitation, and the students refused to send their congratulations to Bonaparte when he crowned himself Emperor Napoléon on 2 December 1804 (we will see that Arago was one of those who organized this refusal). Napoléon then decided to give the school a military regime and to have the students live in the new school building (on the Montagne Sainte-Geneviève in Paris's Left Bank) like soldiers in their barracks. He also decided to stop paying students stipends, in an effort to discourage poor students from applying to the school. This turned out to be detrimental to the quality of French science after 1830 or so.

Napoléon never succeeded in rallying the students to his cause; after his fall in 1815, King Louis XVIII, who replaced him, did no better. He first fired Monge, then, in 1816, temporarily closed the whole school. No surprise that the students were at the forefront of the 1830 revolution against the next king, Charles X. One student, named Vaneau, was killed on the barricades. A new king, Louis-Philippe, was installed after this revolution, but only to confirm the military status of the school, and there were other uproars. The Polytechnique students took part in the revolution of 1848 as well, but with a less revolutionary spirit than in 1830. Later, the political troubles decreased at the school.

Despite all of these problems, Polytechnique remained a very important place for science. It soon developed well-equipped physics and chemistry laboratories, in part with subsidies from Napoléon, who was very interested in science: in particular, he paid for the large electric battery installed in 1811 (see later Fig. 5.3). Professors and assistants received a salary that allowed them to do research even if they had limited means of their own. Many of their results were published in the *Journal de l'École polytechnique*. Together with the observatories, this school was the most active research center during the first half of the nineteenth century.

The École Normale Supérieure[6]

This school was founded on 9 brumaire an III (30 October 1794), soon after the École polytechnique, in order to train the teachers of the *écoles centrales*, the equivalent of present-day high schools. Only a few scientists pursued their studies at the *École normale supérieure*, but some of its former students were excellent scientific collaborators. An example was Antoine Quet (1810–1884), professor at the *Lycée Saint-Louis* in Paris, who helped Foucault with the gyroscope by forming the theory of this apparatus while Foucault performed the experiments.[7] The best-known scientist who went through the school is Joseph Fourier (1768–1830); others were the physicist Claude Pouillet (1790–1868) and the astronomer Jean-Nicolas Nicollet (1786–1843), whom we will encounter later in this book.

[6] For a history of the school, in English, see https://archive.org/details/ecolenormalesup00laddrich

[7] Tobin (2003) p. 166.

There were physics and chemistry laboratories in the school, but only for teaching. It was only in 1851 that a real chemistry laboratory was founded at the school by Henri Sainte-Claire Deville (1818–1881).

The Conditions for Scientific Research

The Nature of Research[8]

The nature of research at the beginning of the nineteenth century did not differ much from that of today's science, but was very different from that in the seventeenth century. At that time, scientists, who were generally called "philosophers," studied nature from their office only. When they were travelling, they gathered information by chance. Science was made more or less at random, without preconceived plans if not with preconceived ideas. However, some scientific enterprises were organized by the Academy of Sciences during the last part of the seventeenth century: the first accurate measurement of an arc of meridian by Jean Picard (1620–1682) from Paris to Amiens in 1669; the first real scientific expedition, that of Jean Richer (1630–1696) to Cayenne in Guyana (1672); and the systematic measurements of the longitude and latitude of the major cities of France and neighbouring countries (1679–1682). There were more and more such enterprises later, culminating with the famous expeditions to measure one degree of meridian in Peru (actually the present Equator, from 1735 to 1743) and in Lapland (1736–1737), and the construction of the very detailed and accurate map of France by Cesar-François Cassini de Thury (1714–1784) and his son Jean-Dominique Cassini IV (1748–1845), which extended over the whole second half of the eighteenth century. Similarly, French and foreign astronomers organized themselves to observe the passages of Venus in front of the Sun in 1761 and 1769 in many places of the world, with the purpose of obtaining better values for the dimensions of the Solar System.

It is clear that these operations required preparations, financing and coordination similar to that of today's large scientific programs. The government by necessity had to intervene in the realization of these major enterprises, which it often ordered itself. Scientists and instrument builders had to work together as a united staff, a novel situation at the time. Theoreticians had to discuss with experimenters or observers for the interpretation of the data, for example in order to determine the shape of the Earth. In short, a degree of maturity similar to the present one was reached in some fields of research, essentially astronomy and geodesy.

However, evolution was slower in other domains of science. It was not easy to get rid of antiquated notions and replace them with new concepts based on observation and experimentation: This would take most of the eighteenth century.

---

[8] This section owes much to Debyser (2007).

Institutions like the Royal Society of London and the Academy of Sciences in France were instrumental in these changes, as testified to by the following text,[9] which dates from 1746:

Observation is the most certain way to succeed in the explanation of natural phenomena, so that the Academy [of Sciences], which has made so much progress in these useful researches, receives more favorably observations than reasonings, over all those reasonings that are not based on observations.

This evolution was essentially finished at the time of the French Revolution. Observations were now quantitative, and began to be formalized using an appropriate scientific language. It was now possible to build tables of data, and even representative graphs (which would however take much more time). Let us cite what Lavoisier said of the specialized scientific languages in his eulogy of Torbem Olaf Bergman (1735–1784), a Swedish chemist who was one of the first to attempt to create such a language[10]:

He had the very natural idea to express by some sorts of formulae all the chemical operations.[…] This sort of new language deserves the attention of philosophers: time is coming when the alphabetic language will not be fast enough, nor rich enough and precise enough to fulfill the requirements of sciences and to follow their progresses; either they will be forced to a stop, or one will have to create a language in which invariably fixed signs will describe the objects of our knowledge, the diverse combinations of our ideas, the operations to which we submit Nature, and those that we execute following our own ideas, a language which will be for all kinds of sciences, but even with more perfection, what the language of algebra is for mathematical analysis.

One has also learned, in order to progress, to renounce provisionally to search for the causes of some natural phenomena, a search which was now often recognized as impossible. Still, it was the only preoccupation, together with the construction of abstract systems, of the "philosophers" or "metaphysicists" of the ancient times. For example, the physicist Charles de Coulomb wrote in 1788[11]:

Whatever the cause of electricity, one will explain all its phenomena, and calculations will agree with experiments, by supposing two electrical fluids, the parts of any one of them repulsing each other proportionally to the inverse square of distance, and attracting the parts of the other fluid according to a similar inverse square law.[…] In the hypothesis of two electric fluids, I have no other intention than to present as simply as possible the results of calculations and of experiments, and not to indicate the true causes of electricity.

---

[9] Malouin, P.J. in **Mémoires de l'Académie royale des sciences* for 1746, p. 151.

[10] *Lavoisier, A.-L. de, in *Histoire de l'Académie royale des sciences* for 1784, p. 31–47, see p. 41–42.

[11] *Coulomb, C., in *Mémoires de l'Académie royale des sciences* for 1788, p. 617–705.

In the same period, communities of scientists formed, such as that which emerged through the spontaneous creation of the *Société d'Arcueil*. Let us cite again Lavoisier in his *Traité élémentaire de Chimie*[12]:

If sometimes I might have adopted without citing them the experiments or the opinions of M. Berthollet, of M. de Fourcroy, of M. de Laplace, of M. Monge and of those who in general have adhered to the same principles as myself, it is because the fact of living together and of communicating to each other our ideas, our observations and our way of thinking has established between us a community of opinions in which it is often difficult for ourselves to discern what belongs more particularly to one of us.

Everything was therefore ready for the harvest of discoveries to occur during the next years.

Science and Society

The reservations of the public with respect to science are a relatively recent phenomenon, although some reasons for these reservations—for example, the progress of more and more destructive weapons—have always been present. There were essentially no such reservations before the twentieth century. However, the attitude with respect to science evolved at the time of the French Revolution. During the eighteenth century, the Encyclopedists created a vast movement to promote science and techniques, but only for a restricted elite, that which attended public experiments in the "physics cabinets". Scientists were forming an isolated social group, and their influence was limited to this enlightened public. Most often they were ignored by the lower classes and also by many aristocrats, for lack of appropriate education and even of general education. The bourgeoisie was more interested in philosophy than science. There was even some vague hostility with respect to scientists, not because of the nature of their activity but because they were considered as belonging to a privileged class.

At the end of the eighteenth century and during the whole nineteenth century, interest in science and technique spread into a much larger public. There are several causes for this. First, the increasing number of *écoles centrales* educated a broad elite able to understand abstract notions. Second, at a higher level, the *grandes écoles* trained a large number of scientists, often of modest origin, whose example impressed the whole nation. Third, the applications of scientific discoveries progressively transformed daily life. The continental blockade imposed on France by Great Britain in 1806–1808 forced the country to be self-sufficient, and then to develop techniques to fabricate products that were imported previously: a famous example is the extraction of sugar from beets, to replace the cane sugar imported from Martinique. Chemistry began to find industrial applications, in particular

[12] Lavoisier (1790), see *Discours préliminaire*, p. 13.

thanks to Berthollet and Jean-Antoine Chaptal (1756–1832), and the steam engine born in England was quietly appearing in France.

The personal role of Napoléon Bonaparte in the scientific and technical development of France cannot be underestimated. He was interested in science, as shown by his organization of the famous expedition in Egypt, which was instrumental for the formation of a broad community of scientists. Bonaparte managed to be elected in 1796 in the mechanics section of the *Première classe de l'Institut* (see Fig. 1.1); although politically imposed, this election was beneficial for the Academy and for science in general. All during his reign, Napoléon distributed subsidies for research and financed costly scientific equipment. The kings who succeeded him were less interested in science, but the precedent was set. Napoléon III, who reigned from 1852 to 1870, had again a direct interest in science, perhaps inspired by the positive attitude of his uncle.

However, many workmen understood from the start that the machines driven by steam engines would replace them and would be a cause of unemployment; also, many manufacturers were initially reticent for various reasons. We will see in Chap. 11 that Arago exerted all his powers of persuasion to promote the use of machines, demonstrating their positive aspects. Indeed, one started to notice that inventions were able to make life more comfortable and to create wealth. Consideration for the work of scientists grew accordingly: the more educated members of the public admired those discoveries they could (or believed they could) understand, for example those of Fresnel or Ampère. There was also much enthusiasm for the discovery of fossils of extinct animals, and interest in the controversies raised by the first evolutionist ideas of Jean-Baptiste de Lamarck (1744–1829) and Étienne Geoffroy Saint-Hilaire (1772–1844), which were opposed by the Academician Cuvier.

There was now public demand for instruction in science and technique. Journalists tried to fulfill this need by writing accounts of the works presented at the Academy of Sciences. There was a weekly scientific *feuilleton* (feature story) on the front page of the daily journal *Le Globe* from 1825, then, after 1829, in the influential *Journal des Débats*, which had already published occasional scientific articles from its foundation in 1789. In 1830, most journals contained articles about science. This popularization was not appreciated by all scientists, for example by Cuvier who, as one of the two Perpetual Secretaries of the Academy, expelled from the meetings some people suspected of having organized leakages of information. These scientists obviously enjoyed staying in their ivory tower. The other Perpetual Secretary, Arago, fought this tendency with success, even during the life of Cuvier.

Financing Research

During the time we are discussing, one had to be rich to do research. This was the case of Laplace, and also of Lavoisier, who equipped at his own expense the superb chemistry laboratory now preserved at the *Musée des Arts et Métiers* in Paris. But his fortune, acquired as a *fermier général* (tax collector), was the cause of his condemnation and execution. Later, Hippolyte Fizeau (1819–1896), using his own money, bought from Paul-Gustave Froment (1815–1865) the apparatus he used to measure the velocity of light in 1849; so did Léon Foucault (1819–1868) for his rotating mirror and his pendulum built by the same Froment in 1850 and 1851 respectively. These were very expensive instruments requiring much work and care.

One could also seek the favor of the king or some wealthy patron. It was in this way that Charles le Monnier (1715–1799) managed to have his own observatory, which could rival the Paris Observatory, financed by King Louis XV. This king also had a private observatory in La Muette, then a suburb of Paris, whose instruments came into the possession of the Board of Longitudes at the Revolution. This kind of sponsorship was continued all during the nineteenth century: it took most often the form of prizes or "encouragements," often quite substantial, generally financed by rich people and distributed through competitions by the Academy. Finally those scientists who had the good fortune of belonging to the staff of the École Polytechnique or to the Board of Longitudes might have their equipment subsidized by these institutions. This was the case for Ampère, who was a *répétiteur* (assistant) at the École Polytechnique, and for Arago himself, who, after being promoted to *Directeur des observations,* could have the Board of Longitudes buy several rotating mirrors from Louis Breguet (1804–1883) for his experiments on the velocity of light.

The scientific institutions paid a salary to their members. We have seen that this was the case for the Board of Longitudes. The teachers and assistants in the grandes écoles were also paid. The National Institute (the Academies) paid an annual stipend of 1200 francs to its members. The members were allowed to accumulate stipends from several organizations. For example, in 1816, Arago received a total of 9700 francs, reduced to 8200 francs after the "dismissal" of the École Polytechnique, and re-established later (Fig. 1.6). After 1830, he received 6500 francs from the Board of Longitudes, 6000 francs from the Academy as Perpetual Secretary, and 3800 francs as an examiner at the School of Artillery in Metz, a total of 16,300 francs per year (equivalent of about US$50,000). At that time, he had renounced to his position of professor at the École Polytechnique in order "not to be accused of accumulating too high salaries,[13]" but more probably because he had no time left for this duties. His functions as a deputy at the Chamber and Paris city councillor were not remunerated, and he refused the fee attached to his short tenure as a minister in 1848. However, not everyone had the same financial opportunities as Arago.

Arago noticed that a number of university professors were not doing any research. He recommended that full-time research positions be created, but with little success. He was probably thinking of his friend Ampère, a bright scientist but a poor teacher. After his great discoveries on electromagnetism, to earn a living Ampère had to take a position of inspector of education, for which he was not suited and which took up time better spent on research.

[13] *Barral (1853).

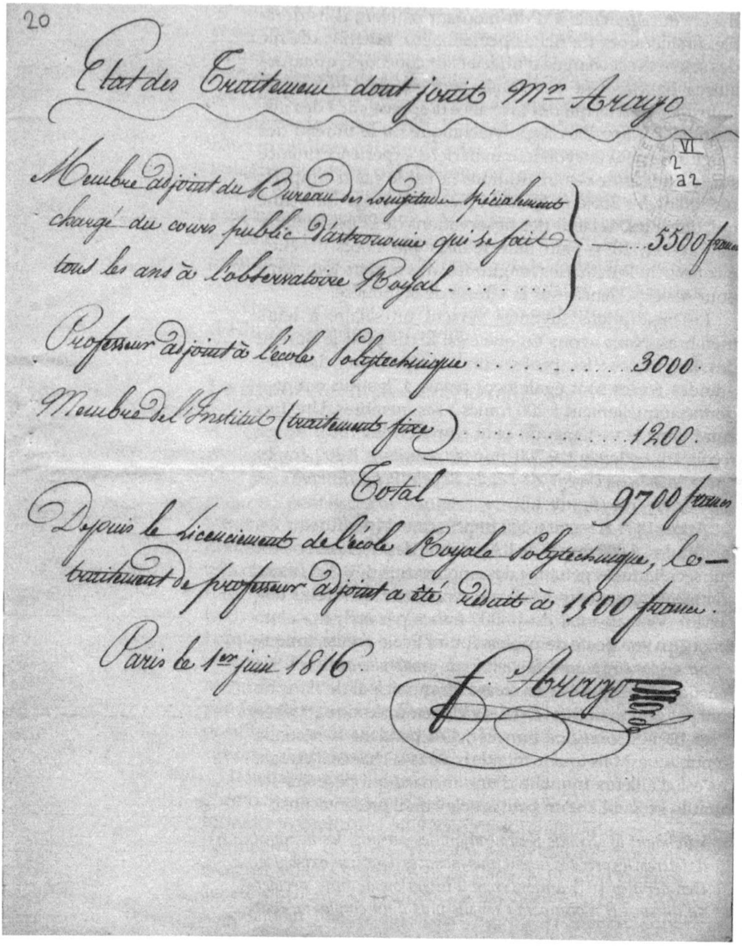
20

État des Traitemens dont jouit Mr Arago

Membre adjoint du Bureau des Longitudes spécialement chargé du cours public d'astronomie qui se fait tous les ans à l'observatoire Royal. . . . . . . ——— 5500 francs

Professeur adjoint à l'école Polytechnique . . . . 3000.

Membre de l'Institut (traitement fixe) . . . . 1200.

Total ——— 9700 francs

Depuis le licenciement de l'école Royale Polytechnique, le traitement de professeur adjoint a été réduit à 1500 francs.

Paris le 1er juin 1816

F. Arago

Fig. 1.6 The salaries of Arago in 1816 (autograph manuscript). Bibliothèque de l'École Polytechnique

Publications

After 1665, the date of the founding of the *Journal des sçavans*, the French scientists could publish their results in specialized journals. The *Journal des sçavans* disappeared in 1789, but it was also possible to publish in the *Mémoires de l'Académie royale des sciences*. The *Connaissance des temps et des mouvements célestes*, an annual ephemeris founded in 1679, also contained scientific papers. This publication became the official organ of the Board of Longitudes at its creation in 1794. Its first editor was Jérôme de Lalande (1732–1807, Fig. 1.7), succeeded by Delambre (Fig. 1.8) in 1808. It contained the observations made at the Paris Observatory (until 1809) and observations made in other observatories as a

Fig. 1.7 Jérôme de Lalande (1732–1807). A contemporary said of this convinced atheist: "That God whose existence he so often denied, by making him ugly deserved his revenge…." Wikimedia Commons, Gdr

Fig. 1.8 Jean-Baptiste Delambre (1749–1822). Wikimedia Commons

supplementary annual issue. This issue also included until 1809 a "Histoire de l'Astronomie," which summarized the work and results obtained during the preceding year, to which Delambre added his own commentaries.

Besides the *Connaissance des temps*, the Board published a small yearly volume entitled *Annuaire du Bureau des longitudes*. It was

> *extracted from the* Connaissance des temps, *contained all that is useful to the public in a small volume, to be understood by anyone, and distributed in sufficient numbers in all parts of the Republic.*

It was in this yearbook that Arago published his numerous scientific notices, which are excellent examples of the popularization of science and sometimes contained original developments. We will examine some of them in Chap. 11.

The *Première classe de l'Institut* created its own publication, *Mémoires,* from 1790. However, these memoirs were only published several years after having been approved by an ad hoc commission. This was one of Arago's motives in persuading the Academy to publish, starting in 1835, the *Comptes rendus hebdomadaires des séances de l'Académie des sciences*, which are still in existence. The Royal Society of London was better organized: from 1665, it published memoirs rapidly in the *Philosophical Transactions of the Royal Society of London*, which were quite popular even outside England.

Aside from these official publications, one found in France publications by the grandes écoles, the most popular being the *Journal de l'École polytechnique*. There were also scientific journals edited by private individuals for their own benefit, for example the *Observations et Mémoires sur la Physique, sur l'Histoire naturelle, et sur les Arts et Métiers*, founded in 1771 by Abbé François Rozier (1734–1793).[14] However, this publication, renamed the *Journal de physique, de chimie, d'histoire naturelle et des arts* during the Revolution, fell into jeopardy and disappeared in 1823. Indeed, a strong competitor was born in 1816: the *Annales de Chimie et de Physique*, founded by Arago and Berthollet. All these journals had a very personal tone: the *Journal de physique, de chimie, d'histoire naturelle et des arts* began every year with a "preliminary discourse" of about 100 pages, where the editor summarized the scientific works of the preceding year. In *Annales de Chimie et de Physique*, we can find the mark of the editors everywhere, including unsigned articles and notes that were clearly selected according to their tastes.

In all of these journals we can observe the popularization of science so typical of the time. It is explicitly for this purpose that the *Bulletin des Sciences de la Société philomatique de Paris* was created during the Revolution. Similar publications existed in foreign countries, for example, in Switzerland the *Bibliothèque universelle de Genève*, which covered various scientific fields with a predilection for physics thanks to his editors, the physicists Charles Gaspard de la Rive (1779–1834) and his son Auguste (1801–1873).

We should not be surprised to find many overlaps between these publications. Every time a discovery was made or some important scientific event occurred, each journal editor accounted for it and competed to obtain a text from the author. Here the *Annales de Chimie et de Physique* were at the forefront, because it was difficult to resist the requests of Arago and Berthollet. We can find in this journal the first articles by Ampère, Fresnel, Dulong, Joseph Fourier, many papers by Jean-Baptiste Biot (1774–1862), Humboldt and the best physicists and astronomers of the time, including, of course, Arago himself. In general, these texts had just been read at the Academy of Sciences or at the Société philomathique. Publication in the *Annales* was also a way to shortcut the long publication delay in the *Mémoires* of the Academy. The creation of the *Comptes rendus* in 1835 resulted in a progressive decrease of the impact

[14] He was succeeded by Abbé Jean-André Mongès le Jeune (1750–1788) in 1780, who left in 1785 to join the ill-fated Pacific Ocean expedition of La Pérouse. The journal was then edited by Jean-Claude Delamétherie (1743–1817), followed by the zoologist Henri-Marie Ducrotay de Blainville (1777–1850).

of the *Annales*; but one can still find there, long after this date, interesting original papers, essentially long memoirs that could not take place in the *Comptes rendus*.

An International Science

A striking feature of scientific journals at the time is the large number of foreign articles translated into French, and of substantial reports on research in other countries. No important results from Great Britain, Germany, Italy, Denmark, Switzerland or Sweden were ignored by French scientists. There was no nationalism in the way foreign colleagues were cited. For example, the work of the British astronomer William Herschel was described in long summaries in several journals, and an original paper by Alessandro Volta (1745–1827) describing his discovery of the electric "pile" (battery) was published by Delamétherie in his *Journal de physique*. The *Connaissance des Temps* published astronomical tables built by Herschel and other foreign scientists and related in a very laudatory way discoveries like that of the minor planet Ceres made in Palermo by Giuseppe Piazzi (1746–1826) in 1801; it also described in detail geodetic surveys made in other countries. The *Annales de Chimie et de Physique* and the *Bibliothèque universelle de Genève* published *in extenso* translations of fundamental papers like that of Christian Œrsted (1777–1851), written in Latin, which describes the action of an electric current on a magnetized needle.[15] Other translated foreign papers were not of similar importance, but in general the editor's insight was quite good.

It is remarkable that the scientific exchanges between France and other countries were completely ignorant of the conflicts and wars. The only mentions we could find about these problems are concerned with the difficulties of communication: it was not easy and sometimes impossible to obtain articles or books written by foreign colleagues, as Delambre regretted in the introduction of his recension of a book on triangulation printed in England.[16]

Others did not have this difficulty or were more persevering, like Delamétherie, who reported on the works of Herschel during the whole of Napoléon's reign, a period when the relations between France and England were very difficult.

[15] One finds as a note to the paper of Œrsted in **Annales de Chimie et de Physique* (1820, vol. 14, p. 417–425) the following note by Arago: "This paper, which will be published in the next issue of the *Bibliothèque universelle*, was communicated to me in Geneva by M. Pictet [Marc-Auguste Pictet, 1752–1825]; since, several scientists of Paris have received it directly from the author. The readers of the Annales have certainly noticed that we do not publish in general too lightly the announcements of extraordinary discoveries; and up to now we can only be pleased with these reservations. But as to the Memoir of M. Œrsted, the results it contains, how strange they could seem, are accompanied by enough details to be exempt of any suspicion of error." It is true that Arago had seen the experiment himself in Geneva.

[16] **Connaissance des temps* pour 1818, p. 243–278.

Chapter 2
The Life of Arago

Portrait of Arago, Wikimedia Commons, Kelson

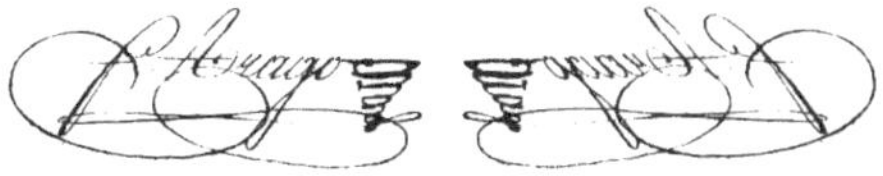

The signature of Arago. As seen in a mirror (*right*) it reproduces vertically the name of Arago. Wikimedia Commons, Skratt

Annex 1 gives a summary of the life and work of Arago, together with a mention of contemporary political events in France.

J. Lequeux, *François Arago*, Astrophysics and Space Science Library 421,
DOI 10.1007/978-3-319-20723-0_2

The Arago Family

Dominique François Arago (he went by François) was born on 26 February 1786 in Estagel, a small city near Perpignan, in the extreme south of France. He was the second of eight living children (three others died very young). His father, François Raymond Bonaventure Arago (1754–1814) was a relatively wealthy and educated countryman who obtained the title of bachelor of law at Perpignan University. He was elected mayor and judge of Estagel in 1790, but abandoned his functions as a judge in 1792 to become a member of the Executive Directory of the Département des Pyrénées-Orientales. The next year, France was at war with Spain, and François Arago's father was at the forefront of resistance against the Spanish troops, which had invaded a substantial part of the region before being defeated the following year. According to a legend, the 7-year-old François attacked a few isolated Spanish soldiers, and was fortunately saved by some countrymen.

In 1796, the family moved to Perpignan, where a *collège* (a middle school) and an *école centrale* had just been created. The young François could then study at these institutions. His father was appointed *Caissier de la Monnaie* (treasurer) of Perpignan, an important but poorly paid position that he kept until his retirement in 1813. In the city, he was at the centre of the intellectual life. He received Pierre-André Méchain, who was on his way to Spain to continue the measurement of the Paris meridian, and the "geometer" (mathematician) Jean-Nicolas Hachette, who was giving lectures on hydrography in Port-Vendres and in Collioure, near Perpignan. Méchain probably made some impression on the young François. On the other hand, the friendship of his father with Hachette was useful to his son: when in Paris later, he lodged at Hachette's home. Alas, Arago's father did not have much time to enjoy his retirement: he died at the end of 1814.

The father had married Marie Roig (1755–1845) in 1778. She was the daughter of a well-to-do peasant. She was a strong, affable and generous woman, very attached to religion; but she did not succeed to impose it on several of her children, including François. After the death of her husband, she retired to Estagel, where she lived to 90, her life illuminated by the visits of her children and long stays of several prestigious friends, such as the sculptor Pierre-Jean David d'Angers, who made her bust and that of François, and the physicist Félix Savary, who retired to her house (and where he subsequently died).

It is not without interest to say a few words about François Arago's siblings, several of whom played an important role in France and elsewhere and exerted some influence over him. I will add short biographies of the two sons of François, Emmanuel and Alfred.

Marie-Rose Emeranciane Arago (1782–1832) was the eldest surviving child. She remained at the family's home until 1812, when she married a wealthy countryman, with whom she had four daughters.

Jean Martin Arago (1788–1836) had a military career, first in France, then in Mexico, where he emigrated in 1816. He was lucky enough to survive through a very tumultuous time, and eventually rose to the rank of general in 1832. He died from malaria in 1836, alone and poor, without descendants.

Jacques Étienne Victor Arago (1790–1854) was bright and eccentric. Married at age 19 and soon father of two children, he abandoned his family to travel the world, pencil in hand. He became blind due to congenital diabetes, which also affected his brothers François and Joseph and his sister Marguerite. Even so, he embarked to Brazil in 1849, where he died after several travels back to France.

Pierre Victor Jean Arago (1792–1867) entered the École Polytechnique, like his brother François, and took up a military career, marked by a bright feat during the siege of Anvers in 1832. He ended his life in a discreet and modest way.

Joseph Honoré Arago (1796–1860) also served in the military. He was attracted to Mexico by his brother Jean. Blind due to diabetes, he finished his career as a colonel.

Marie Thérèse Gaudérique Marguerite Arago (1798–1859) played a larger role in the life of François. She was called in 1820 by her brother to live in the Paris Observatory, as well as their younger brother Étienne. This was to alleviate temporary financial difficulties of the family in Perpignan. At the Observatory, she met an already well-known astronomer named Claude-Louis Mathieu. They married in 1824. After the death of François's wife in 1829, Marguerite took care of his three children. Marguerite and Mathieu had two children. Their daughter Lucie married another astronomer of the Observatory, Ernest Laugier. It seems that Marguerite and Lucie, who were both ardent republicans, had some influence over François. Lucie served as his secretary when he became almost blind and could no longer write.

Étienne Vincent Arago (1802–1892, Fig. 2.1) was certainly, together with François, the most remarkable member of the family. He was an important actor of the political and cultural life of the nineteenth century. After unfinished studies in Perpignan, he joined François at the Observatory in 1820, as I just said. Thanks to his older brother, Etienne found a position of *préparateur* of Joseph Gay-Lussac in

Fig. 2.1 Étienne Arago (1802–1892), brother of François Arago. Wikimedia Commons, Paris 16

chemistry at the École Polytechnique. He soon resigned to begin a politico-literary career. He was a passionate republican and revolutionary, and took a major part in the 1830 revolution against King Charles X. At that time, he was the director of the Théâtre du Vaudeville, which was, according to him, the starting point of the revolt. One says that he is the man depicted by Eugène Delacroix with a top hat, standing on a barricade, in the famous painting *La liberté guidant le monde* (Liberty guiding the world).

Etienne was quick to denounce the excesses of the next king, Louis-Philippe, and to proclaim his republican faith, while brother François waited until the 1848 revolution to do so. Not surprisingly Étienne took an active part in the 1834 riots in Paris and Lyons, then in the 1848 uprising in Paris. He proclaimed himself minister of the postal service of the short-lived Second Republic that followed this revolution, and reorganized it to make it more efficient. He then introduced to France the postage stamp, which had been invented in Britain. Étienne was elected to the Chamber of Deputies at the same time as François and his son Emmanuel, Mathieu and the poet Alphonse de Lamartine. After Louis-Napoléon Bonaparte's election as president of the republic (later, to proclaim himself Emperor Napoléon III), Étienne fled to Belgium. He remained there until 1859, the date of a general amnesty. In 1870, he served as the mayor of Paris during the siege by the Prussians. He resigned during the Commune, criticizing its excesses. Then he made his living for some time from the sale of paintings and drawings he had accumulated, and ended his long life as the curator of the Luxembourg Museum in Paris. He was co-founder of the famous newspaper Le Figaro.

François Victor Emmanuel Arago (1812–1896, Fig. 2.2) was the eldest of François's three sons. After an untroubled childhood, he took to writing poems and theatre plays, under the influence of his uncle Étienne. He eventually became a barrister, specialising in invention patents (a domain familiar to his father). But he was more keen on defending his political friends like Armand Barbès (1809–1870) or Alexandre Auguste Ledru-Rollin (1807–1874). He joined the 1848 Revolution and was sent to Lyons as the superintendent of the young Republic. However, he was overwhelmed with the events that were particularly violent there, and was summoned back to Paris. This did not prevent his election as deputy. Then he was sent to Berlin as ambassador. There, he met alexander von Humboldt, who wrote to his father "that he [Emmanuel] was behaving in a much seemly way." After the election of Louis-Napoléon Bonaparte as president of the republic, he resigned from this function but was immediately re-elected deputy until 1851. Then he resumed his work as a barrister with some success. In 1870, he was named Minister of Justice, then Minister of Interior. He was again elected deputy in 1871, and senator 4 years later. He was ambassador to Switzerland from 1880 to 1894, and died in 1896, much esteemed by the republicans.

Alfred Arago (1815–1892), the second son of François, was a painter appreciated by his contemporaries. He was appointed General Inspector of Arts from 1852 to 1870, after which he found a position of director of a tobacco factory.

The third son of François, Gabriel, died in 1832 aged 15.

Fig. 2.2 Emmanuel Arago (1812–1896), son of François Arago, caricature by Daumier. The caption translates: "Fierce leftish, but too fat". Wikimedia Commons, Fabricio Cardenas

Youth (1756–1809)

Arago tells us himself what happened during his youth in a charming posthumous work entitled *Histoire de ma jeunesse* (History of my youth), written at the end of his life and inserted by the editor Jean-Augustin Barral (1819–1884) at the beginning of Arago's *Œuvres complètes*.[1] Knowing Arago's honesty and his excellent memory, which that he kept lasted until his death, we can believe most of what he wrote in this text, although events have probably been embellished in places.

Born on 26 February in 1786, Arago was too young to have participated in what he called "the excesses of our first Revolution." There was nothing important to mention for the first years: average studies, except in mathematics with the help of an amateur mathematician named Hippolyte Raynal, and meeting with a young

[1] *Arago, F. *Œuvres complètes,* t. 1, p. 1–102.

officer in a bright uniform, Jacques François Célini de Creyssac (1778–1856). Arago asked the latter about the way to become an officer, and decided to postulate for the École Polytechnique. For the entrance examination, which consisted only of an oral test in mathematics, Arago studied "the newest books that I obtained from Paris," in particular one by Adrien-Marie Legendre (1752–1833). The first trial was canceled due to the illness of the examiner who should have come to Montpellier to interview the candidates. The following year, 1803, when he took the test again, Arago had time to study the works of Leonhard Euler (1707–1783) and of Louis de Lagrange (1736–1813), as well as the *Mécanique céleste* of Laplace. The examiner, in Toulouse this time, was Legendre himself. Arago passed the exam, coming in sixth overall and first of those applicants who intended to take up artillery in the school.

Arago then moved to Paris and lived in the home of the mathematician Hachette, a friend of his father, professor of geometry at the École Polytechnique and an ardent republican. He met another mathematician, Siméon-Denis Poisson. It was no surprise that Arago would be one of the leaders of the students who refused to congratulate Napoléon after his crowning as emperor on 2 December 1804. General Jean-Gérard Lacuée, who had just been named governor of the school, was asked to send to the emperor a list of names of the agitators, which began with Arago. Napoléon was embarrassed and said:

I do not exclude the firsts of the promotion. Ah, if he would have been at the end... Mr Lacuée, stop at that point!

Soon after, the position of *secrétaire-bibiothécaire* (secretary-librarian) of the Bureau of Longitudes, located in the Paris Observatory, became vacant. Poisson and Laplace urged Arago to take the job, which he did on the condition that he could come back to artillery if he felt he was more suited to it. This was accepted, and although he was appointed at the Observatory on 25 January 1805 with a salary of 1800 francs per year (equivalent to some 7500$), he was still on the list of the students of the school, and again when he was sent to Spain the following year. However, he would not return to the school as a student, but as professor.

At the Observatory, he would often meet Laplace; he had much admiration for his science but was disappointed by his pettiness in daily life. Immediately after entering the place, he started astronomical observations and worked with Jean-Baptiste Biot (Fig. 2.3) on the refraction of light in gases, a program begun long before by Charles de Borda (1733–1799). He did not neglect his duties as librarian: 2 months after his arrival, the Board of Longitudes decided, most probably at his request, that "no book from the library could be borrowed by any member without a receipt." Although Arago was only librarian for 1 year, he would always take care of preserving and enriching the patrimony of the Observatory: in 1844 he offered to the Observatory 44 manuscript registers by Lalande that were given to him personally, and in 1848 he obtained from the Academy of Sciences the complete collection of the *Comptes Rendus* and *Mémoires* of this institution.

Fig. 2.3 Jean-Baptiste Biot (1774–1862). Wikimedia Commons, Scewing

The Catalan Odyssey

Arago writes in his *Histoire de ma jeunesse* that he and Biot proposed to Laplace resuming the triangulation of the Paris meridian, interrupted in 1804 by the death of Méchain, and to extend it to the Baleares islands. They would also add measurements of gravity with a pendulum. This project was accepted with enthusiasm, and Laplace obtained the necessary funds from Napoléon. The two scientists departed for Catalonia on 3 September 1806, together with a Spanish commissioner named Jose Rodriguez, who was a student in Paris, and a man named Chaix. The observations began badly: the parabolic mirror of an oil lamp located on the continent and which was supposed to be seen from the summit of Ibiza was not well oriented and thus very difficult to see at night. Arago had to stay for 6 months at the Desierto de las Palmas in Spain before succeeding in the observations. It is true that the distance to Ibiza is no less than 170 km! The rest of the measurements were less problematic, despite political difficulties and troubles with the local populations.

Biot had to return to France in May 1807, leaving Arago alone with his two assistants. He came back for a short stay at the end of the year, bringing a new instrument to replace a damaged one. Arago finished the work himself, extending the triangulation to Formentera then Majorca. In August 1807, Arago was named *in absentia* astronomer-adjunct by the Bureau des Longitudes, replacing Michel Lefrançois de Lalande (1766–1839), promoted titular. The report for this nomination states:

The astronomical knowledge that distinguishes M. Arago, and the services he renders presently in the important operation he is entrusted with, justify the choice of the Board.

In the spring of 1808, France declared war on Spain and the troops started to invade Catalonia, so that the Arago's situation became increasingly uncomfortable. Then followed an odyssey told with considerable humour in *Histoire de ma jeunesse,* and with fewer details in the *Discourse on Algeria* pronounced on 19 April 1833 at the Chamber of Deputies.[2]

Taken prisoner by the Spanish, Arago succeeded in escaping from jail, helped by his knowledge of the Catalan language (which was actually his mother tongue). He found a boat to reach Algiers. There, he embarked for Marseilles on a ship carrying two lions and several monkeys the Dey of Algiers offered to Napoléon. However, the ship was intercepted by a Spanish corsair, which towed it to Catalonia. Arago managed to discreetly inform the Dey of this adventure and of the death of one of the lions. Furious, the Dey demanded the Spanish to release his ship, which resumed its trip to Marseilles after being blocked for 3 months. This was not the end, however. A strong *mistral*, the northern wind that can blow so violently in the Mediterranean Sea, pushed the ship to the Algerian coast. It landed in Béjaïa, 180 km to the east of Algiers, with the surviving lion and the monkeys, which helped the unfortunate passengers to delouse themselves. There was no hope of reaching France from Béjaïa, so Arago had to travel by land to Algiers, which was not without danger. There he could find a ship to Marseilles, which he reached on 2 July 1809, almost 3 years after leaving Paris. All along this odyssey, he managed to keep his instruments and the notes with his results.

As usual for travellers by sea, Arago was kept in quarantine at the lazaret of Pomègue island, which faces Marseilles. During this time, he met Jean-Louis Pons (1761–1831), a great astronomer of comets who was then at the Marseilles observatory. Pons was initially the *concierge* (keeper) of the observatory, but he did so well in observing that he was appointed director of the Lucques observatory in Italy in 1819, and director of the Florence observatory in 1825.

The Friendship of Humboldt

While in quarantine, Arago received a letter from the already famous scientist Alexander von Humboldt (Box 2.1), who used to call himself Alexandre de Humboldt when he was in France.[3] Enthusiastic about Arago's adventures, he congratulated the Frenchman and offered him his friendship. This friendship was to last for 44 years. They shared the same scientific and political ideals. For Humboldt (Fig. 2.4), Arago was a genius whom he considered (wrongly) as intellectually superior to him. In fact, Humboldt was probably homosexual and was attracted to Arago, as we can understand from many of his letters, for example[4]:

Since I have not had the good fortune to listen to you and to strain you with my questions, there is no hour of my life when I am not thinking of you as the person I admire most in the world.

[2]* Arago, F. *Œuvres complètes*, t. 12, p. 548–533.

[3] This letter has unfortunately disappeared.

[4]*Hamy (1907), Lettre XI, p. 29–36.

Or reproaching Arago for not writing to him more often[5]:

I am asking from you only two or three lines from your hand as I have had the luck to receive in the past for twenty years. After fifteen months of complete silence, the request of a friend who has always been devoted to you as much as and perhaps more than any member of your family, should not be importunate.

Actually, Arago did not write often to Humboldt, probably to distance himself from the lovelorn German. His letters are warm but without excess. For example, he wrote to him in 1841, learning of his visit to Paris[6]:

Box 2.1 Alexander von Humboldt[7]

Humboldt was born in 1769 in an aristocratic Prussian family at the Tegel castle, near Berlin. He spent there the end of his long life and died there. When quite young he showed a great interest in natural history. Traveling through Europe led him to Paris in 1790, and he was passionate for the French Revolution: His advanced social ideas contrasted with those of his family and surroundings. He entered at the School of Mines of Freiberg, and after obtaining his degree was named General Inspector of mines in 1792. He was instrumental in ameliorating the conditions of miners, while taking part in various diplomatic missions for the Prussian state. In 1798, with the French botanist Aimé Bonpland (1773–1858), he left for a 5-year, 5000 km trip through the "equinoctial America." They discovered the junction between the Orinoco and Amazon rivers, described the cold current along the west coast of South America and gathered an unrivalled amount of observations. Humboldt settled in Paris in 1808, and remained until 1827, taking an active part in the French scientific life and working often with his great friend Arago. He wrote there in French his *Voyage aux regions équinoxiales du Nouveau Continent* (Travel in the equinoctial regions of the New Continent), in 30 volumes. He was called back to Prussia by the king, who gave Humboldt several missions and made him his friend in spite of his leftish opinions. He gave a large number of popular lectures, like Arago was doing in Paris. From 1829, he made a trip of 19,000 km in Russia and Siberia, then other less important ones including a new one to Paris in 1841. The last years of his life were devoted to writing his great synthesis work, this time in German, entitled *Kosmos*, to unceasing visits at Tegel and to an enormous exchange of letters: He received 3000 letters a year and answered 2000. He died in 1759, aged 90.

Humboldt was the last encyclopedic scientist, whose culture and activity were astonishing. His influence on his contemporaries was immense, in particular on Arago. He was one of the greatest explorers of all times. His work on geophysics and oceanography is particularly remarkable.

---

[5] *Ibid.*, Lettre XXXVII (1833), p. 126–133.

[6] *Ibid.*, Note after Lettre LXX (1841).

[7] From *Humboldt (1846–1851).

Fig. 2.4 Alexander von Humboldt (1759–1859), portrait by Joseph Karl Stieler. Wikimedia Commons, Ben-nb

Apart from my family, you are without comparison the person I like the most fondly in the world. But you have also to resign yourself: you are the only friend on whom I would rely in difficult circumstances. I am really happy to think that I am going to spend some evenings with the person to who I owe my taste for meteorology and terrestrial physics. There will be a bed for you at the Observatory.

Whatever the nature of their relationship, the two men were close and saw each other almost every day when they were both in Paris. We will see that they made together a variety of researches at the Observatory, and the text just cited shows what Arago owes to Humboldt scientifically.

The Time of Major Scientific Activity (1809–1830)

Elected to the Academy of Sciences

A few days after his arrival in Paris, on 18 September 1809, Arago was elected as the *Première classe de l'Institut*, the provisional name of the Academy of Sciences. He was only 23, and was the youngest scientist ever elected in this assembly. It is obvious that his tribulations were quite decisive in his election, but he was also already an established scientist: He succeeded in bringing back to Paris all his geodesic observations, which turned out to be excellent, and the work he had done in 20 months before his departure is impressive by its quality and its originality. However, the election was not without difficulties: Laplace pushed Poisson, who was older than Arago and was already professor of geometry, while Biot, Delambre, Lagrange and Legendre supported Arago. It is amusing to report on the exchanges that took place before the election, as remembered by Arago at the end of his life when dictating his *Histoire de ma jeunesse*:

M. Laplace, without denying the importance and the usefulness of [my] works and of [my] researches, saw only a promise in them: then M. Lagrange said to him in strong terms:

"Yourself, M. de Laplace, when you entered Academy you had not yet done anything remarkable: you gave only hopes. Your great discoveries came only afterwards.[8]*"*

Lagrange was the only man in Europe who could make such observations to Laplace. The latter did not reply on the personal fact, but added: "I maintain that it is useful to show to the young scientists that a seat at the Institut could be a reward to excite their zeal."

"You remember me, replied M. Hallé [the academician Jean-Noël Hallé, 1754–1822], of this cab-driver who attached a bundle of hay to the pole of his carriage. The poor horses made more and more efforts, but the hay bundle kept fleeing in front of them. Ultimately, this produced their decline and their death"

The Laplace's manoeuvres did not succeed, and Arago was elected by 47 votes against four for Poisson and one for Nicolas-Antoine Nouet (1740–1811), an astronomer who had taken part with Méchain to the expedition in Egypt. On 4 October 1809, Napoléon approved the election of his ancient enemy; he had no choice.

Arago told other stories about the elections at the Academy, where he attempted to give to himself the good parts, although this was not always justified. From the testimonies of his contemporaries, in particular of Laplace who called him "the great elector," he played a major role in these elections. In 1822, after the death of Delambre, another Perpetual Secretary had to be elected. Biot and Joseph Fourier were both candidates, and Arago was pushed by some colleagues to be a candidate himself. He declined, probably because he considered himself too young to occupy this position, claiming that "he had already as many positions as he could fill." He said that Biot was in the same situation. By doing so, Arago undermined Biot, and this did not help the already poor relations between the two men. Fourier was thus elected. After his death in 1830, Fourier was replaced by Arago; Biot was definitively eliminated.

Arago's election at the Academy opened many doors for him. He frequented bright circles, where he met Lucie Carrier-Besombes (1788–1829). They married on 11 September 1811. As a familiar of Laplace, he entered the *Société d'Arcueil*, of which Humboldt already was a member: he made friends there with Gay-Lussac

[8]Laplace was elected at the Academy in 1773 as "adjoint mécanicien", aged 24. He became a member of the Institut de France at its creation in 1795.

Fig. 2.5 Joseph Gay-Lussac (1778–1850), lithography by François Séraphin Delpech. Wikimedia Commons, Materialscientist

(Fig. 2.5) and with another former member of the Egypt expedition, Étienne-Louis Malus (1775–1812), who was elected to the Institut in 1810 thanks to him. He also met Gaspard Monge, the founder of the École Polytechnique: this was to be the beginning of his teaching carrier.

Professor at the École Polytechnique

Monge, professor of "analysis applied to geometry" at the École Polytechnique, was 64 and not in good health. He was looking for a substitute. He first considered his *répétiteur*, Jacques Binet. But the Emperor, contacted by Monge for the necessary approbation, had in hand a police report that denounced Binet as a "disciple of Loyola" (a Jesuit). Monge had to find someone else: He chose Arago.

Arago delivered his first lecture on 26 October 1810 "with the highest clarity" as a contemporary said. In May 1812 he was promoted to adjunct professor at Monge's request. He had to teach not only "analysis applied to geometry," but geodesy, a subject with which he was fortunately very familiar. When named a full professor in 1816, he added a course in "social arithmetic," a new discipline that applied the probabilities developed by Laplace to demography and economics. The time was indeed at statistics of all kinds using this new tool, which raised enthusiasm amongst the scientists. Arago's teaching was rather elementary, but he gained knowledge that turned out to be very useful for his future career in the Chamber of Deputies.

Arago also gave a "course of machines."[9] Hachette had taught this course until his dismissal in 1816 for political reasons. After some delay, the course was given in 1819 to Alexis Petit (1791–1820), who fell ill and died the following year. Arago acted as a substitute for 15 lectures, but asked to be replaced because he was overwhelmed with work. Andre-Marie Ampère, a professor of mechanics, was then named to the position, but he fell ill himself after a few lectures. Arago had no choice but to resume teaching the course, which he gave from 1821 to 1830. It consisted of 24 lectures each year, in which he described various machines and explained how they worked. The students also had to make finished drawings of a hydraulic press, a flour mill and a Watt steam engine, as well as a "fire engine used to move a boat". Several beautiful examples of such drawings survive (see Fig. 11.4). For those students who really wanted to understand how these machines worked, Arago wrote his remarkable *Notice historique sur les machines à vapeur* (Historical notice on steam engines).[10]

After the 1830 Revolution, Arago was too busy to continue his teaching, and so he resigned. His successor was Félix Savary, who had been his *répétiteur* for 2 years. Savary taught until 1841. However, as he fell very ill, he was replaced temporarily by Urbain Le Verrier, the future discoverer of Neptune. Le Verrier did not obtain the position at Savary's death and was quite frustrated. Was this due to some intervention by Arago? We do not know; in any case this did not make for better the relations between the two men, which were already poor and became increasingly strained.

Arago, who temporarily had a good relationship with King Louis-Philippe, was put in charge of a reorganization of the École Polytechnique. Curiously, he was very conservative in doing this: For example, he recommended that its military status be preserved. This caused controversies that lasted at least until 1836. Later, Arago, who was a member of *Conseil de perfectionnement* (Council of Improvements) of the school from 1832 to 1844, had to intervene again in the functioning of the school. He was again quite conservative in this. However, the school keeps a good remembrance of him and has recently given his name to an amphitheatre.

The Scientific Work

Arago was overall an experimentalist and an observer who left theory to others; he had original ideas, especially at the beginning of his career, but he could also work with others, acknowledging their contributions; and he was enthusiastic about emerging domains of science and rushed into them, even if this meant abandoning his current research.

---

[9] Dupont (2000) p. 1–101.

[10] Several editions in **Annuaire du Bureau des longitudes* for 1829, 1830 and 1837. The last one is reproduced in *Arago, F. *Œuvres complètes*, t. 5, p. 1–116.

In 1810, Arago was already an accomplished scientist. He could well have given up research after coming back from his odyssey and being elected at the Academy, but he decided to resume the work he had to abandon to go to Spain. First, he attempted anew to determine if the light coming from various celestial bodies was always deviated in the same way by a prism, hoping in this way to measure possible changes of the velocity of light. He took an interest in the polarization of light, which had just been discovered by his friend Malus. At the end of 1810, he made an important discovery, that of chromatic rotational polarization. Using this property of crystalline quartz, he built a *polarimetre* that he used to discover the polarization of the Moon and comets, and to show that the emitting surface of the Sun and (later) of the stars was made of an incandescent gas. It was the first time that something could be said about the physical nature of celestial bodies: thus, these discoveries marked the beginning of astrophysics.

But Arago was already driven toward other subjects, leaving to others like Biot the development of practical applications of rotational polarization. However, he used his knowledge of polarization to conceive various astronomical and other instruments during the rest of his life. For the moment, he was busy measuring the refraction and dispersion of light in various gases. This was certainly less exciting than polarization, but useful, and the study extended over several years. At the same time, Arago had to teach at the École Polytechnique and to prepare the public lectures in astronomy required by the statutes of the Board of Longitudes, which had charged him to deliver them: the first lecture took place on 3 February 1813, and the success was such that Arago continued to teach the course of "popular astronomy" until 1846.

By 1815, Arago was famous. A testimony of this fame is the following anecdote reported in Arago's necrology of Monge.[11] The scene takes place after the second abdication of Napoléon, in 1815.

The vanquished of Waterloo was living in the Élysée palace. During one of his private talks with Monge, Napoléon discussed his projects: he wanted to sail to America, and thought that the English would let him go [which did not materialize: they kept him prisoner on the island of Saint Helena]... "Since I would no more be allowed to command armies, I see that only science would be able to take strongly possession of my soul and of my mind. To learn that others have done would not be enough for me. I want, in this new career, to do works and make discoveries worthy of me. I need a companion who would first instruct me rapidly of the present status of science. Then we would travel together through the New World, from Canada to Cape Horn; during this immense trip, we would study all the great phenomena of the physics of the Earth that the scientific world has not yet described." Monge, enthusiastic, exclaimed: "Sire, your collaborator is found: I will go with you." Napoléon thanked his friend with affection. But he made him understand, not without difficulty, that a septuagenarian would not withstand such a perilous venture.

---

[11] **Mémoires de l'Académie des sciences* 24, p. I–CLVII, also in *Arago, F. *Œuvres complètes*, t. 2, p. 427–592.

One then called a much younger scientist [who evidently was Arago]. Monge exposed to his colleague with considerable persuasion how glorious the proposition was by its purpose, and even more because of the position of the illustrious person who made it. A large amount of money would compensate the loss of his positions by the young academician; another sum would be used to buy a complete collection of instruments of astronomy, physics and meteorology. But the negotiation failed.

Of course, Napoléon wanted to follow in the footsteps of Humboldt. As for Arago, he detested Napoléon, and he was not keen on embarking on such an uncertain venture.

The same year, 1815, marked the encounter with Augustin-Jean Fresnel: a real scientific "bolt from the blue" that drove Arago to defend Fresnel who was just constructing his wave theory of light, and also to perform with him several very interesting experiments. Their collaboration extended in 1819 to lighthouses for navigation. The deep knowledge that Arago had of interferences allowed him to explain for the first time in 1814 the scintillation of stars, and later to realise an instrument able to measure accurately by interferences the refraction index of gases.

In 1820, Arago saw in Geneva the experiment of Hans Christian Œrsted, which showed the deviation of a magnetized needle by an electric current. He reproduced it in front of the Academy of Sciences, raising considerable interest among his colleagues. Arago behaved with Ampère in the same way as with Fresnel. While Ampère, starting from Œrsted's experiment, was discovering in a few months the laws of electromagnetism, Arago collaborated with him and discovered on his side that the electric current can magnetize a piece of iron. The two men built the first electromagnet. Then Arago discovered in 1823 the "magnetism of rotation," which is the action on a magnetized needle of a rotating copper disk. This observation had a significant impact and in 1827 Arago won the prestigious Copley medal of the Royal Society of London. However, neither Arago nor any of his contemporaries could find the explanation for this phenomenon. It was finally understood in 1831 by Michael Faraday (1791–1867), after his discovery of electromagnetic induction, to which Arago's observation contributed strongly.

The discovery of the magnetism of rotation ended the most fecund period of Arago's scientific activity. Very busy with his teaching, with his functions at the Board of Longitudes, at the Paris Observatory and at the Academy, then with his political activities, he had little time left for research and could only make sporadic incursions in scientific experimentation. The most interesting ones are his measurement with Pierre Louis Dulong, between 1825 and 1829, of the "elastic force of water vapor" (the saturation vapor pressure) as a function of temperature, and his aborted essays to compare the velocity of light in air and in water, an experiment finally done with success in 1850 by Léon Foucault and by Hippolyte Fizeau. After 1848, when he was almost blind but had more time for himself, Arago had novel and interesting ideas on a variety of topics.

At the Observatory

Despite his work in physics, Arago was basically an astronomer. He lived at the Paris Observatory (Fig. 2.6), and took an active part in the observations. He was also charged by the Board of Longitudes of various tasks that were usually done by astronomers, for example the geodetic connection of France and England in 1821, and a new measurement of the velocity of sound in 1822. Arago had already visited England with Gay-Lussac in 1816, meeting Thomas Young (1773–1829), then in 1817 with Humboldt and Biot. In 1818, he travelled to Dunkirk with Biot and the Englishman William Mudge for a new measurement of its latitude. His last important trip was to Scotland in 1834.[12] Arago had thus frequent relationships with Britain, and understood its language. He also travelled several times to Geneva to visit his friend Charles Gaspard de la Rive, at least in 1812, where he could see a demonstration of the electric arc invented by Humphry Davy (1778–1829),[13] and in 1820 to attend Œrsted's experiment.

The Board of Longitudes controlled the Observatory with a strong hand, but in a rather erratic way. Arago certainly had an important role because of his scientific reputation, justified by his efficacious, and abundant, activity. It is however difficult to assert what was his real weight on the equipment and management of the Observatory. One might think that his influence increased progressively while that of Laplace declined, and that the most eminent astronomer of the Observatory, Alexis Bouvard (Fig. 2.7) was busy with routine observations, endless calculations and administrative duties. At the death of Laplace in 1827, Arago had become the preeminent French astronomer.

The year 1820 saw the beginnings of the "Arago clan," which was to dominate French astronomy for more than 30 years. His sister Marguerite had married in 1821 the astronomer Mathieu, while Arago's sister-in-law had married the physicist Alexis Petit. Otherwise the 1820s did not see major events in Arago's life, with the exception of the premature death of his wife Lucie in 1829. She was a very beautiful and pleasant woman, but rather discreet; this is understandable given the flamboyant nature of her husband. After her death, their three children were raised by their aunt Marguerite. François would not marry again, and I do not know of any love affair with him, with the possible but unproved exception of Le Verrier's wife. The Observatory continued to be a meeting place, with receptions on Mondays, but one may imagine that they were less bright than before Lucie's death. Humboldt had left for Berlin in 1826, and his closest friends were dead: Berthollet in 1822 and Fresnel in 1827. As for Gay-Lussac, he was now running after honours and money,

[12] For the travels in 1816 and 1834, see **Mémoires de l'Académie des sciences* 13 (1832) p. LVII–CV, also *Arago, F. *Œuvres complètes*, t. 1 p. 241–294. In Scotland, Arago met Lord Henry Brougham (1778–1868), a conservative amateur scientist, specialist of light, with whom he had stormy relations about scientific priority.

[13] See *Arago, F. *Œuvres complètes*, t. 7 p. 530–547.

Fig. 2.6 View of the Observatory taken from the north-east. Arago lived in this part of the building. Private collection

which did not please Arago. Arago took refuge in work, accepting in spite of his heavy teaching and observations to write many articles and reports in his journal, the *Annales de Chimie et de Physique*, and his famous Scientific Notices in the *Annuaire du Bureau des Longitudes*. One should not forget also the many reports required by the Academy. He was now one of its most influential members, as confirmed by his election on 7 June 1830 to its most prestigious position: that of Perpetual Secretary for the mathematical sciences.

Fig. 2.7 Alexis Bouvard (1767–1843), lithography by Boilly, 1821. Wikimedia Commons, Gdr

The Politician (1830–1838)

Although Arago had claimed at several occasions his opposition to Napoléon and had rallied for the constitutional monarchy in 1815, after the fall of the Emperor, he only entered active political life in 1830. On 25 July of that year, after a dispute with the Chamber of Deputies, King Charles X signed ordinances that restrained the freedom of the press, dissolved the just-elected Chamber and restricted the conditions to be an elector of the next Chamber. All this was contrary to the 1814 charter that defined the new constitutional monarchy. The next day, Arago came to the full Academy to inaugurate his new functions as perpetual secretary, and to read his obituary of Fresnel.[14] He discovered there the ordinances. An emotional Arago said he would not give his speech, citing the new limits on freedom of expression. This threat provoked a strong reaction from the other perpetual secretary, the conservative Georges Cuvier, who feared that the Academy would be suppressed. Finally a compromise was found: Arago would present his eulogy of Fresnel, but inserted in it some noticeable allusions to the restrictions of liberty, for example:

Fresnel, as so many well-disposed persons, shared in 1814 the hopes that produced the return of the Bourbon kings. The 1814 charter, written with an open mind, seemed to him to contain all the premises of a wise liberty. He saw in it the dawn of a political regeneration that would extend smoothly from France to the whole Europe.

These words caused "frantic applause," including that of Étienne Arago and of the writer Alexandre Dumas. The latter wrote with enthusiasm[15]:

It was not a simple success that was obtained by Arago: it was a triumph. It is impossible to be more original, more frank, more beautiful even than François

[14] *Arago, F. *Œuvres complètes*, t. 1, p. 107–185.

[15] Cited by Sarda (2000) p. 112.

Arago when he is passionate.... He is not only the science, but the conscience; not only the genius, but the probity; I want to be amongst those who claim it.

The following day, the insurrection of the *Trois Glorieuses* (the three glorious days) broke out. Étienne Arago was one of its leaders.

The Deputy

During these 3 days, France went from the tyranny of Charles X, the last Bourbon king, to the milder regime of a new king, Louis-Philippe, that was very close to being a republic. François did not share the revolutionary views of Étienne, and he rallied behind the new regime. He was an officer of the *Garde Nationale*, a militia of 60,000 citizens commanded by the old Lafayette (1757–1834) and was even promoted to colonel; but the new ministers did not ratify this promotion, and Arago would later complain about it to the king.

However, there were other troubles, and the regime became increasingly repressive. Arago turned to the opposition, but his complete motivations are not known. He announced on 7 April 1831 that he was seeking election to the Chamber of Deputies, both from Paris and in his birthplace, the *Département des Pyrénées-orientales* (this kind of duplication was allowed at the time). He was elected to both on 6 July, and chose the *Pyrénées-orientales*. He had already been elected the previous year to the Paris city council, so that he could be influential in both government bodies. He would serve in the Chamber of Deputies and the Paris city council for 20 years, with some interruptions.

The year 1832 was marked by tragedy: the death of his son Gabriel and of his sister Marie-Rose. It is also the year when he suffered the first symptoms of diabetes, an incurable illness in this epoch. But the political situation required that he overcome his personal problems, and his political activities continued to growing. The many speeches he gave at the Chamber and at the council about Paris (such as fortifications, navigation of the Seine, the water supply) showed a deep understanding of the city. Arago, who was a great orator, also covered education, technical progress and social problems. He also intervened in order to limit the waste of public funds.

The most remarkable of his speeches is probably the one he delivered at the Chamber on 16 May 1840, calling for electoral reform.[16] He described the sufferings of the workers, the need for better working conditions, and he maintained that social progress would benefit from replacing the present electoral system, which required that the voters had a significant income, by the universal franchise for men. (No one would consider giving women the vote.) Box 2.2 gives some extracts of this speech.

[16] *Arago, F. *Œuvres complètes*, t. 12, p. 589–614.

Box 2.2 Extracts of Arago's Discourse of 16 May 1840 at the Chamber of Deputies

"Messieurs,

…I owe it to you to be frank right from the beginning. I declare that I will support the broadest requests…in favor of a deep change of the electoral system. I am aware that this opinion has few supporters on these benches, even within my political friends….

I maintain that the principle of political representation is an empty thing in a country where, only one man in forty can vote.

I maintain that the part of the population that is deprived of political rights is not only the most numerous, but also that which contributes the most important part of the income of the State (*protestations, interruptions*)….

I now come to the major difficulty. One pretends that the citizens for whom we give the right to vote would not possess sufficient capacities to exert it. What capacity? Should they pass an examination?…The capacity a voter should possess is to distinguish an honest man from a crook, a good citizen from an egoist, an unselfish man from an ambitious one….

One also pretends that the people, if asked to compose the Chamber, would always elect illiterate men.…History gives a denial to this: It shows that the masses know how to find merit where it is.

One of our past assemblies was elected by the people: the *Convention*[17] (*Ah! Ah! Murmurs*). During its reign, there were deplorable excesses.…But, don't forget it, the Convention has saved our country, our territory, our nationality. It did not let foreign armies, enemy armies to progress to our capital[18]; it brought our frontiers to the natural limits of France; it created most of the beautiful institutions that made the glory of our nation for half a century.…

There is, gentlemen, in our country an important part of the population that suffers much. It suffers so much that it is undermined, tortured by misery and hunger (*brisk interruption*).…This part of the population is particularly made of workmen.

This problem grows everyday. The small capital, in industry, will not be able to stand for long in front of the great capital. That industry that uses machines will win over the industry that uses only the natural human power. The industry that uses powerful machines will always win over that which uses small machines.

In the next years, the whole population of workmen will be dominated by a few capitalists.

(continued)

[17] The *Convention nationale* governed France from September 1792 to October 1795, during the French Revolution. Elected by universal franchise, it proposed a very democratic constitution, but this constitution was never applied. From October 1793 to July 1794, a regime of terror took place, followed by a strong reaction. Nevertheless, the Convention decided, especially during this last period, of the creation of several great schools, of the Bureau des Longitudes, of the Museum of Natural History, of the Conservatory of Music, and created the metric system of units.

[18] Allusion to what occurred in France in 1815 during the fall of Napoléon.

Box 2.2 (continued)

Gentlemen, I have just pointed out a real evil.... I will raise murmurs if I derive from these facts their logical consequence, if I proclaim the need to organize work.... If you find this idea extraordinary, I will say that you already adopted it when you examined a law limiting the work of children in factories.

There are disciples of Fourier, of Saint-Simon, of Babeuf[19] (*various exclamations*) who believe that they have found a remedy for this great social problem.... I would like the Chamber of Deputies to substitute itself for these audacious empirical thinkers, who would kill the patient with the illness. I would like that the Chamber of Deputies deserves the full confidence of these interesting classes of society that one confuses, that one deceives, that one blinds....

If one may judge from some symptoms, the people are presently convinced that their participation in politics would hasten the ameliorations they hope for, and will not give up easily their legitimate claims. It would be, I believe, an enormous mistake to take away any hope from the people. Given the extreme tension in the society, the reform looks to me like a safety valve.

The 1830 revolution was made by the people. By examining their requests, let us close the mouths of those who say that it was not made for the people."

Eight days after this address, ten thousand workmen came to the Observatory to thank Arago. This would be one of his dearest memories. His speech was still remembered 40 years later.

Of course, in spite of his fame but also because of it, Arago was violently attacked by his political adversaries. After 1846, the government became increasingly rigid and attempted to neutralize him, but in vain: Arago was overwhelmingly re-elected in Pyrénées-Orientales on 2 August of that year.

Arago was also a victim of jealousy. A particularly rancorous paper was published in March 1840, 2 months before the famous speech, in the *Revue des Deux-Mondes*. It was not so much against Arago the politician than against the Arago the academician. The *National* defended Arago, but *La Presse*, a conservative journal, tried to destroy him by minimizing his scientific work: "None of his works took him more time than needed for a report." Arago soon detected the author of these diatribes. It was Guglielmo Libri, an Italian refugee banished because of his political ideas (Box 2.3). And yet, Arago had welcomed him in Paris, had helped him to obtain French citizenship in 1833, and had supported him for his election to the Academy of Sciences. He also managed to obtain for him appointments to lecture at the *College de France* and the *Sorbonne* to make his living. But in spite of his

[19] Charles Fourier (1772–1837), not to be confused with the physicist Joseph Fourier, was a utopian. Claude Henri de Rouvroy, Comte de Saint-Simon (1760–1825), was an economist who was very influential for the development of the industrial society in France. Gracchus Babeuf (1760–1797) was the first revolutionary communist.

Box 2.3 Guglielmo Libri[20]

Born in Italy in an allegedly aristocratic family, Libri soon distinguished himself by his mathematical work. He was pursued because of his revolutionary activities and took refuge in France in 1830. Arago protected and helped him. He was named General Inspector of public instruction and was in charge of inspecting France's major libraries. He took this opportunity to steal many precious books and manuscripts. He was finally denounced in 1846 and 1847. Police deposited on the desk of Prime Minister François Guizot a secret report signalling Libri's misappropriations, amounting to an estimated 500,000 francs. This report would have probably been forgotten had Guizot not been fired during the 1848 Revolution. The report was made public, and Libri fled to England with 30,000 books and many manuscripts; only six cases of books were intercepted by the customs. He sold the rest to rich Englishmen. In France, he was sentenced in absentia in 1850 to 10 years of prison, although he was defended by several notables, such as Prosper Mérimée (1803–1810). Libri died in England in 1869. From time to time, items stolen by this exceptional crook reappeared for sale and could be handed back to the libraries from which they were taken.

real scientific qualities, Libri was a detestable character and was often unfair, so that the two men quarrelled violently at the Academy. We can find traces of these fights in the *Comptes Rendus*.

Arago also had problems with a few astronomers. One of them, Gustave de Pontécoulant, was a mediocre researcher. He attacked Arago in his *Précis d'Astronomie*, and even wondered in a public letter "whether Arago [was] in possession of all his faculties." Arago, who never let an attack go unanswered, wrote a long open letter, first published in the *National,* then as a separate leaflet. Even worse was the attitude of Le Verrier, who attempted to take Arago's place at the Observatory, and almost succeeded thanks to his connections to Prime Minister Guizot; but the plot failed thanks to the 1848 Revolution. We will come back to this story in Chap. 7.

Fortunately Arago had many more friends than enemies. His friends were scientists like Humboldt, Malus, Fresnel, Gay-Lussac and Ampère, artists like the sculptor David d'Angers, writers like Victor Hugo, François-René de Chateaubriand and Jules Michelet and the close relatives of famous dead people like Lavoisier, Condorcet[21] and Laplace. Always generous, Arago supported those men whose value he acknowledged, even the detestable Le Verrier, without any jealousy or condescension. He was friendly with upper-class people and appreciated their praise, but even then he spoke his mind. He was acquainted with bankers, as we can see from his speeches in the Chamber of Deputies, but he was not interested in money and left only a small estate upon his death.

[20] See the biography of Libri by Mostert (1995).

[21] Arago and the daughter of Condorcet together published the works of the latter.

The 1848 Revolution[22]

A revolution was inevitable after Guizot's opposition to any reform, in particular to the right to vote. Meetings being forbidden, the opposition used banquets as substitutes (every attendant paying for his meal), which were generally authorized by the police. From 1847, about 70 banquets were organized in France, gathering some 22,000 subscribers. This displeased the king and Guizot, who forbade on 14 January 1848 a republican banquet in the 12th *arrondissement* (district) of Paris, then that of the Observatory. A part of the opposition decided to go ahead with the banquet, on 22 February. At this date, a violent confrontation with police began. Realizing the gravity of the situation, the king asked for the dismissal of Guizot, to replace him with Louis-Mathieu Molé (1781–1855), who was the right-hand man of the king. Molé was unpopular, but no matter: Paris celebrated Guizot's dismissal with demonstrations, which soon degenerated. Soldiers shot demonstrators, the fight extended to the whole city. King Louis Philippe found it safer to abdicate in favour of his grandson, the regency being insured by his mother, the duchess of Orléans. She showed up at the Chamber of Deputies, but the room was invaded by a mob calling for the republic. The royal family, as well as Guizot, were sent into exile in England. Republican politicians Emmanuel Arago, Lamartine and Ledru-Rollin proposed a provisional government, with François Arago's endorsement (Fig. 2.8). The republic was proclaimed on 24 February by Lamartine from the balcony of the town hall, then officially on 27 February by François Arago and two others from the base of the column of the *Place de la Bastille*.

The provisional government named several ministers. They were rather moderate, but they were soon confronted by the extremists. Arago was appointed minister of the Navy and Colonies, According to Lamartine, this was "his natural place… given his science, his knowledge of sophisticated weapons, and his fame as large as the globe over which his name was going to float" [!]. He was also named minister of war a few days later. We have seen that his son Emmanuel promoted himself as the Minister of Postal Office, which gave him the means to announce rapidly the creation of the Republic to the whole nation.

François Arago quickly promulgated some liberal and humanistic measures: On 2 March, he signed a decree limiting the working day to 10 h. On 12 March he issued a decree forbidding corporal punishment in the Navy, and on 28 March still another decree to facilitate the naturalization of foreigners. But above all, on 27 April he signed the decree abolishing slavery. Victor Schœlcher (1804–1893, Fig. 2.9) is generally credited with the abolition (which allowed his remains to be transferred to the Panthéon in 1948). But we will see that this credit was in fact shared with Arago.

---

[22] The history of the 1848 Revolution is the subject of a book by Daniel Stern, pseudonym of Marie d'Agoult (1805–1876), a mistress of Liszt and mother of Cosima Wagner: *Histoire de la revolution de 1848* (1850–1853), Paris, Sandré, 3 vol; the 1869 illustrated edition is accessible via http://gallica.bnf.fr

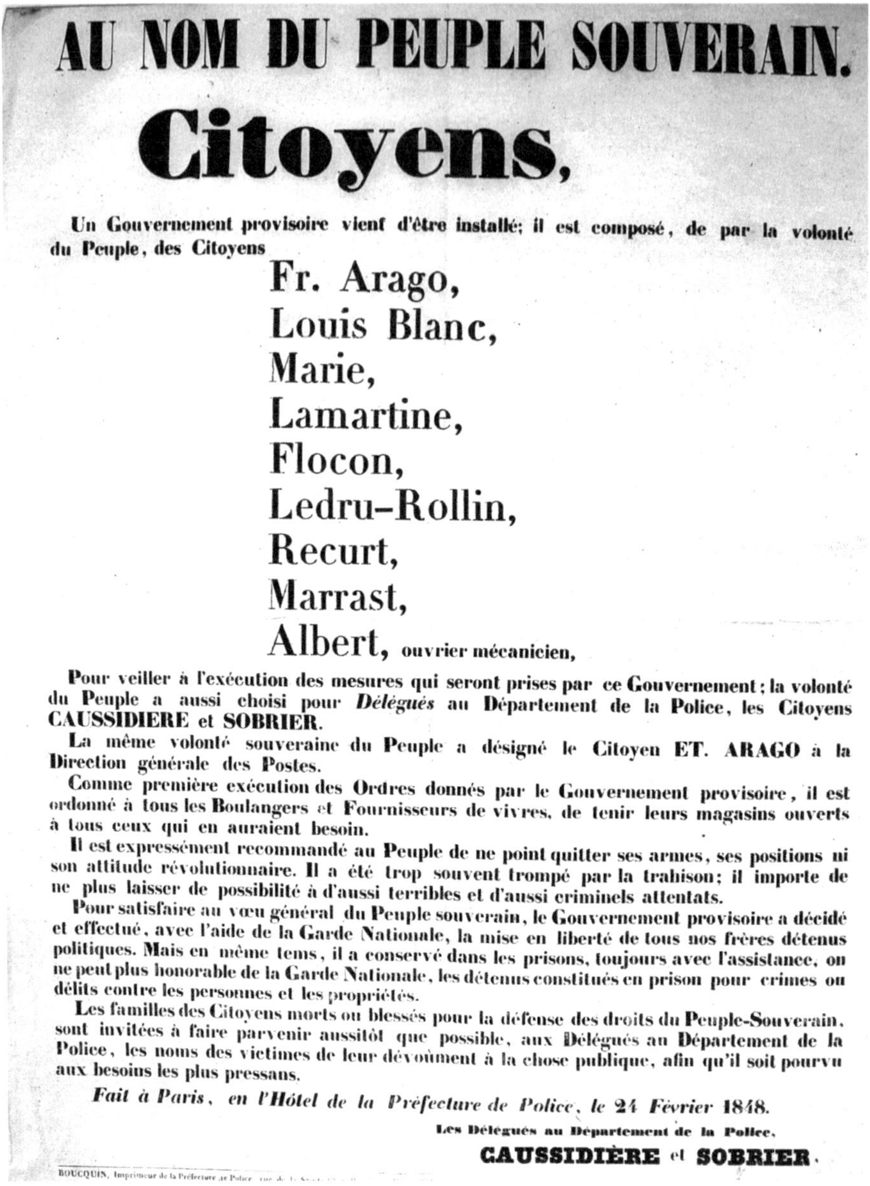

AU NOM DU PEUPLE SOUVERAIN.

Citoyens,

Un Gouvernement provisoire vient d'être installé; il est composé, de par la volonté du Peuple, des Citoyens

Fr. Arago,
Louis Blanc,
Marie,
Lamartine,
Flocon,
Ledru-Rollin,
Recurt,
Marrast,
Albert, ouvrier mécanicien,

Pour veiller à l'exécution des mesures qui seront prises par ce Gouvernement; la volonté du Peuple a aussi choisi pour *Délégués* au Département de la Police, les Citoyens CAUSSIDIERE et SOBRIER.

La même volonté souveraine du Peuple a désigné le Citoyen ET. ARAGO à la Direction générale des Postes.

Comme première exécution des Ordres donnés par le Gouvernement provisoire, il est ordonné à tous les Boulangers et Fournisseurs de vivres, de tenir leurs magasins ouverts à tous ceux qui en auraient besoin.

Il est expressément recommandé au Peuple de ne point quitter ses armes, ses positions ni son attitude révolutionnaire. Il a été trop souvent trompé par la trahison; il importe de ne plus laisser de possibilité à d'aussi terribles et d'aussi criminels attentats.

Pour satisfaire au vœu général du Peuple souverain, le Gouvernement provisoire a décidé et effectué, avec l'aide de la Garde Nationale, la mise en liberté de tous nos frères détenus politiques. Mais en même tems, il a conservé dans les prisons, toujours avec l'assistance, on ne peut plus honorable de la Garde Nationale, les détenus constitués en prison pour crimes ou délits contre les personnes et les propriétés.

Les familles des Citoyens morts ou blessés pour la défense des droits du Peuple-Souverain, sont invitées à faire parvenir aussitôt que possible, aux Délégués au Département de la Police, les noms des victimes de leur dévoûment à la chose publique, afin qu'il soit pourvu aux besoins les plus pressans.

Fait à Paris, en l'Hôtel de la Préfecture de Police, le 24 Février 1848.

Les Délégués au Département de la Police,

CAUSSIDIÈRE et SOBRIER.

BOUCQUIN, Imprimeur de la Préfecture de Police

Fig. 2.8 Poster proclaiming the provisional government on 24 February 1848. Private collection

Fig. 2.9 Victor Schœlcher (1804–1893) in 1885. Wikimedia Commons, Raoli

The Abolition of Slavery

Slavery was initially abolished in the French colonies by the Convention in 1794, as proposed by Condorcet and a few others. Condorcet had already written against slavery in 1776 in his eulogy of Pascal, and then at several other occasions. It is remarkable that several of his proposals can be found in the English laws of abolition (that of 1807 forbidding the commerce of slaves, and that of 1833 abolishing slavery itself). However, Napoléon re-established slavery in 1802 to please his wife Joséphine de Beauharnais (1763–1814), who belonged to a family of colonists in the Caribbean island of Martinique. After a few isolated attempts, the fight for abolition resumed in 1830, led by Schœlcher. He discussed this question with Arago, the workers' advocate Louis Blanc (1811–1882), and the artist David d'Angers, among others.

In February 1848, when Arago became a minister, Schœlcher was in Senegal. Ledru-Rollin and Arago were both opposed to slavery, but the colonists reacted angrily and the politicians hesitated for fear of bloody conflicts. Schœlcher was back in Paris on 3 March, and pushed Arago to act as quickly as possible. Arago was easily convinced, and immediately dictated to Schoelcher the following text:

The provisional government of the Republic, considering that no slave should exist on its territory, decides:

A commission is created to the provisional ministry of navy and colonies in order to prepare in the shortest possible delay the immediate freedom of slaves in all the colonies of the Republic.

The minister of navy will be in charge of the execution of the present decree.

Paris, 4 March 1848

Another decree designated the members of the commission, headed by Schœlcher. The decree of abolition they prepared was ready on 23 April and signed on 27 by Arago. This is a superb text in its concision[23]:

The provisional government, considering that slavery is outrageous to the human dignity; that, in destructing the free will of men, it suppresses the natural principle of right and duty; that it is in flagrant violation of the republican dogma "Liberty, Equality, Fraternity[24]"; considering that, if efficient measures do not follow closely the principle of abolition, deplorable disorders could occur in colonies, decrees: ... the abolition two months after the promulgation [of the present decree], ... amnesty, ... indemnification, ... interdiction of any trading or exploitation of slaves on pain of loss of French citizenship and punitive sanctions.

Surprisingly, abolition was easily accepted in spite of some incidents. It is true that the trade of blacks was abandoned in practice about 30 years before, and that colonists were ready to accept abolition. The 2-month delay was even shortened in Martinique, following a bloody revolt of slaves in the colony.

Thus, Schœlcher and Arago shared the credit of one of the most emblematic decisions of the ephemeral Second Republic: Schœlcher was the inspiration, and Arago used all his power for the realization. They should not be separated.

The End of a Dream

In spite of his permanent illness, Arago was efficient in accomplishing his ministerial duties during the short time they were to last. As Minister of the Navy between 24 February and 10 May 1848, he favoured the transition between sail and steam navigation that took place from 1830 to 1850, he forbade corporal punishment, took some measures to insure better food for the crews and introduced several other reforms that were to last. As Minister of War between 19 March and 17 May[25] he decided on limited purges in the army, accompanied by compensation, reinforced the *garde nationale* and ameliorated the transportation of troops by the purchase of 30,000 horses. His competence in military topics was acknowledged by contemporaries like Admiral Charles Baudin, who wrote to him on 25 November 1848 that his short tenure would have left only good memories if, "having put too much trust in M. Schoelcher, he had not allowed him to upset our colonies[26]".

[23] Cited by Sarda (2000) p. 286.

[24] Still the motto of the French Republic.

[25] He was then replaced by General Cavaignac.

[26] Cited by Sarda (2000) p. 287.

The greatest satisfaction for Arago during this period was to have seen his 1840 dream realized: the universal franchise. The elections for the new Constituent Assembly took place in this way on 23 April 1848. They saw the triumph of the Arago family: François was elected both in Pyrénées-Orientales and in Paris, in the latter case with almost as many votes as Lamartine but far ahead of the other members of the government. This time, he chose to represent Paris. Étienne and Emmanuel were also elected, as was Mathieu, who shared with Lamartine the department of Saône-et-Loire. The Arago dynasty was found "invasive," and not without reason, and the word *aragocratie* appeared. And this was not the end. Another honor, redoubtable this time, was waiting for Arago.

The Constituent Assembly now had to form a government for France. It chose to elect an executive commission of five members, who would nominate new ministers. The election of 10 May gave an unexpected result: one was expecting Lamartine to come in first, but this time it was Arago who had the most votes, 725, followed by Louis-Antoine Garnier-Pagès (715 votes), Pierre Marie (702 votes) and Lamartine (643 votes). The fifth member was Ledru-Rollin, the most left-wing of all, who obtained only 458 votes. Arago became the president of the Executive Commission that ruled France… for only 7 weeks. He held all the winning cards: his support for the Republic, his social preoccupations, the results of his actions as a minister and, of course, his prestige. But difficulties were soon to occur, due to dissention in the government and to the action of revolutionaries who had not been admitted in this government. Several of them organized an insurrection that some historians consider a provocation fomented by the conservatives. Elections on 4 and 5 June to fill 40 empty seats saw the victory of the extremists and the defeat of the moderate republicans. Even Louis-Napoléon Bonaparte, the future emperor Napoléon III, was elected. Lamartine wanted to have him sent into exile, but the Assembly accepted his election as valid. Bonaparte preferred to withdraw, biding his time.

Then, the Executive Commission made a fatal error: its decree of 21 June dissolving the Parisian *Ateliers Nationaux* (national factories). These factories were created some time before to reduce unemployment. The workers there were relatively well paid but had little to do. But the cost of *Ateliers Nationaux* was so high that they led the State to bankruptcy. The workmen of the suppressed ateliers were asked to join the army or to leave Paris. Their displeasure was the reason for another insurrection. Some historians maintain that Arago was responsible for it. Actually, the relevant testimonies are contradictory. What is certain is that Arago lived through this ordeal confused and powerless, while trying to calm the insurrectionists on the barricades. General Louis-Eugène Cavaignac (1802–1857) was charged by the Commission to restore order, which he did with brutality. Overwhelmed by the events, the Executive Commission collapsed and resigned (Fig. 2.10). Cavaignac took power provisionally. The Assembly approved a new constitution; the election of the president of the Republic took place on 10 December. Cavaignac was a candidate, but the winner was Louis-Napoléon Bonaparte. Étienne and Emmanuel Arago resigned from the Assembly. As for François, he retired at the Observatory,

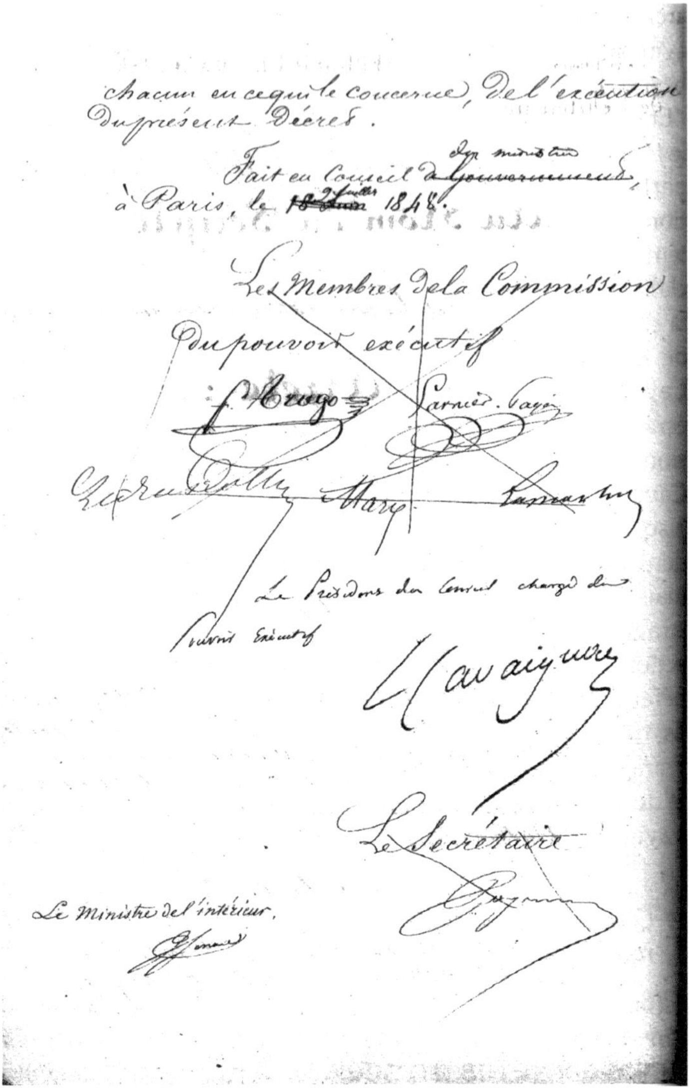
chacun en ce qui le concerne, de l'exécution du présent Décret.

Fait en Conseil de Gouvernement, à Paris, le [illegible] 1848.

Les Membres de la Commission du pouvoir exécutif

Le Président du Conseil chargé du Pouvoir Exécutif

Le Secrétaire

Le Ministre de l'intérieur,

Fig. 2.10 The end of the Executive Commission. Cavaignac signed a decree prepared by the members of this commission, whose signatures have been struck out. Private collection

exhausted by his failure and increasingly ill. He attended fewer and fewer meetings of the Assembly. Victor Hugo wrote about him[27]:

Arago does not show up at the Assembly.

When one has the two specialities of looking at the sky or staring at the ground, I understand that one prefers the first.

[27] Hugo (1848) p. 582.

The Last Years

Disappointed by politics and seeing the degradation of his health, Arago took refuge in science. He was however elected deputy for the last time in May 1849. Mathieu was eliminated in Saône-et-Loire, but the left wing resisted in the south of France. Emmanuel was also re-elected, as were several political friends. Arago would now only take part to some important votes. His only visit to his electoral district of Pyrénées-Orientales took place in 1853, in the terminal phase of his illness.

However, his scientific activity remained important. Resuming his functions of Perpetual Secretary, he was as assiduous as possible at the Academy meetings, and continued to make reports and to read biographic notices of deceased academicians in the solemn meeting of the end of year. He read the necrology of his former friend Poisson on 16 December 1850, but his sight was so bad on 20 December 1852 that his notice of Gay-Lussac had to be read by Ernest Laugier. As to that of Malus that he prepared for the meeting of December 1854, it had to be read after his death, by special decision of the Academy, by the new Perpetual Secretary, Léonce Élie de Beaumont (1798–1879).

Humboldt implored Arago to gather his scientific works and to realize with his own help a "general rescue" as a book that would have as the title: *Optical Astronomy, Physics of the Earth: fragments gathered by François Arago*. Arago made an attempt in this direction, but alone because Humboldt was in Berlin. He was now almost completely blind, and had to be helped by his niece Lucie Laugier and above all by young colleagues of the Observatory. Jean-Augustin Barral (1819–1884, Fig. 2.11), the future editor of Arago's *Œuvres complètes*, wrote[28]:

In 1849, M. Arago made me the honor to tell me of his intention to publish his works… Soon he started to dictate, first to members of his family, then to someone who could devote all his time to the work [he was initially Albert Terrien, then Jean-Jacques-Émile Goujon (1823–1856), one of the élèves-astronomes*]. A large part of the manuscripts that were used for the printing of the Œuvres of M. Arago are from the hand of M. Goujon who, during three years, wrote almost every day during about ten hours per day from the dictation of the learned astronomer.*

Barral himself did the bibliographic researches and wrote down from memory the texts that Arago improvised in front of him. The result is a large number of articles, some new and some already published in *Annuaire du Bureau des Longitudes*, but updated. They are published in *Œuvres complètes*. Arago also dictated some parts he had not yet written of what would become his best-known work: the *Astronomie populaire*, which gathers most of what he taught throughout his career. Its voluntarily modest title hides a real compendium of the astronomical knowledge of the time.

---

[28] *Arago, F. *Œuvres complètes*, t. Tables, p. VII–CCLXIII.

Fig. 2.11 Jean-Augustin Barral (1819–1884), editor of Arago's *Œuvres complètes*. Wikimedia Commons, Infofiltrage

Amongst the new texts, that of the first *Memoir on photometry*[29] begins with the following sad introduction:

My poor health and the deep alteration that my sight has met almost suddenly have led to the desire, or rather the duty to publish promptly some results that were lying in my drawers for a long time. I decided to begin with photometry.

Rules required that the Memoirs be read at the Academy. Six of the seven memoirs on photometry were thus read in succession from 18 March to 17 June 1850. It seems that only the first one could be read entirely by Arago, but he had to interrupt his reading for the three next ones, while the fifth and sixth memoirs probably had to be read by someone else. As to the seventh, it was only printed after his death in his *Œuvres complètes*. Although an important part of the material of these memoirs had actually been "lying in drawers," they also contained the results of recent experiments and observations that were realized at his request by colleagues at the Observatory. Jules-Célestin Jamin (1818–1886), Laugier, Frédéric Petit, Goujon and Mathieu were also in charge of carrying on various experiments started by Arago, or whose principle he described. The results were published by Barral in *Œuvres complètes*.

In 1853, found in Arago's papers was a memoir written in 1810 that he decided to publish in the *Comptes rendus* "without changing a single word," although the subject had evolved a lot in the meantime.[30] This text dealt with the velocity of light,

[29] *Arago, F. *Œuvres complètes*, t. 10, p. 168–183.

[30] **Comptes Rendus de l'Académie des sciences*, 36 (1853) p. 38–49. Also *Arago, F. *Œuvres complètes*, t. 7, p. 548–568.

a topic that preoccupied Arago all his adult life. He would have the satisfaction of seeing the experiment on the comparison of the velocity of light in air and in water realized successively by Foucault and by Fizeau in 1850. Arago also attended with admiration the pendulum experiment that Foucault presented for the first time to the public at the Observatory in February 1851.

Arago continued to exchange letters with scientists in Europe and even America, like Samuel Morse (1791–1872), and, of course, with Humboldt. The latter, always perspicacious, asked Arago some embarrassing questions in 1849, about an error that he had made concerning the velocity of light.[31]

The Arago family was spreading far and wide. His brother Étienne fled to exile in 1849, and his son Emmanuel married the following year in the south of France. The political situation deteriorated. Louis-Napoléon Bonaparte led a coup d'état on 2 December 1852 and became Emperor Napoléon III. Arago tried to provoke a reaction from the Academy, but without success. Now, the Emperor decided that all civil servants should take an oath to him on pain of dismissal. Arago was a civil servant, as the director of observations, but he refused to take the oath and presented his resignation letter, as he did years earlier with the previous emperor, Napoléon I. His letter[32] was noble, but also very clever: he warned Napoléon III that "he would have to inform from his decision the foreign academies and eminent correspondents." One reads also in this letter:

It is not without a deep affliction that I would abandon those beautiful instruments to which I have contributed in various degrees; it is not without fear that I would see some means of researches that I created fall into malevolent or enemy hands. But my conscience spoke, and I had to go on.

These "malevolent or enemy hands" were obviously those of Le Verrier. Victor Hugo rejoiced at Arago's attitude and wrote in 1863 from his exile in Guernsey[33]:

M. Bonaparte wanted that Arago would swear. You have to know that astronomy must take an oath… A free astronomy is almost as dangerous as a free press. Do you know what could happen in the nocturnal meetings between Arago and Jupiter? If it was Leverrier, no problem, but with a member of the provisional government… It is necessary that the Board of Longitudes swears not to conspire with celestial bodies, in particular with those mad ones that make celestial coups d'état, the comets… And also, … one becomes fatalist when one is called Bonaparte. The great Napoléon had a star, the small one[34] should probably have a nebula; astronomers are certainly somewhat astrologers. Take an oath, Messieurs. Of course Arago refused… One may laugh, but this is a heart-rending thing… Take an oath or starve. They refuse. The silence and the darkness into which they stoically proceed know the rest.

[31] *Hamy (1907), lettre CVII p. 294–306 and CVIII, p. 306–308.

[32] Cited from Daumas (1987) p. 275 and Sarda (2000) p. 320.

[33] *Hugo (1863a) p. 206–208.

[34] Hugo generally called Napoléon III, that he hated, « Napoléon le petit », by contrast with Napoléon I.

However, in the case of Arago, Napoléon III judged that it would be politically wise to dispense with the oath for such a famous man "who has illustrated France by his works and whose life the government does not want to sadden." Le Verrier would have to wait for a few more months! Emmanuel was very pleased with the attitude of his father and with the happy conclusion of the affair. From his exile in Brussels, Étienne exulted when learning the news via the electric telegraph, and sent to his brother congratulations in trashy verse.

Arago's health deteriorated. A sojourn in Vichy in 1851 did nothing. The doctors pushed Lucie to go with him to take the waters in Amélie-les-Bains, near Perpignan in his native Catalonia. He went for the last time to the Academy on 22 August 1853, then travelled to Perpignan with Lucie. The trip was very arduous because the train went only as far as Dijon, and because Arago suffered from various diseases, in particular from colon disorder. The new sojourn brought no improvement, as could be expected in the case of diabetes; but Arago could see for the last time his beloved native region.

He began to realize how bad his condition was (Fig. 2.12). He despaired that he was unable to carry on his functions as Perpetual Secretary, and proposed resigning. Biot, reconciled with him and feeling that the end was near, came to see him and tried to dissuade him. Arago died on 2 October 1853 at the Observatory. Humboldt, who could not come from Berlin to see Arago, who he knew to be dying, wrote to Étienne on 6 October[35]:

Many thanks for all the details you give me: this scrupulousness, this oversight of himself until the last moment of his life, these words to Biot, and this certainty that he would not be conscious of his death and that his end was without suffering, this does not console me, but this individualizes, this relieves because it makes me almost believe that I was attending myself the death bed... I grieve and I fear to see M. and Mrs Neptune [Le Verrier] invade the place where we have lived and to upset the interior peace.

Victor Hugo also wrote to Étienne, but his style is quite different[36]:

The death of Arago is a decrease of light. He illuminates from his grave. Seeing that all was dead in the nation, he said: "Let me die." He went to bed in the shroud next to France, and France quivered. Your mourning is that of the Republic.

[35] *Hamy (1907), lettre CXX p. 341–343.

[36] Cited by Sarda (2000) p. 324.

Fig. 2.12 Arago 2 weeks before his death and in very poor condition. Curiously, this photograph seems to be the only one in existence of this pioneer of daguerreotype. Bibliothèque de l'Observatoire de Paris

Chapter 3
The Nature of Light

The Precursors

In Arago's time, there were two competing ideas about the nature of light. Both were present, although in a confusing way, in the *Dioptrique* of René Descartes (1596–1650): Either light was made of luminous, massive corpuscles,[1] an idea adopted and developed by Isaac Newton (1642–1727, Fig. 3.1), or it was the propagation of the vibration of an ether supposed to fill the whole space. The latter idea was stated more precisely by Father Francesco Maria Grimaldi (1618–1663) and fully developed by Christiaan Huygens (1629–1695) in his *Traité de la lumière* (Treatise of light) published in French in 1690.[2]

Newton's results are scattered in his *Philosophiae naturalis principia mathematica* (Mathematical principles of natural philosophy) of 1687 and in his *Optice sive de reflexionibus, inflexionibus et coloribus* (Optics, or of reflection, refraction and colors) of 1704. There are many editions and translations of these works, which are often difficult to read and somewhat obscure. The experimental part of Newton's optics is admirable, superior to the theoretical one. Newton discovered the dispersion of light by a prism into its different colors, described the colors of thin plates and films, and the rings that bear his name (Fig. 3.2), and reproduced the experiments of Grimaldi, who discovered diffraction; but the explanations he gave of these phenomena are confused. His basic idea was that light was made of particles launched by the luminous body with a high velocity, and whose mass depended on the color. He explained refraction as due to an acceleration of those particles when they penetrate into a denser material (Fig. 3.3). In this theory, often called the "emission theory," the velocity of light was thus greater in a "denser" (actually:

[1] These corpuscles have nothing to do with our modern photons, which have no mass.

[2] *Huygens (1690).

J. Lequeux, *François Arago*, Astrophysics and Space Science Library 421,
DOI 10.1007/978-3-319-20723-0_3

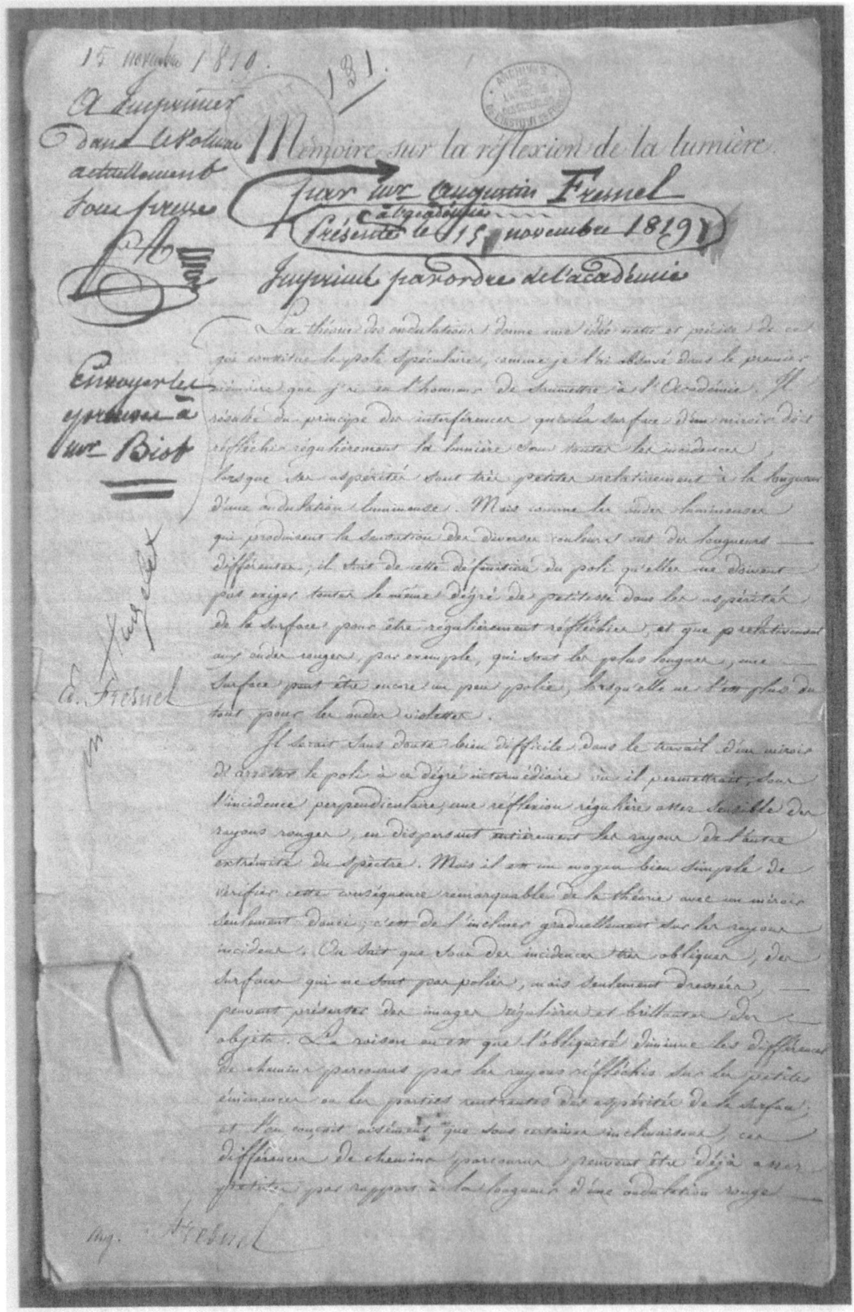

Autographed Memoir of Fresnel with the approbation of Arago, for publication by Academy of sciences. Archives de l'Académie des sciences

Fig. 3.1 Isaac Newton, portrait by Sir Godfrey Kneller, 1689. Wikimedia Commons Soerfm

more refractive) medium, contrary to what we know now. The ratio of the sine of the angle of incident ray with that of the refracted ray was equal to the *inverse* of the ratio of the velocities of the corpuscles before and after they have crossed the surface of separation. For Newton, the dispersion of light was supposed to be due to the different masses of the particles: Those particles with a higher mass would be more attracted by the denser medium and thus more refracted. If this were true, the dispersion would be proportional to the refraction index, but we know now that this is wrong. The partial reflection on a transparent surface was not really explained in this theory.

Given the triumph of his theory of universal gravitation, few people dared to criticize Newton's ideas on the nature of light. Among them was Leonhard Euler (1707–1783), who produced a *Nova theoria lucis and colorum* (New theory of light and colors)[3] in 1746. This was a slight improvement of the wave theory of Huygens. Euler wrote[4]:

Newton was without doubt one of the greatest geniuses who ever existed; and his profound science and his penetration inside the most hidden mysteries of nature will always be the most brilliant subject of our admiration and of that of our posterity; but the aberrations of this great man should serve to our humiliation, and to

[3] *Euler (1746).

[4] *Euler (1761) Letters XVIII and XXXI.

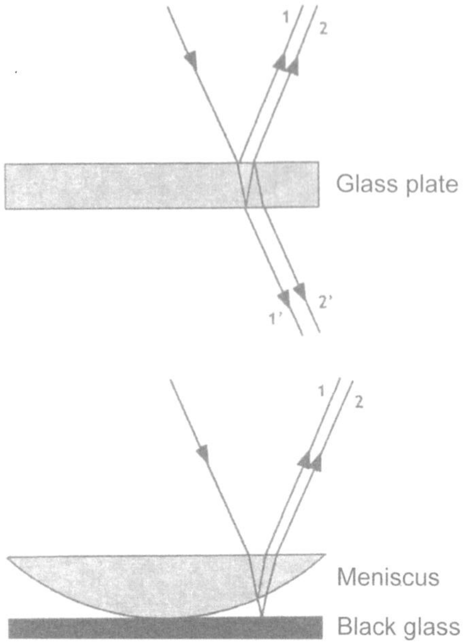

Fig. 3.2 Colors of a thin transparent plate (*top*) and Newton's rings (*bottom*). (*Top*) The light that falls on a thin plate is partially reflected by the upper surface as *1*, and also by the lower surface as *2*. Interferences between *1* and *2* produce a color, because the phase difference between *1* and *2* depends on wavelength. There are also colors by transmission due to interferences between *1′* and *2′*. (*Bottom*) A meniscus lying on a glass plate (preferably *black*) produces interferences between the reflected rays *1* and *2*. The places for which the phase difference between *1* and *2* is the same are concentric rings. In monochromatic light, the rings for which this difference is an entire multiple of the wavelength are bright; they are separated by dark rings. If the light is not monochromatic, the rings are colored and disappear at large radii. The central spot is *black* because of a phase change of π at the reflection from air on glass. If the glass plate is transparent, it is possible to observe rings in transmission

acknowledge the weakness of the human spirit that, after rising to the highest possible capabilities, risks nevertheless to rush into the worse errors.

This was going too far for the contemporaries, and Euler did not have much success. Even more courage was necessary for Thomas Young (1773–1829, Fig. 3.4, Box 3.1) to criticize Newton's ideas, because he was living in England. Desirous of convincing an English readership that was fully Newtonian, he remarked that Newton himself acknowledged having met difficulties in explaining some phenomena with his theory, and that he was less certain about it than one would have imagined[5]:

The optical observations of NEWTON are yet unrivalled; and, excepting some casual inaccuracies, they only rise in our estimation, as we compare them with later attempts to improve them. A further consideration of the colours of thin plates, as

[5] Young, T. (1802) *Philosophical Transactions of the Royal Society of London* 92, p. 12–48, accessible as http://www.jstor.org/stable/107113

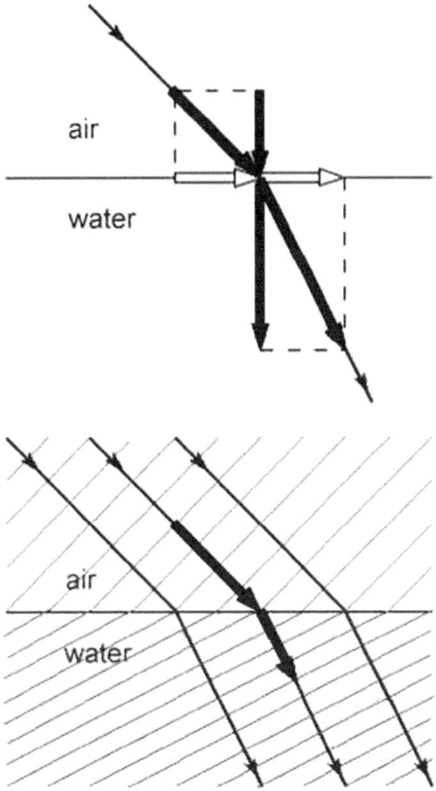

Fig. 3.3 Refraction. (*Top*) In order to explain refraction, the "emission theory" of Newton assumed that when the light particles penetrated a denser transparent body like water, the component of their velocity perpendicular to the surface of separation increased, while the parallel component was unchanged. (*Bottom*) In the modern wave theory, one assumes the continuity of the wave planes, which are perpendicular to the direction of propagation, when one goes from one medium to the other. Because the velocity of light is lower in a medium like water that has a larger refraction index than air, the wave planes are closer to each other, producing the refraction. Note that Huygens produced in 1690 a drawing very similar to the present one

they are described in the second book of NEWTON's optics, has converted that prepossession which I before entertained for the undulatory system of light, into a very strong conviction of its truth and sufficiency. ...

A more extensive examination of NEWTON's various writings has shown me that he was in reality the first that suggested such a theory as I shall endeavor to maintain; that his own opinions varied less from this theory than is now almost universally supposed; and that a variety of arguments have been advanced, as if to confute him, which may be found nearly in a similar form in his own works; and this by no less a mathematician than LEONARD EULER, whose system of light, as far as it is

Fig. 3.4 Thomas Young (1773–1829), engraving by Henry Adlard, after Sir Thomas Lawrence (1769–1830). Wikimedia Commons, Materialscientist

Box 3.1 Thomas Young[6]

Born in 1773 in Milverton in England, Young was an infant prodigy: It has been said that he could read at age 2 and spoke ten languages at 16. At 19, he started medical studies. His work on vision was so remarkable that he entered the Royal Society 2 years later. He became its Foreign Secretary in 1802, a position he held until his death in 1829. In 1800, he was named Professor of Natural Philosophy (physics) at the Royal Institution of London, where he gave lectures for a large audience; but it is likely that only a few people could understand him. Young was a universal scientist, whose contributions to medicine, Egyptology and philology are impressive. In physics, he was the first to call *energy* the quantity mv^2, where m and v are respectively the mass and the velocity of a body, and called *work done* the quantity Fd, where F is a force and d the displacement of its application point. He is famous for his discovery of light interferences and his other works on light. Unfortunately these works did not impress his contemporaries who were all Newtonians, and raised violent criticisms, in particular from Lord Brougham.

[6] For an analysis of the works of Young in optics, see Mollon, J.D. (2002) *Philosophical transactions* A, 360, p. 807–819, accessible via http://www.jstor.org/stable/3066507

worthy of notice, either was, or might have been, wholly borrowed from NEWTON, HOOKE, HUYGENS, and MALEBRANCHE.

This is going too far in the other direction, and the reference to Euler is rather inaccurate: He did not borrow anything from Newton!

In the same article, Young enounces the following propositions, and curiously cites Newton for the first three:

HYPOTHESIS I. A luminiferous Ether pervades the Universe, rare and elastic in a high degree. [Actually Newton did not need the ether in his theory, his light particles being bullets that could move in a vacuum, but the idea of ether was so deep-rooted in the minds that he assumed its existence.]
HYPOTHESIS II. Undulations are excited in this Ether whenever a Body becomes luminous.
HYPOTHESIS III. The Sensation of different Colours depends on the different frequencies of Vibrations, excited by Light in the Retina.
HYPOTHESIS IV. All material Bodies have an Attraction for the ethereal Medium, by means of which it is accumulated within their Substance, and for a small Distance around them, in a State of greater Density, but not of greater Elasticity.

The last hypothesis is quite obscure, but fortunately was not crucial in the demonstrations.

The works of Young were a great improvement over those of his predecessors. He was the first to explain correctly not only the colors of thin plates and the Newton rings, but also the diffraction by the edge of a screen, which was, according to him, "the most difficult part of the undulatory system." He understood completely how the combination of waves of the same period produce interferences,[7] and even produced a mechanical model for demonstration[8] (Fig. 3.5). Of course, Young's best-known interference experiment is his famous "Young's double hole," or "Young's double slit" (Fig. 3.6).

However, something is missing in Young's work: the polarization of light, which was only discovered in 1808.

The Polarization of Light

What Is Polarization?

We know today that light is a transverse electromagnetic wave (Fig. 3.7). Transverse, because the wave vibration is perpendicular to the direction of propagation, contrary to what occurs for sound waves, which are longitudinal compression

---

[7] This was suspected by Newton who heard about the combination of tide waves in the region of Haiphong in Vietnam, but not extended by him to sound or light waves.

[8] Young (1807).

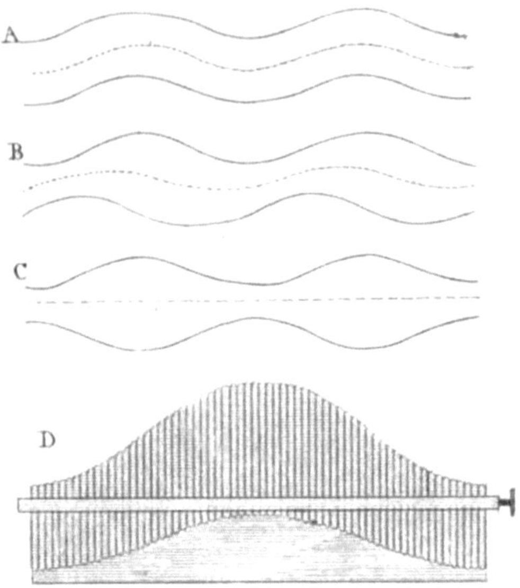

Fig. 3.5 A mechanical model used by Young for his demonstrations of interference. In this model (D), vertical rods whose lengths follow a sinusoid slid on another sinusoid of same period and amplitude. The top of the rods gave the sum. Here, the amplitude was at its maximum. Examples are given in A, B and C

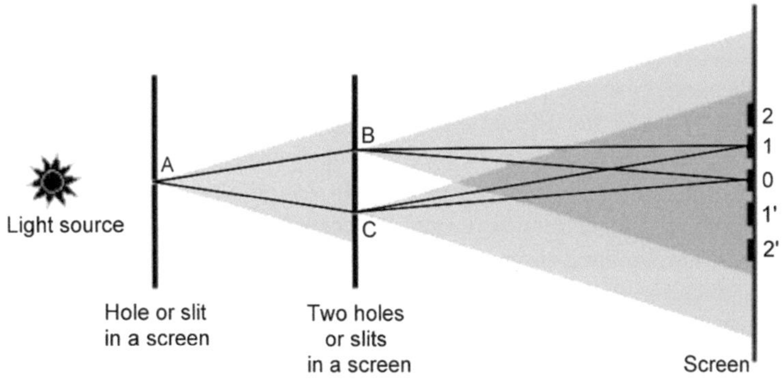

Fig. 3.6 The Young's double-hole experiment. A light source illuminates a small hole (or a narrow slit) A, which by diffraction illuminates two other holes (or slits) B and C. The light beams from these holes interfere in their common part, producing fringes on the screen. These fringes are bright when the paths ABD and ACD are equal (order 0) or differ by an entire multiple of the wavelength (order *1*, *2*, etc.). They are *dark* in between. In *white light* the fringes of order *1* and more are colored and vanish for high orders

waves. Electromagnetic, because light is formed of an electric field and a magnetic field perpendicular to each other and varying together with a frequency that characterizes the color.

In natural light, both fields have a randomly varying orientation around the direction of propagation (Fig. 3.7c). The instruments named *polarizers* transmit only the component of the electric field along some orientation (Fig. 3.7d). If one looks at a light polarized in this way through another polarizer (named in this case *analyzer*),

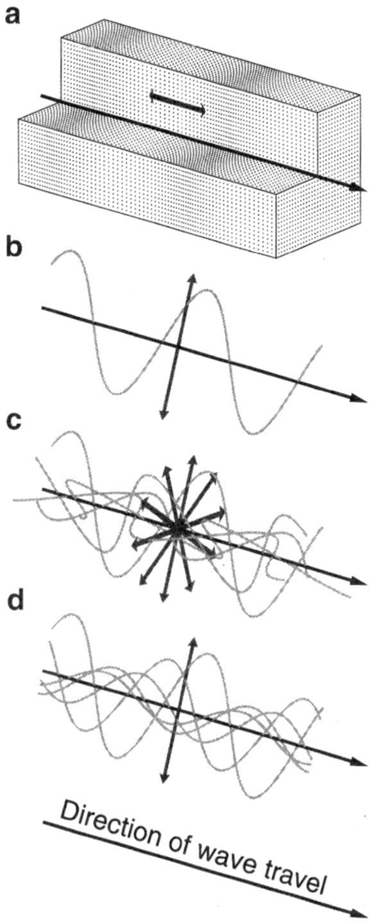

Fig. 3.7 Longitudinal and transverse waves. (**a**) A longitudinal wave is formed of a series of compressions and rarefactions of a medium that progresses in the direction of the wave. (**b**) In a transverse wave, the vibrations are perpendicular to the direction of propagation. (**c**) A beam of natural light (not polarized) is made of vibrations of various amplitudes and random orientations (five such vibrations are represented). (**d**) A beam of linearly polarized light obtained by filtering the preceding natural light with a polarizer; it contains the components of all the vibrations projected on the direction of polarization. From Tobin (2003), with thanks

the light is more or less attenuated according to the angle between the privileged planes of the polarizer and the analyzer, and one sees nothing if these directions are perpendicular to each other (Fig. 3.8).

Some natural crystals are polarizers. The best known is Iceland spar, a natural, transparent crystal of calcium carbonate (Fig. 3.9), whose strange properties have been described and used by many physicists of the seventeenth and eighteenth centuries. The first one was Erasmus Bartholin (1625–1698), professor of mathematics and medicine at the University of Copenhagen, who received from Danish merchants a sample of this material.[9] He observed the double refraction by this crystal, which produces two images of an object seen through it. He noticed that, when he rotated the crystal around an axis parallel to the direction of arrival of light, one of these images did not move while the other did, and "sheweth an *Unusual* kind of refraction hitherto un-observed by *Dioptricians*."

[9] Bartholin (1669).

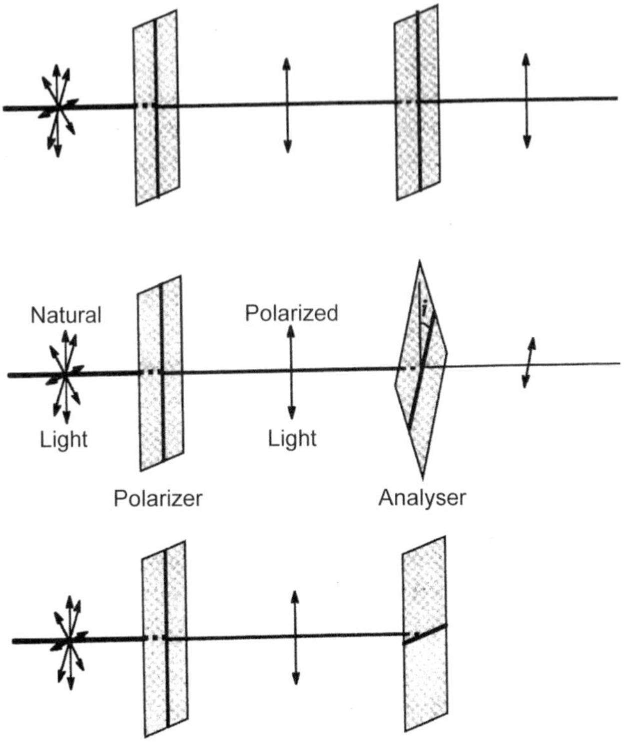

Fig. 3.8 Polarizer and analyzer of light. A polarizer filters from natural light the components of the wave electric field in the plane of polarization: the wave is linearly polarized (see Fig. 3.7d). If this polarized beam crosses another polarizer, called the analyzer, only its component in the plane of polarization of the latter is transmitted. If the two planes of polarization are the same there is no attenuation (*top*). If they are perpendicular, no light gets through (*bottom*). In the intermediate case, the light is attenuated by a factor $\cos^2_i$, i being the angle between the two planes of polarization (*middle*)

Iceland spar is a polarizer: It splits a beam of natural light into two beams polarized in directions orthogonal to each other, as indicated on Fig. 3.10. Huygens noticed in 1678 that when one observes a source through two successive spar crystals, one sees four images, which cancel two by two when rotating one of the crystals. He was not able to understand that the first crystal acts as a double polarizer and the second one as a double analyzer. His observation remained unexplained for a long time.

While the physics of crystals of spar or quartz were not understood, they were used for various purposes. For example, in 1777 the astronomer and physicist Alexis Marie de Rochon (1741–1817) built a "double prism," which divided a light beam into two rays with an angle defined by construction (Fig. 3.11).[10] Rochon used this

[10] Many details on the trials and realizations of Rochon can be found in **Journal de physique, de chimie et d'histoire naturelle* (1812) 74, p. 321–340.

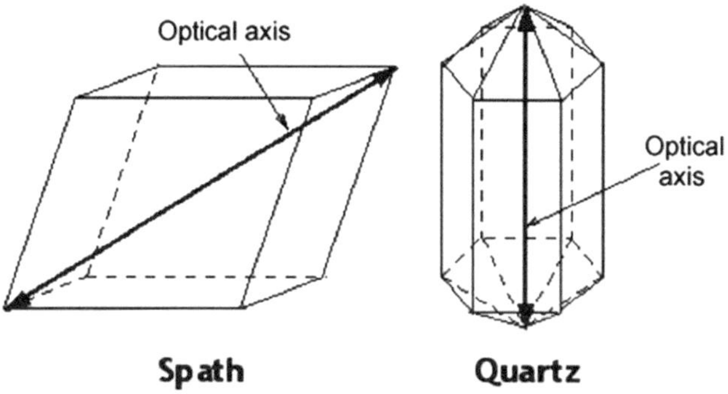

Fig. 3.9 Crystals of spar (*left*) and quartz (*right*). The very interesting optical properties of these two crystals made them very useful until the middle of the nineteenth century. They are somewhat different, but in both cases there is symmetry around the optical axis indicated in the drawings

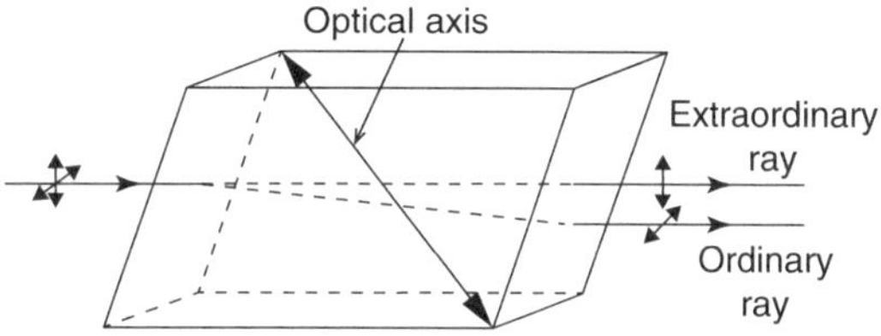

Fig. 3.10 Iceland spar, as all birefringent crystals, splits a beam of natural light into two separate rays polarized perpendicularly to each other

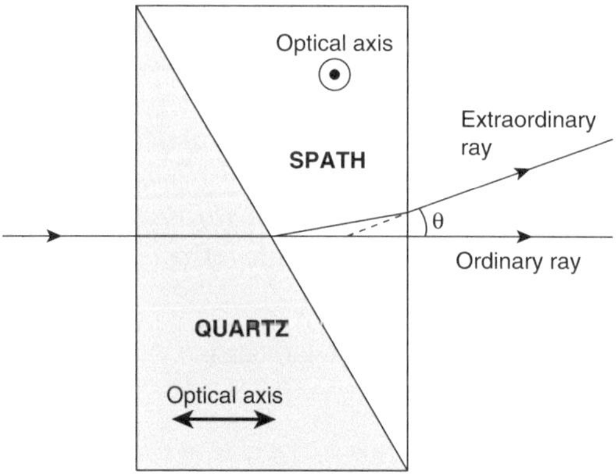

Fig. 3.11 Rochon's double prism. It was made of a quartz prism cut perpendicularly to its optical axis, which does not produce double refraction, and of a similar quartz prism, but cut parallel to its optical axis, which gives double refraction. The two rays produced in this way were polarized perpendicularly to each other and separated by an angle θ that could be chosen between certain limits by setting the angle of the prisms. For an angle of 30°, θ was equal to 20 arc min

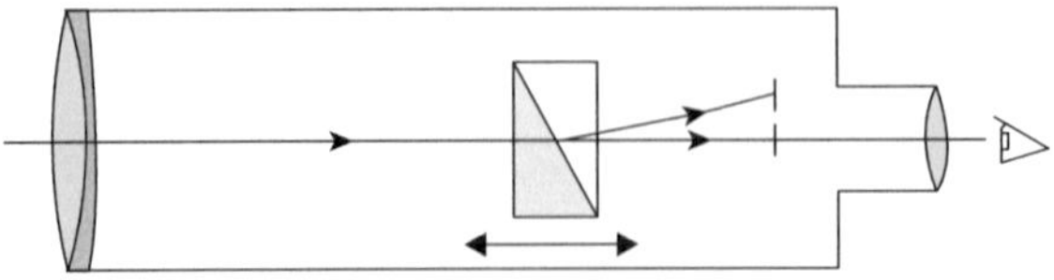

Fig. 3.12 Rochon's telescope. This was an ordinary refractor inside which a Rochon's double prism (see Fig. 3.11) was inserted. It gave a double image of an objet. By displacing the double prism along the axis of the telescope one could change the angular distance between the two images, which remained in focus. For measuring the diameter of a planet, one set the prism so that the upper edge of the lower image coincided with the lower edge of the upper one. After calibration, the diameter could be read on a scale

double prism inside a telescope to measure angular distances between stars or the angular diameter of the Sun, the Moon or planets (Fig. 3.12).

Malus and the Discovery of Polarization

In 1808, Étienne-Louis Malus (1775–1812, Fig. 3.13, Box 3.2) discovered polarization by chance.[11] Arago tells this discovery as follows[12]:

Some day, from his house of Rue d'Enfer, Malus looked through a crystal doted of the double refraction the sun reflected by the glass of the windows of the Luxembourg [palace]. Instead of the two intense images he was expecting to see, there was only a single one, the ordinary or the extraordinary one according to the position of the crystal in front of his eye[13]*... During the night, he looked at the reflection of the light falling from a candle under the angle of 36°, and noticed, using a birefringent crystal, that the reflected light was polarized as if it was coming from an Iceland's crystal. A similar experiment with a glass mirror under an angle of 35° gave the same result. It was thus proved that the double refraction was not the only means of polarizing light.*

It is Malus who coined the term of polarization.

[11] The works of Malus on polarization can be found in *Mémoires de la Société d'Arcueil* (without date) t. II p. 145–158, accessible via http://commons.wikimedia.org/wiki/Category:Mémoires_de_Physique_et_de_Chimie,_de_la_Société_d'Arcueil?uselang=fr and in *Mémoires de l'Académie des sciences* pour 1810, p. 105 and 112.

[12] *Arago, F. *Œuvres complètes* t. 3, p. 141–142.

[13] This is inexact and was rectified elsewhere by Arago: the polarization was only partial, and the images did not disappear completely.

Fig. 3.13 Étienne-Louis Malus (1775–1812), engraving by Ambroise Tardieu. Wikimedia Commons Gdr

Box 3.2 Étienne-Louis Malus
Malus was born in 1775 in a wealthy family. He studied at home, then at the Royal School of Military Engineering in Mézières. He attracted the attention of his teacher, who was none other than Monge. He started a military career, interrupted by studies at the École Polytechnique. He was a member of the scientific expedition to Egypt, and was probably one of the "12 mathematician members" of the temporary Cairo Institute. His work on light, begun before 1803, was recorded in various publications. Malus was elected to the *Première classe de l'Institut.* in 1810 and, in spite of the war between Great Britain and France, the Royal Society of London gave him the Rumford medal. He died from tuberculosis in 1812.

Arago's Contribution

Arago was very interested by Malus's discovery, and they soon became friends. In 1810, he backed him for his successful election to the Academy of Sciences. Unfortunately, Malus died 2 years later. Arago and Jean-Baptiste Biot then became the specialists of polarization, without rivals, except for David Brewster in Scotland. Arago was asked to write an article on polarization in the *Encyclopaedia Britannica*: It was written in French in 1824 and reproduced as such in the *Œuvres complètes*,[14]

[14] *Arago, F. *Œuvres complètes* t. 7, p. 291–454.

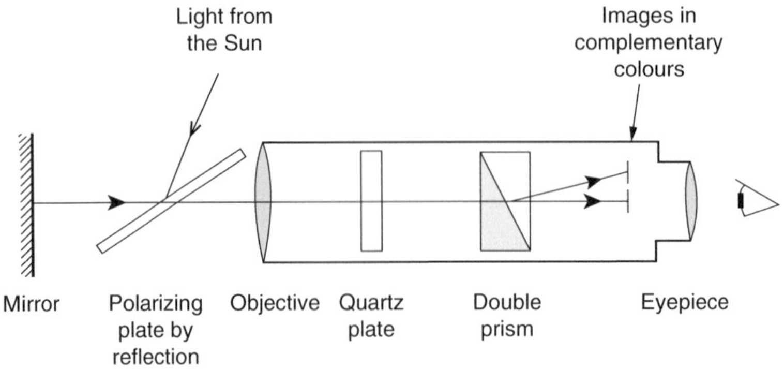

Fig. 3.14 Principle of Arago's experiment to observe rotational chromatic polarization. Explanations in the text

and translated by Young himself. This paper is very clear and shows that most of our knowledge of polarization was already obtained. It acknowledges the dominant contributions of Malus and Augustin Fresnel, but Arago's were far from negligible.

His major discovery, which dates from 1811, is that of chromatic and rotational polarizations. Chromatic polarization can be observed by placing between a polarizer and a crossed analyzer a plate of birefringent material, for example a foil of mica. While light would be completely extinguished without the plate, it is now more or less vividly colored.[15] More interesting is what occurs when the plate is a slab of quartz cut perpendicularly to the optical axis. Then one speaks of rotational chromatic polarization. Figure 3.14 shows how Arago observed it, and its principle is explained in Box 3.3.

The genesis of this experiment is detailed in Arago's paper in the *Encyclopedia Britannica*. The light, probably that of the Sun sent by a heliostat, was reflected by a glass plate under the Brewster angle and was thus completely polarized. After reflection on a mirror the light entered a telescope that produced an image of the Sun observed through an eyepiece. The quartz plate was placed inside the telescope, followed by a Rochon double prism (see Fig. 3.11). This prism produced a double image. The quartz plate possesses the property, discovered on this occasion, of rotating the plane of polarization of the light that crosses it: this is the rotational polarization. If one observed in monochromatic light by inserting a colored filter on the optical path, one saw two images of the Sun with different intensities depending on the rotation angle. For example, if this angle were 90°, one of the images was completely extinguished. But Arago observed in white light. In this case, the images

[15] The physicist Jacques Babinet (1794–1872) wrote in **Comptes Rendus de l'Académie des sciences*, 11, p. 618–620: "It is when observing the colors of mica foils lit by the light of a clear sky that M. Arago discovered chromatic polarization." The light of the blue sky is indeed polarized, but an analyzer must be placed between the foil and the eye to see the phenomenon.

Box 3.3 Rotational Chromatic Polarization

A quartz plate cut perpendicular to the optical axis is not birefringent, but has the property of rotating the plane of polarization of the light that crosses it. The rotation angle α depends on the thickness e of the plate and on the wavelength λ of the light. It is proportional to e/λ^2 (Biot's law). In Fig. 3.15, the rotations of the red, yellow and violet components of the incident light are represented schematically for a plate with of 3.75 mm thickness. If the incident light is polarized, these components are also polarized. An analyzer placed behind the plate does not attenuate the color for which the direction of polarization is parallel to that of the analyzer, but extinguishes more or less the other wavelengths, so that the transmitted light is colored. In the position "analyser 1" the transmitted light is orange-red, and in the position "analyser 2" it is blue-green. These colors are complementary if the two positions of the analyzer are perpendicular to each other. This is the case for a Rochon double prism as in Fig. 3.11, which produces two distinct images with complementary colors. One can change these colors by rotating the crystal.

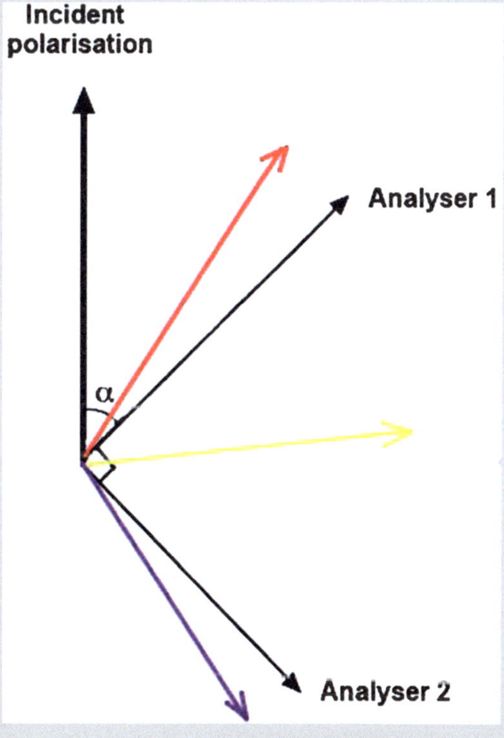

Fig. 3.15 Explanation of the rotational chromatic polarization, see text

Fig. 3.16 Arago's polariscope-polarimeter. The polariscope consists of a Rochon telescope (see Fig. 3.12) into which a plate of quartz has been inserted. If the light of the object is polarized, the two images are colored. By placing in front an inclined stack of glass plates in order to compensate the polarization of the source, as on the figure, it becomes a polarimeter giving the fraction of polarized light of this source. Bibliothèque de l'Observatoire de Paris

are colored with complementary colors, as explained in Box 3.3. If the incident light was not polarized, they were both white. If the polarization was only partial, they showed colors but less vivid because they were mixed with white light.

Arago made using this principle a very sensitive instrument to detect the polarization of light, provided it covered a large spectrum of wavelengths. He placed in front of a Rochon telescope (Fig. 3.12), or inside, a quartz plate cut perpendicular to its optical axis and called this set up a *polariscope*, or *polarimeter.* If one points this instrument to a luminous object, diaphragmed if too extended so that the two images do not overlap, one immediately sees from the presence of colors on the images whether its light is polarized. An elaborated version of this instrument (Fig. 3.16) allowed it to measure the degree of polarization of a light that was only partially polarized. For this, Arago placed in front of his apparatus a stack of glass plates separated by thin chocks. This stack polarizes more or less the light according to its inclination, as shown and studied by Fresnel. It allows equalizing the intensities of the polarized components parallel and perpendicular to it. The two images given by the instrument are then white for some inclination of the stack, whose measurement allows calculating the degree of polarization of the incident light. Until the end of his life, Arago improved his polarizer, which was commercialized.

With this instrument, which was both sensitive and easy to use, Arago made important discoveries that will be described in the following chapters. Various other instruments derived from his polarimeter: Biot's *colorigrade* allowed a quantitative measurement of colors, and the *cyanometer* of Biot and Arago was used to measure the blue color of the sky.

Polarization was very popular during all of the nineteenth century. The physics "cabinets" contained many relevant instruments. Biot discovered the rotational polarization produced by some liquids (initially turpentine).[16] This allowed the building of *saccharimeters* to titrate solutions of sugar. The rotational chromatic polarization is still in use in mineralogy to identify various crystals.

Arago and Fresnel

Fresnel's Come Out

Thomas Young's ideas were criticized by his contemporaries, but anyhow they were launched. However, it was a young engineer, Augustin Fresnel (Fig. 3.17, Box 3.4), who got the credit for re-discovering and improving them, and for formulating the first complete and satisfying theory of light. We will see that Arago played a prominent role in this story.

Fig. 3.17 Augustin Fresnel, engraving by Ambroise Tardieu, 1825. Wikimedia Commons, Tholme

[16] For a history of this discovery see Biot in **Comptes rendus de l'Académie des sciences* (1842) 15, p. 962–975.

Box 3.4 Augustin Fresnel[17]

Fresnel was born in Broglie, in Normandy, from relatively wealthy parents. His uncle was Léonor Mérimée, a good painter, whose son was Prosper Mérimée, a well known writer and archaeologist who directed the restoration of many monuments in France. Fresnel was not fit for classical learning, but soon showed excellence in observations, experiments and mathematics. He entered the École Polytechnique in 1804 at age 16½. His older brother Louis was already a student in the school, and his younger brother Léonor, the future editor of his *Œuvres complètes*, was admitted a few years later. The teaching of mathematics was excellent, but that of physics rather bad, so that Fresnel left the school with only rudimentary notions in this subject. In 1806 he entered the École nationale de ponts et chausées (National School of Bridges and Roads), and after he got his diploma was sent to build roads in several regions of France. As a hobby, he worked by himself on various problems of chemistry and physics, and in 1814 began to be interested in light. His research was of such quality that he was elected to the Academy of Sciences in 1823, and in 1825 to the Royal Society of London as an associate member. He was attached to the *Service des phares* (Lighthouse Service) in 1819, and improved their utility enormously thanks to his lenses. Unfortunately his health deteriorated from 1823, and he died of tuberculosis in 1827, when his friend Arago was bringing to him the Rumford medal of the Royal Society, that he had accepted on his behalf.

While studying the theories on the nature of light, Fresnel became convinced that light could not be made of material corpuscles as proposed by Newton, but that one had to go back to the waves of Huygens. His limited knowledge of physics and the fact that he was a complete stranger to the scientific world, which was entirely Newtonian, were great advantages, as he had no preconceived ideas. Only in 1814, did he hear about polarization, which had been discovered 6 years before. But his progress was rapid.

In April 1815, he publicly declared his opposition to Napoléon, who had just escaped from the island of Elba, and was given a house arrest in his family home, in the small village of Mathieu, near Caen, in Normandy. He could then work full-time on physical optics. He made contact with Arago, who was well known due to his work on polarization. On 23 September, he sent Arago a letter[18] from Mathieu, of which there is an extract:

---

[17] For more details, see *Arago, F., *Œuvres complètes* t. 1 p. 107–185, and the introduction by Verdet to the *Œuvres complètes d'Augustin Fresnel* (Fresnel 1866, 1868, 1870).

[18] Fresnel (1866, 1868, 1870) t. 1, p. 5.

Monsieur,
I believe that I have found the explanation and the law of the colored fringes that are seen on the edges of the shadows of bodies illuminated by a point source. The results I got from calculations are confirmed by observation. But I could not reach in my observations the degree of exactness that is necessary to be ascertained of the correctness of my formula. For this, I would need instruments that I can only buy in Paris. Before doing that, I would like to know if this would not be useless, and if no one has already determined the law of diffraction by sufficiently exact experiments.

On 26 October, Fresnel wrote to Arago another letter[19] in which he said:

Monsieur,
A few days after I have told you that I believed to have found the explanation of diffraction, I built a micrometre with which I could perform observations sufficiently exact to be certain of the correctness of the formulae I obtained in the theory of undulations.

A very simple experiment proved to me that the light rays can act upon each other, and can weaken and even disappear when their vibrations are opposed; they can also add and strengthen each other when they vibrate together. It is on this principle that I based my theory of diffraction.

By extending this theory of undulations and of the mutual influence of light rays to reflection and refraction, I have found the reason for the laws of propagation of the light in these two phenomena.

I exposed this theory and the experiments that lead to it in a Memoir I have sent to my uncle [Léonor Mérimée] on 16 of this month, asking him to show it to the Perpetual secretary of the first class of the Institute [meaning Arago]. You have perhaps already read it. I would like to know your judgment on this work. Your suffrage is the one I would appreciate most.

Then Fresnel wrote that he had also explained Newton's rings and even the diffraction gratings with his theory. He also told Arago that he was working on double refraction and polarization, which, he hoped, could also be explained.

Arago answered on 8 November[20]:

Monsieur
I have been asked by the Institute to examine your Memoir on the diffraction of light: I have studied it carefully, and have found in it many interesting experiments, several of which were already performed by Dr. Thomas Young who, in general, considers this phenomenon in a rather similar way to yours.

---

[19] Fresnel (1866, 1868, 1870) t. 1, p. 35–37. Also *Arago, F. *Œuvres complètes*, t. Tables, p. XLII–XLV. Original at the Archives of the Academy of sciences, file Fresnel.

[20] Fresnel (1866, 1868, 1870) t. 1, p. 38–39.

The Memoir of Fresnel[21] is truly remarkable, and Arago was impressed. It began by a deep criticism of Newton. Then Fresnel described an interference experiment that consisted in observing the shadow of a narrow strip or of a wire lit by a point source: one saw diffraction fringes on the edges and interference fringes in the middle. He made measurements on a screen placed at different distances from the wire, and concluded that the locus of a fringe of a given order is a hyperbola whose foci are the light source and one of the edges of the wire. This property particularly impressed Louis Poinsot (1777–1859) and Arago, who were the referees of the Memoir. They wrote in their report[22]:

The fact discovered by Fresnel, of the propagation of the bands [sic] as hyperbolas, looks to us as one of the most curious results of optics… We do not understand how this singular fact could be explained by the emission theory. The simplest circumstances of the formation of the interior bands are not explicable, or at least not explained in the ordinary theory… If we add that there is no diffraction experiment known today that cannot be, we would not say explained, but even calculated, one cannot but recognize, whatever opinion one can have on the root of the question, that the hypothesis of M. Fresnel deserves to be followed and to fix the attention of the physicists.

Then Fresnel explained refraction and reflection in the same way as Huygens. He used the principle of Huygens, for which he gave a better form so that it is often called the principle of Huygens–Fresnel[23]:

The vibrations of a luminous wave at each of its points are equal to the sum of all the elementary motions that would send, at the same instant, as acting individually, any part of this wave considered at any of its previous positions.

Fresnel wrote also in his Memoir of 1815:

This explanation of refraction does not need to assume… that the surface of polished surfaces is without asperities; it suffices to admit that these asperities are very small compared to the wavelength… With these considerations, it seems easy to me to explain the colored images reflected by ruled surfaces, a curious phenomenon described for me by M. Arago.

We find here the first mention that the optical quality of a mirror must be better than a small fraction of the wavelength in order to give good images, a point developed later by Fresnel, and the announcement of the explanation of diffraction by gratings, with its different orders. Fresnel gave a more detailed explanation in a complement[24]; here is an extract of this text:

> *If ones engraves on a polished surface parallel, equally distant lines… it will reflect, together with the usual image of a luminous object, another image that is colored, and even*

---

[21] Fresnel (1866, 1868, 1870) t. 1, p. 9–33. This Memoir was not published by the Academy.

[22] Fresnel (1866, 1868, 1870) t. 1, p. 79–87. See p. 86–87.

[23] Fresnel (1866, 1868, 1870) t. 1, p. 174.

[24] Fresnel (1866, 1868, 1870) t. 1, p. 41–60.

several, if the ruled surface is large enough. The colors will be ordered as in the solar spectrum or in colored rings: the violet is the closest to the uncolored image, and the red the most distant.

This is a diffraction grating, for which Fresnel gives the fundamental formula.[25] He also studied Newton's rings by transmission and by reflection, but did not succeed initially to explain why the central point is black (this is due to a phase shift by half a wavelength in the reflection glass-air).

In his Memoir on diffraction published in 1816,[26] Fresnel made for the first time reference to Young. Before this date, he did not know the works of Young, and he asked Arago to send him information on these works. Young was informed by Arago of the works of Fresnel, and thought initially that they contained nothing new. However, he recognized later the genius of Fresnel, as can be seen in the exchange of letters between Arago, Fresnel and Young, reproduced in Fresnel's *Œuvres complètes*. These letters are generally friendly, but each of the three men tried to ascertain his priority. A letter of Young dated 18 June 1827 informed Fresnel that the Royal Society of London had given him its Rumford medal "for his works on the application of the undulatory theory to polarization." Arago wrote to Young on 6 August that he was elected as one of the eight foreign associates of the Academy of Sciences, and announced the death of Fresnel, "who was already half dead when I gave him your medals."

Fresnel and Arago in Paris

After the second abdication of Napoléon, Arago wrote on 19 December 1815 to Gaspard, baron Riche de Prony (1755–1839), the director of the École nationale de ponts et chausées, hence Fresnel's superior who now had reintegrated the Corps of bridges and roads. He asked Prony to let Fresnel visit the Paris Observatory, to show him his Fresnel's experiments that he found "very remarkable by their novelty, I would even say their boldness, which would only be admitted after a serious examination." This permission was granted. This was the beginning of a collaboration that lasted until 1819. Fresnel, who resumed his duties as an engineer, could nevertheless come to Paris rather frequently. He was posted to Paris in 1818, and could then benefit from a laboratory that Arago had installed for him at the Observatory. An optical bench that Fresnel probably used can still be found in the Observatory's collections.

[25] The invention of the diffraction grating is often attributed to Joseph von Fraunhofer (1787–1826). However, its properties were first described at the end of the eighteenth century by the American astronomer David Rittenhouse (1732–1796; see Cope T.D., 1932, *Journal of the Franklin Institute* 214, p. 99–104). Young mentioned and explained it in 1802. Fresnel, who always cited his sources, clearly ignored these works.

[26] **Annales de chimie et de physique* 1, p. 239–281.

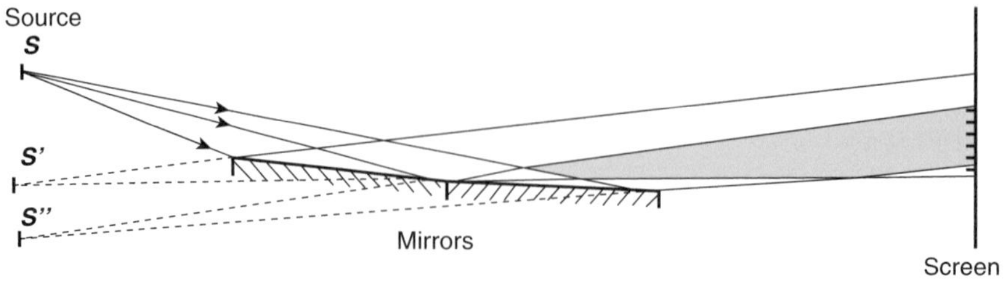

Fig. 3.18 Fresnel's mirrors. The two mirrors, almost in the same plane, give of a point source or a narrow slit S two virtual images S′ and S″ which play the role of the Young's holes. Interference fringes are observed in the common part of the two reflected beams (*grey*). This device is much more luminous than the holes or slits of Young

In 1819, Fresnel was appointed to the Lighthouse Service and continued to work with Arago in this new domain; we will detail this in Chap. 10. Although Arago was very famous, he was only 2 years older than Fresnel, and the two men got on very well. Both had independent minds, and they were not crippled by the weight of tradition. Fresnel carried out experiments and theory. Arago, who was an excellent experimentalist but was not very keen on theory, took part in Fresnel's experiments. For example, Fresnel wrote on 4 March 1816 to his brother Léonor[27]:

I am satisfied with the verification that Arago made of my theory... He carried out recently a new experiment that I had not imagined, whose result is another confirmation of my theory. Instead of intercepting the light of one of the edges of the wire by an opaque body [to make the fringes disappear], he placed there a glass plate and the fringes disappeared [also]. We came back home to search for the reason: I showed him that this came from the delay that the light suffered when crossing the glass on one side, so that the fringes of order 1, 2, 3 and 4, the only ones that one can see, were outside the shadow. I told him that if one put a very thin sheet of mica instead of the glass, or one of these foils of blown glass, it might be that the inner fringes would stay inside the shadow, and would be displaced on the side of the transparent foil. We carried out this experiment the next day, and everything occurred as I had foreseen. He was very pleased, and presented an account of the work the last Monday at the Institute, in a note where he said that my Memoir is such as to make a revolution in science.

In a supplement to his second Memoir on diffraction,[28] Fresnel described an interference experiment with the mirrors to which his name was given later (Fig. 3.18):

In order to eliminate the effect of the edges of the body, of the screen or of the small holes in the formation or disappearance of the inner fringes, I searched to produce

[27] Fresnel (1866, 1868, 1870) t. 1, p. 75.

[28] Fresnel (1866, 1868, 1870) t. 1, p. 129–170.

similar ones by crossing the beams reflected by two mirrors, and I succeeded after a few trials… M. Arago replaced them by platinum mirrors, and obtained similar fringes.

Then he introduced the "integral mechanism" that allowed him to completely calculate the diffraction fringes by the edge of a screen. The theory of diffraction and interference was then fully established. However, the Newtonians, who were many at the Academy, alarmed by the progresses of Fresnel, attempted to take their revenge by organizing in February 1817 a competition whose subject was the explanation of diffraction by the edge of an obstacle. Fresnel, and an unknown person who would not be taken into consideration, were the only ones to answer. The commission, composed of Biot, Arago, Laplace, Gay-Lussac and Poisson, named Arago as chairman. This was to be the triumph of Fresnel, who received the prize in March 1819. His Memoir,[29] crowned by the Academy, is actually a treatise of wave optics; however, it was only published in 1826.

The battle was over, but not the war: the Newtonians did not surrender despite the evidence.[30] In a famous controversy, Poisson tried to corner Fresnel by predicting from calculations that, if the undulation theory was exact, one should see a bright point at the center of the shadow of a circular screen illuminated by a point source: this looked absurd to him. However, the phenomenon was actually observed[31]! This did not prevent Poisson from remaining a Newtonian until his death in 1840. Even Biot, in spite of his numerous works in optics, was not yet converted to the wave theory in 1820.

What had still to be found was the nature of the light waves: were they longitudinal or transverse? The answer cannot be obtained from the classical experiments of diffraction and interference. At the time, it was natural to admit that the light waves were longitudinal like the sound waves; the analogy between light and sound had been emphasized during all of the eighteenth century and the beginning of the nineteenth century, and the propagation of light in ether was understood as a compression wave in this ether. So, no one really asked the question. However, in 1816, Arago had the idea for a new experiment involving polarization[32]:

One of us (M. Arago) imagined that it would be curious to search if the actions that the light rays exert on each other would not be modified if one interfered two light beams after having polarized them.

In a first experiment, the two scientists attempted to produce interference fringes between the ordinary and extraordinary rays of a spar crystal, but could not see any.

---

[29] Fresnel (1866, 1868, 1870) t. 1, p. 247–382.

[30] It is true that the arguments of the supporters of the wave theory were not always excellent. As an example, Arago cites in his biographic notice of Fresnel (*Arago, F. *Œuvres complètes* t. 1, p. 107–185) some fruitless experiments to detect what we now call the radiation pressure (which really exists), and concluded that its absence was an argument against the emission theory; but he noticed, judiciously in the context of the time, that "these sorts of experiments and of arguments that can be traced back to Franklin will never lead to anything decisive."

[31] Young had already observed this phenomenon by looking at some observational artifacts produced by diffraction in a microscope. But this was probably not known in France.

[32] Fresnel (1866, 1868, 1870) t. 1, p. 509–522.

Then they carried out Young's slit experiment, placing in front of each slit a polarizer made of an inclined stack of mica foils. When the beams were polarized perpendicularly to each other no fringe could be seen. They concluded that "the production of fringes is impossible if the two light beams … are polarized in opposite senses."

It took several years for Fresnel to understand why: He only gave the interpretation in 1821.[33]

M. Ampère, to whom I communicated the results of our experiments … and I understood that these phenomena could be explained very simply if the oscillatory movements of the polarized waves would only take place in the plane of the wave. But what about the longitudinal oscillations along the paths of the rays? … These difficulties looked so embarrassing that I abandoned our initial idea, and continued to assume longitudinal oscillations in polarized rays, but supposing at the same time transverse motions without which I always found impossible to understand polarization and the absence of mutual influence between rays polarized at right angle. It is only a few months ago that, thinking with more attention at the subject, I recognized that it is very probable that the oscillatory motions of the light waves are only along the plane of these waves, for direct as well as polarized light. [Then follows a mechanistic explanation in terms of vibrations of ether].

I must say at this point that a letter of M. Young dated 29 April 1818, communicated to me by M. Arago, contributed to my doubts about the existence of longitudinal oscillations. M. Young … concluded that the undulations of ether could well be similar to those of an infinitely long string, and could propagate in the same way.

But if Fresnel and Ampère took the plunge, Arago had a hard time accepting their hypothesis. Arago was even reluctant to present a report to the Academy on Fresnel's memoir, for which he was a referee with Ampère and Joseph Fourier. He wrote in effect:

The Memoir of which we just presented an analysis contains a very extended chapter, in which M. Fresnel exposes his theoretical ideas on the particular kind of undulations that, according to him, constitute the light; time has lacked to us to examine it with all the necessary attention.

The corresponding "very extended" passage was only a single page! Arago did not mention transverse vibrations in his review paper for *Encyclopaedia Britannica*, nor in his biographic notice on Fresnel written in 1830.[34] In an interesting testimony, William Whewell (1794–1866) wrote[35]:

When [Arago] and Fresnel had obtained their joint experimental result of the non-interference of oppositely-polarized pencils, and when Fresnel pointed out that

---

[33] Fresnel (1866, 1868, 1870) t. 1, p. 609–653, see p. 629–630.

[34] Curiously, Arago placed himself sometimes in the frame of the corpuscular theory, at least until 1842. Was this for the purpose of avoiding conflicts at the Academy when this theory did not play a crucial role in the analyses?

[35] Whewell (1866) vol. 2, p. 101 and p. 114–115.

transverse vibrations were the only possible translation of this fact into the undulatory theory, he himself protested that he had not courage to publish such a conception; and accordingly, the second part of the Memoir was published in Fresnel's name alone…

M. Arago would perhaps have at once adopted the concept of transverse vibrations, when it was suggested by his fellow-laborer, Fresnel, if he had not been that he was a member of the Institute, and had to bear the brunt of the war, in the frequent discussions on the undulatory theory; to which theory Laplace, and other leading members, were so vehemently opposed, that they would not even listen with toleration to the arguments in its favor.

There was perhaps a more scientific reason to Arago's perplexity. A transverse wave would require a change in the notion of ether, which should now be rigid, thus crystalline. This meant coming back to the crystalline ether of Aristotle, to Arago's dismay.[36] Arago would finally accept the idea of transverse vibrations in 1852, in a Memoir that summarizes his joint research with Fresnel.[37] Barral wrote about this[38]:

M. Arago took care to warn the reader that the undulations of light waves are perpendicular to the direction of propagation.

Given the reluctance of many of his contemporaries to accept the wave theory, we should not be surprised that Arago decided to carry out a "crucial experiment" that would definitively eliminate the emission theory. We will discuss this in the next chapter. In the meantime, Fresnel finished off his *œuvre* by giving a complete and quantitatively exact theory of vitreous reflection, of the double refraction of spar, of the rotational chromatic polarization discovered by Arago, of circular and elliptical polarizations, and, in general, of all the phenomena known at this time that involved the wave nature of light. I cannot comment in detail this magnificent work, but it is worth mentioning that all its explanations and calculations rest upon a mechanical reasoning that uses the conservation of momentum, a reasoning that takes its origin from Young (Box 3.5). This did not convince everybody, as can be seen by this extract of the *Treatise on light* of John Herschel (1792–1871)[39]:

The fact is, that neither the corpuscular nor the undulatory, nor any other system which has yet been devised, will furnish that complete and satisfactory explanation of all the phenomena of light which is desirable. Certain admissions must be made at every step, as to modes of mechanical action, where we are in total ignorance of the acting forces; and we are called on, where reasoning fails us, occasionally for an exercise of faith.

[36] Cawood (1974) discusses this point in detail p. 136–149. For him, a mathematically oriented physicist like Fresnel could accept a transverse wave, which was only a model that could allow progress in the interpretation and prediction of phenomena, while for a physicist of Nature like Arago, this hypothesis would imply an unacceptable description of the Universe.

[37] *Arago, F. *Œuvres complètes* t. 10, p. 132–149.

[38] *Arago, F. *Œuvres complètes* t.Tables, p. CV.

[39] Herschel (s.d), p. 450.

The situation did not change until the beginning of the twentieth century, when the electromagnetic theory of James Clerk Maxwell (1831–1879, who still believed in the existence of ether), then the withdrawal of the idea of propagation of light in ether and the acceptance of its propagation in a vacuum would at last give a satisfactory solution.

After Young, Fresnel, Arago and Biot there was little progress in physical optics until Maxwell. Augustin Cauchy (1789–1857) encumbered the *Comptes rendus* with scores of long mathematical papers based on Fresnel's type of reasoning; they did not report much that was new, except for a study of reflection on metals, confirmed by the experiments of Jules-Célestin Jamin (1818–1886).[40] Some innovations were due to George Biddell Airy (1801–1892), who studied the diffraction by a circular aperture in 1835, and to Jacques Babinet (1794–1872). The latter produced the principle that bears his name: the diffraction produced by the edges of an aperture is complementary to that produced by a screen with the same shape. He also showed that the vibrations of a polarized light are in the plane of polarization, and not in the perpendicular plane as assumed by Fresnel.[41] Finally, one can cite a curiosity discovered and explained in 1832 in Dublin by Sir William Rowan Hamilton (1805–1865) and Humphrey Lloyd (1800–1881): the conical refraction produced in bi-axial crystals, in which an incident ray is refracted in a luminous cone or cylinder.[42] But all this is not very important compared to the revolution brought by Maxwell's electromagnetic theory.

Box 3.5 Fresnel's Mechanistic Explanation of the Reflection of Light

The reasoning of Fresnel was "rigorously verified" by Poisson, according to him (Fig. 3.19). It rests on the conservation of momentum. Fresnel proposed in 1815 "that in the [vitreous] inflection [refraction] and reflection the molecules of matter reproduce the vibratory movements induced by the incident rays." He first considered what occurred for the component polarized in the plane of incidence and reflection. He assumed that a small element of mass m of the first medium where the light arrives takes a velocity proportional to the displacement of the ether produced by the incident ray, then drags along, in moving over the surface, a contiguous element of mass m' of the second medium. Then, the conservation of momentum implies that their common velocity is $mv/(m+m')$, v being the initial velocity of m, to that the velocity lost by m is $v-mv/(m+m')=m'v(m+m')$. After the two masses separate from each other, the mass m, which now insures the reflection, "must receive, in the inverse direction, all the part of the initial velocity it has lost": the final velocity of m corresponding to the reflection is thus $mv/(m+m')-m'v(m+m')=v(m-m')/(m+m')$. The reflected wave is thus produced by the mass m with an

(continued)

[40] **Comptes Rendus* (1846) 23, p. 1103–1105.

[41] **Comptes Rendus* (1849) 29, p. 514–515.

[42] See. e.g. http://www.vct-ag.com/conical_refraction.pdf

Box 3.5 (continued)

Fig. 3.19 Reflection and refraction between two transparent media. (*Bottom*) The polarization is in the plane of incidence and reflection, the case studied by Fresnel. (*Top*) The polarization is perpendicular to this plane

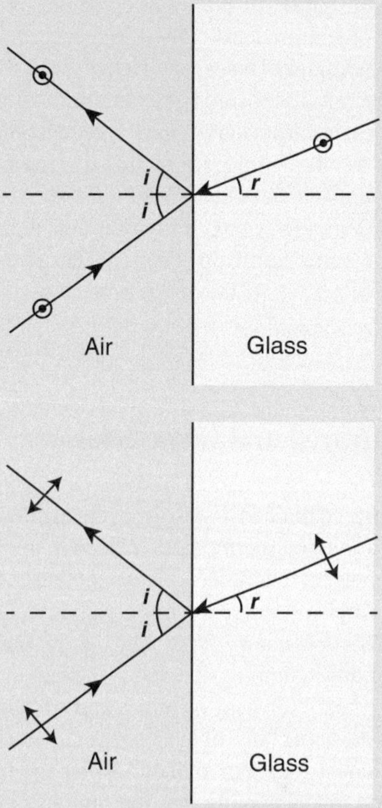

amplitude proportional to its velocity: thus, the ratio of the amplitudes of the reflected and the incident waves is $(m-m')/(m+m')$, and the ratio of their intensities is $[(m-m')/(m+m')]^2$. As to the refracted wave, it is produced by the motion of m', with the complementary intensity. One has then to know the ratio of the masses m and m'. Fresnel assumed that it is that of the masses contained in the light beam on either side of the separating surface, with thicknesses proportional to the wavelength on each side. He also assumed that the specific masses (densities) of these elements are inversely proportional to the square of the velocity of light in each medium, which is quite approximate. He then obtained $m'/m = \tan r/\tan i$, i and r being respectively the angles of incidence and of reflection, such that $\sin i = (v_1/v_2) \sin r$, v_1 and v_2 being respectively the velocities of light in the first and in the second medium. After some calculations, Fresnel found the exact result for the reflection coefficient of the component polarized parallel to the plane of incidence and reflection: $R_{//} = [\tan(i-r)/\tan(i+r)]^2$. Then, using the deviation "observed by Arago" of

(continued)

Box 3.5 (continued)
the plane of polarization of the reflected light, he obtained empirically the reflection coefficient of the other polarized component, that he found equal to $[\sin(i-r)/\sin(i+r)]^2$. Hence, by a disputable mechanical reasoning combined to an observation, he obtained the exact formulae for the vitreous reflection, including polarization! This reasoning would be taught during the whole nineteenth century. It should be remarked that it is not very different from some present demonstrations that avoid using the Maxwell equations[43]: In these demonstrations, reflection and refraction are explained by the action of the electric field of the wave on the electrons of the media, then by the re-emission of a wave by these electrons, that thus play the role of the differential elements of Fresnel.

Infrared and Ultraviolet

Infrared radiation was discovered by William Herschel around 1800, by observing the heating of a thermometer placed in the spectrum of the Sun beyond the color red. But nobody at that time, with the notable exception of Young, understood that infrared was a simple prolongation of optical light. Delamétherie, when reporting on this discovery in his *Journal de Physique*,[44] wrote that Herschel had concluded that "light and heat are different things", what is inexact to some extent. One spoke in general of "radiating heat" for near infrared, and of "culinary heat" for mid-infrared. Joseph Fourier wrote about "invisible heat radiation mixed with the solar light".[45] In 1804, the German scientist Johannes Wilhelm Ritter and the English physicist Francis Wollaston discovered independently that the Sun also emits "invisible rays" that blacken silver chloride: this is ultraviolet radiation, which prolongs the visible spectrum beyond the violet. These invisible rays were often called "chemical rays." But all this was very confused. For example Gay-Lussac and Thenard wrote in 1811[46]:

Although it seems demonstrated by the various experiments that we have described that, each time light produces chemical changes in bodies, one can also produce them by heat, it is very difficult to decide, given our present knowledge, whether this is effectively because light produces heat in the bodies that absorb it, or if it acts chemically upon them. It is according to us what is more probable; but one has discovered recently properties of light so singular, that it is not yet possible to reconcile the effects it produces by its different rays. [And in note:] In effect, one has

[43] For example in Feynman et al. (1964), chapter 31.

[44] *Journal de Physique*… (nivôse an IX = 1801), vol. 52, p. 7, accessible via http://www.biodiversitylibrary.org/bibliography/6637#/summary

[45] **Mémoires de l'Académie des sciences* (1823) t. 6, p. LXI–LXXXI, see p. LXXV.

[46] Gay-Lussac and Thenard (1811) t. 2, p. 205–206.

found that the chemical action of the violet rays is larger than that of the other rays, even the red ones; however, one knows that the violet rays heat less the globe of the thermoscope than the red rays.

After Fourier's death, Arago read at the Academy on 18 November 1833 his eulogy,[47] which contains interesting historical information on *calorique rayonnant* (radiating heat or caloric):

The famous academicians of the Academia del Cimento [in Florence] found, almost two centuries ago, that this caloric can be reflected like light; like light, it can be concentrated at the focus of a concave mirror. Replacing hot bodies by snow balls, they even proved that one can form frigorific focuses by reflection.

A few years later, Mariotte [Edme, 1620–1684], member of this Academy [the French Academy of Sciences] discovered that there are different kinds of radiating caloric: that which accompanies the solar rays [near infrared] traverses all transparent media as easily as light; that which emanates from a heated, but still dark material, as well as that which is mixed with light rays from a slightly incandescent body [mean and far infrared] are stopped almost entirely by the most transparent glass plate.

The examination of the original texts shows that Arago had somewhat over-interpreted them.[48]

Things then accelerated. In 1835, Ampère, with his usual insight, exposed his concept as follows[49]:

It consists in considering radiant heat as a series of undulations excited in ether by the vibrations of the hot bodies. These undulations would be longer than the waves that make light if the calorific source is dark. But in the case of those sources that are at the same time calorific and luminous, there would always be a set of waves that possess simultaneously the two properties of heating and illuminating. Thus, in this view, no essential difference would exist between radiant heat and light.

Arago nearly agreed with Ampère. In 1836, he spoke "of the three properties that possesses the light emanating from the Sun: 1° that to illuminate; 2° that to heat: 3° that to destroy or determinate chemical combinations." However, in spite of his numerous experiments on reflection, refraction and polarization of the radiating heat, Macedonio Melloni (1798–1854, Fig. 3.20), a specialist of infrared (Box 3.6), did not believe in Ampère's explanations. In 1837, he was still convinced that light and radiating heat were different things, in spite of their common features. For him, like most of his contemporaries, radiating heat should be more similar to the heat

47 *Arago, F. *Œuvres complètes* t. 1, p. 295–369.

48 See Lequeux, J. (2009) Early Infrared Astronomy, *Journal of Astronomical History and Heritage* 12, p. 125–140, accessible freely via http://adsabs.harvard.edu

49 Cited by *Melloni, *Comptes rendus* (1835) 1, p. 503–509. But Melloni did not yet agree with Ampère's concept.

Fig. 3.20 Macedonio Melloni, caricature by Michael Sobkowski, ca. 1830. Wikimedia Commons

Box 3.6 Macedonio Melloni

Born in Parma in 1798, Melloni was appointed professor of physics in the university of this city in 1824. But he was banished in 1831 and had to go to Geneva, then to Paris, because he had publicly expressed his satisfaction at the fall of the king of France, Charles X. He was helped by Arago and by Humboldt, who interceded for him in 1839 with Metternich. He was allowed to return to Italy. He was then named Director of the Conservatory of Arts and Crafts of Naples, and also of the meteorological observatory of Vesuvius. Fired anew in 1849 because of his political convictions, he retired at Portici, where he died of cholera in 1854. Melloni is known for his work on the infrared radiation that he studied with his thermopile. He is also the founder of the paleomagnetism; he discovered that volcanic lava is magnetized, and that this magnetization was acquired in the Earth's magnetic field during cooling; the lava preserves the memory of the direction of the terrestrial magnetic field at the time of this cooling.

that propagates slowly inside the matter, so well studied by Fourier. Finally, in 1842, Melloni changed his mind and rallied in support of Ampère[50]:

Light, heat and chemical reactions [in fact ultraviolet that produces these reactions] are three manifestations of the undulations of ether that make solar radiations. The dark undulations responsible for chemical or calorific actions are perfectly similar to luminous undulations; they only differ from them by wavelength.

[50] *Melloni, *Comptes rendus* (1842) 15, p. 454–460.

This was confirmed for infrared 5 years later by the superb experiments of interference by Fizeau and Foucault.[51] However, these experiments seem to have been largely ignored by the contemporaries, for example by Jamin, who still wrote in 1859, in his course of physics at the École Polytechnique, on the "probable identity of heat and light".[52]

Ultraviolet resisted less than infrared. Young, then Arago in 1821, made interference experiments on the "chemical rays" that impressed a paper impregnated with silver chloride.[53] In 1842 Edmond Becquerel (1820–1891), using photography, discovered in the ultraviolet spectrum of the Sun, Fraunhofer lines similar to those already observed in the visible.[54] Everyone was then convinced that chemical rays were similar to light and only differed by their shorter wavelengths.

[51] See Tobin (2003) chapter 6, or Lequeux (2014) chapter 2.

[52] Jamin (1858–1866), t. 2, p. 247–249.

[53] *Arago, F. *Œuvres complètes* t. 10, p. 484–492.

[54] *Arago, F. *Œuvres completes* t. 7, p. 530–547, see p. 540.

Chapter 4
The Velocity of Light

Rotating mirror and its mechanism built by Louis Breguet in 1844 for Arago. After a few modifications, it was used by Fizeau in 1850 for the comparison of the velocity of light in air and in water. Bibliothèque de l'Observatoire de Paris

J. Lequeux, *François Arago*, Astrophysics and Space Science Library 421,
DOI 10.1007/978-3-319-20723-0_4

The Constancy of the Velocity of Light

In his biography of Arago,[1] Maurice Daumas (1910–1984) wrote that Arago carried out an experiment that announced the theory of relativity:

Arago demonstrated that light propagates with the same velocity whatever the body that emits it… [He also showed] that it is impossible to measure the absolute motion of the Earth. The question never ceased to bother physicists. At the end of the 19th century, Albert Michelson attempted to solve it by a famous experiment that also failed. It is by considering the results of these experiments that Einstein was led to extend the principle of mechanical relativity to electromagnetic and optical phenomena. "It is impossible, according to this principle, by experiments of physics carried out inside some system, to find out a rectilinear and uniform motion of the ensemble of this system."

The young historian of science was somewhat carried away by his subject (he was later much more rigorous). We will see what really occurred: Arago's experiment is interesting, but did not at all demonstrate the properties indicated by Daumas. However, it gave strong arguments against the corpuscular theory of Newton. It must be placed in its context for its significance and importance to be appreciated correctly.[2] But we must first understand what the aberration of light is.

The Aberration of Light

Aberration is a common observed phenomenon: For example the sound of an airplane in flight seems to come from a direction behind its visible position, because the plane has advanced during the time the sound took to arrive to us. Another manifestation is the fact that a vertical rain seems to fall obliquely when we are moving. Aberration is thus a purely geometrical phenomenon that has nothing to do with any composition of velocities.

For astronomers, the aberration of light, which is due to its finite velocity, is such that all stars make an apparent annual displacement in the sky, with total amplitude of about 40 arc sec. Several astronomers saw this displacement during the seventeenth century, but they did not understand its cause. It was explained for the first time in 1727 by James Bradley (1693–1762).[3] Its explanation is rather simple (Fig. 4.1). The result is that the maximum deviation of the star with respect of its mean position, when expressed in radians, is the ratio of the orbital velocity of the Earth (about 30 km/s) to the velocity of light (about 300,000 km/s): this corresponds to 20.4 arc sec.

---

[1] Daumas (1987) p. 77–78; first edition in 1943.

[2] See for this e.g. Eisenstaedt (2005).

[3] *Philosophical transactions* (1728) 35, p. 637–661, accessible via http://www.jstor.org/stable/view/103725

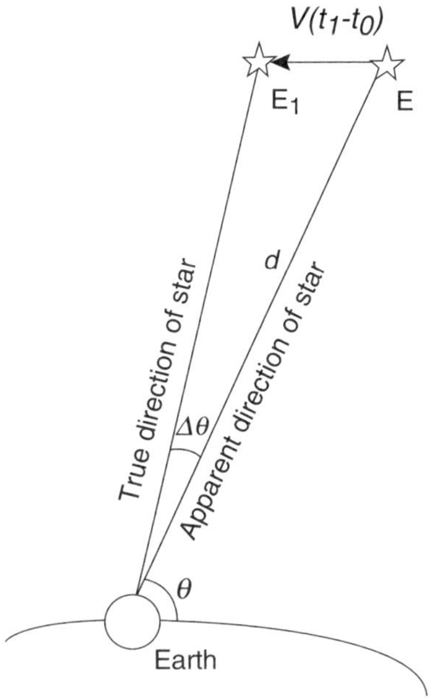

Fig. 4.1 The aberration of stars. It is convenient to place oneself in the system of the Earth. The observer O is then fixed and the star has a velocity V opposed to that of the Earth. At time t_0 the star is in E; its light reaches O at time t_1. The interval $t_1 - t_0$ is the time taken for the light to travel through the distance d with velocity c: thus $d = c(t_1 - t_0)$. At time t_1 the star has moved to E_1 such that $EE_1 = V(t_1 - t_0)$. The true direction of the star is OE_1 while its apparent direction is OE. The angle between these directions is $\Delta\theta = (EE_1/d)\sin\theta = (V/c)\sin\theta$. Time is eliminated, and the result is independent of the distance of the star. The direction of the motion of the Earth changes during the year, so that the apparent position E describes an ellipse around E_1, with a major axis V/c. The angle $\Delta\theta$ is very exaggerated in the figure

Bradley wondered if the velocity of light depended on the emitting star, in which case the numerical value of the aberration would have been different from star to star. He saw nothing, but his measurements were only accurate to 1 arc sec, so that he could only see variations larger than 5 % in the velocity of light.

John Michell and the Variations of the Velocity of Light

However, many expected the existence of such variations. John Michell (ca. 1724–1783) thought that, in the frame of the Newtonian theory of light, the corpuscles emitted by a star would be slowed down by the gravitational attraction of this star: the velocity of the emitted light should then vary according to the mass and the radius of the star. Nothing was known about the latter quantities, but he imagined that the effect could be large: after all, Bradley had only observed 12 stars.

Michell even thought that gravity could be so large as to prevent the light to leave the star. In 1783, he wrote in a paper[4] that should be read to understand what Arago had in mind:

If there should really exist in nature any bodies whose density is not less than that of the Sun, and whose diameters are more than 500 times that of the Sun, since their light could not arrive at us; or if there should exist any other bodies of a somewhat smaller size, which are not naturally luminous; of the existence of bodies under either of these circumstances, we could have no information from sight; yet, if any other luminous bodies should happen to revolve about them we might still perhaps from the motions of these revolving bodies infer the existence of the central ones with some degree of probability…[5]

Michell considered next what occurs when the slowed light originating from a star penetrates a dense transparent medium like glass. In Newton's corpuscular theory, refraction corresponds to an *increase* of the velocity of light in the denser medium (see Fig. 3.3). Michell wrote (p. 51):

Let us suppose with Sir ISAAC NEWTON (see his Optics, prop. VI, paragr. 4 and 5) that the refraction of light is occasioned by a certain force impelling it towards the refracting medium, an hypothesis which perfectly accounts for all the appearances [sic!]. Upon this hypothesis the velocity of light in any medium, in whatever direction it falls upon it, and the sines of incidence and refraction will, in consequence of this, bear the same ratio to each other with these velocities inversely.

The increase of velocity in the denser medium is supposed to be due to a "force" produced by this medium (in modern terms, an increase of kinetic energy of the corpuscles, which were supposed to have a mass). Michell made the logical, but arbitrary, hypothesis that this "force" is a property of the medium that does not depend on the velocity of these corpuscles. He made a little calculation, that I reproduce here in a modern form, that showed that the refraction angle depended on the velocity of light.

Let v_1 be the incident velocity of the light, and v_2 its velocity inside the denser medium. The relation between the incidence angle i and the refraction angle r with these velocities is, in the Newtonian theory:

$$\sin i / \sin r = v_2 / v_1$$

---

[4] *Philosophical transactions* (1784) 74, p. 35–57, accessible as http://www.jstor.org/stable/106576; see p. 50. I cannot resist the pleasure to reproduce the full title: "*On the Means of discovering the Distance, Magnitude, etc. of the fixed Stars, in consequence of the diminution of the Velocity of their Light, in case such a diminution should be found to take place in any of them, and such other Data should be procured from Observations, as would be farther necessary for that Purpose.*"

[5] Such objects really exist, but their properties are quite different from those imagined by Michell: these are the black holes. The reason why light does not escape from them is due to their high gravitation, but this is a general relativity effect: the space-time is locally very deformed, the velocity of light is unchanged but its redshift is infinite. As far as "normal" stars are concerned, the effect of gravitation on light is generally small, even for the very compact white dwarf stars: for them, the gravitational redshift is conspicuous, but the velocity of light is unchanged.

The ratio between v_1 and v_2 writes:

$$\tfrac{1}{2} m v_2^2 = \tfrac{1}{2} m v_1^2 + A$$

where m is the mass of the particle and A is a constant characteristic of the dense medium. One can also write the latter equation as:

$$v_2 = \left(v_1^2 + 2A/m\right)^{1/2}$$

This relation is not linear, so that v_2 is not proportional to v_1: their ratio, hence the refraction angle r, depends on the arrival velocity of the light. One can then, in this theory, search if v_1 varies for different stars by observing the corresponding changes in refraction. Michell wrote (p. 52):

From hence a prism, with a small refracting angle [read: angle between the faces], might perhaps be found to be no very inconvenient instrument for this purpose: for by such a prism, whose refracting angle was of one minute, for instance, the light with its velocity undiminished would be turned out of its way 33", and with the diminished velocity 35", 88 nearly, the difference between which being almost 2".53"'... [In order to ameliorate the sensitivity] a prism... should have a much larger refracting angle than that we have proposed, especially if it was constructed in the achromatic way, according to Mr. DOLLOND's principles[6]*... We might probably be able to discover considerably less diminutions in the velocity of light, as perhaps a hundredth, a two-hundredth, a five-hundredth, or even a thousandth part of the whole.*

The stake of these measurements was of course very great. Michell hoped that one could measure with a sufficiently magnifying telescope the apparent diameter of some stars. If one observed the decrease in velocity of the light coming from them, one could estimate their linear diameter, assuming that their density is similar to that of the Sun. Then one would get their distance by comparing the apparent and the linear diameters. Needless to say that this never succeeded for two reasons: the linear diameters were too small, and there was no change in the velocity of light.

Arago's Experiment

Appointed secretary-librarian of the Paris Observatory in 1805, Arago immediately started experiments on the refraction index of gases, under Biot's direction, and participated in the astronomical activities of the Observatory: measurements of stellar positions for an accurate determination of the latitude, and determination of the orbits of comets.

---

[6] John Dollond (1706–1761) in 1757 built the first achromatic objectives for telescopes, by associating a converging lens of crown glass with a diverging lens of flint glass. He also built achromatic prisms (see Fig. 4.2).

He also took an interest in the velocity of light, perhaps on the instigation of Laplace. He wanted to know whether the velocity of light differed from one star to another. Also—and this is different from Michell's propositions—he would try to detect a possible variation of the velocity of light when the Earth drew nearer or farther from a star. In the latter case, he assumed naturally that the velocity of the Earth would add or subtract to the velocity of light (we know now that this is wrong). In the first case, he had to look for a possible change in the deviation by a prism of the light from several stars, and in the second case from the same star observed at different times of the year.

His work is described in a Memoir presented at the Institute on 10 December 1810, but unpublished. He found it in his papers shortly before his death in 1853 and had it published by the Academy of Sciences "without changing a single word".[7]

In 1805–1806, he used a simple prism of angle 45′, whose dispersion was sufficiently small to allow to work with white light. He observed with and without the prism artificial light sources as well as astronomical ones: stars, the Sun, the Moon and several planets. He found that the deviation of their light by the prism was always equal to 25′, with variations of 5 arc sec at most, that he considered rightly to correspond to errors in the measurements. He concluded that the relative variations of the velocity of light could not be larger than 1/480. This result made much noise because it seemed to contradict the corpuscular theory.

Arago was then sent to Spain for measuring the extension of the Paris meridian, and nothing happened until his return in 1809. Then, the Memoir he had written in 1806, in which he described his first observations, had not yet been examined by a referee. Delambre hastened to present a report on this at the First Class of the Institute, on 4 September 1809. This was certainly one of the decisive elements for Arago's election 2 weeks later. Arago decided to carry out new observations with another prism giving a larger deviation; these measurements took place mostly in 1810. To continue to work with white light, Arago had an achromatic prism built, with a negligible dispersion and a deviation of about 10° (Fig. 4.2). It is easy to see that in the frame of the corpuscular theory, following Michell's reasoning, this improvement would not have perturbed the result of the measurements.

It is curious that in his approach Arago followed Michell exactly, using a small-angle prism, then an achromatic prism. This is probably not by chance. Arago might have seen Michell's paper because the *Philosophical transactions of the Royal Society of London*, where it was published, were in the library of the Observatory, of which he was the librarian. However, Arago wrote in his 1810 Memoir:

My experiments were almost completed, when the reading of one of the beautiful Memoirs that doctor Young inserted on the Philosophical transactions told me that M. Robisson [John Robison (1739–1805)], professor of physics in Edinburgh, had considered theoretically this question of the velocity of light: I have found since, in various works, that it was examined under different points of view by Boscowich, Michell, Wilson and Blair.[8]

[7] **Comptes rendus* (1853) 36, p. 38–49; *Arago, F. *Œuvres complètes* t. 7, p. 548–568.

[8] A possible path from Michell to Arago, via Blair and Robison, is described by Eisenstaedt (2005). Robert Blair (1748–1828) had built a series of 12 achromatic prisms to make the observations, but this project did not succeed because of excessive absorption of light and insufficient correction for dispersion by the prisms.

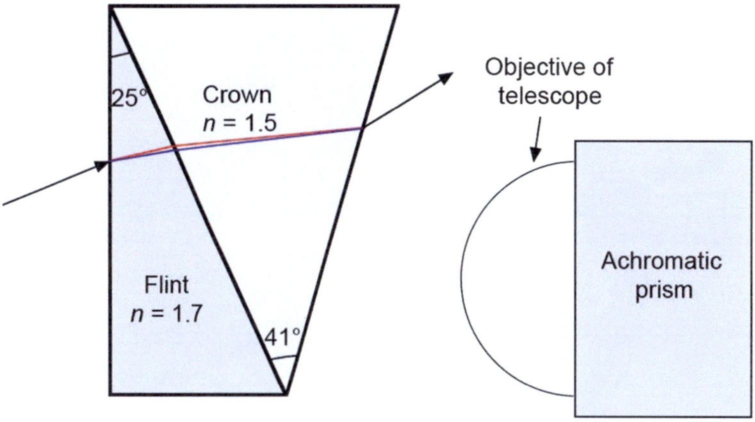

Fig. 4.2 Arago's achromatic prism. The light entered a prism of flint glass with an angle of 25°, where it was dispersed in a different way according to its color, then in an inverted prism of crown glass with an angle of 41°, which corrected the dispersion of the flint. (I assume for this figure that Arago used prisms with the same angles as obtained empirically by Dollond: *Philosophical transactions* (1758) p. 733–743, accessible as http://www.jstor.org/stable/105323: see p. 740. This is difficult to check because the refraction indices of the glasses are not known sufficiently well.) The overall deviation was about 10°. In his last series of observations, Arago put a series of two achromatic prisms glued together in front of one half of the objective of a telescope, as indicated on the right. The overall deviation was then about 22°

Was it Laplace, and not Arago, who had read Michell and suggested the experiment to him?[9]

Let us come back to the observations. The achromatic prism, which gave a deviation of about 10°, was fixed in front of the objective of a mural meridian circle, perhaps the large one built by John Bird (1709–1776), so that the stars were observed at their meridian crossing. It could be rotated upon itself and moved laterally. Arago measured the angular distance of the stars to the zenith with and without the prism. He repeated these measurements at different times of the year. He found essentially no difference in the deviations by the prism measured in this way, within some small changes that he attributed rightly to measurement errors. To improve the accuracy, he had a second achromatic prism constructed and resumed his measurements with the two prisms in series, with a total deviation of 22° 25′. This time, he put this device in front of a *cercle répétiteur* (repeating circle), presumably the one that is presently in the collections of the Observatory (Fig. 4.3). He had probably already used this instrument for geodesy in Spain. The prism was set in front of half of the objective of the telescope, so that it sufficed to successively observe the star with the part containing the prism and with the free one, and to read the difference on the graduated circle.

This time again, no significant difference was seen from star to star. The velocity of light did not seem to depend on the star. Also, if the orbital velocity of the Earth

[9] One often attributes to Laplace the idea that the gravitational field of some stars could slow or stop their light, a prefiguration of the black holes. But the idea is Michell's; Laplace, who rarely cited his sources, had perhaps borrowed it from him.

Fig. 4.3 The repeating circle built by Jean-Nicolas Fortin (1750–1831) and used by Arago for his measurements. From Arago's *Astronomie populaire*, author's collection

was added or subtracted from the velocity of light, Arago would have easily detected the effect, which would have produced for two opposite times of the year a deviation of 6″ for the single achromatic prism and 12″ for the two prisms, at least for stars close to the ecliptic. Nothing like this was seen.

Arago and Laplace were puzzled by these results. They hesitated between two explanations: Either the Newtonian theory was wrong and the velocity of light was constant, or the eye could only see the light corpuscles with a given velocity. They adopted the second hypothesis, and drew from this a rather unexpected consequence:

It would perhaps be useful to note that the observations I just described and the hypothesis that explained them agree very remarkably with those of Herschel, Wollaston and Ritter... The [chemical rays] might be those to which only a small increase in velocity is lacking to become visible, and the calorific rays might be those that an excessive velocity has deprived from the property of illuminating.

In his great work, the *Exposition du système du monde*, Laplace reproduces this conclusion, almost in the same terms.[10] This explanation is of course completely ad hoc and looks to us as given to keep up appearances. However, it shows that Arago and Laplace were beginning to realize that the infrared and ultraviolet radiations are of the same nature as the visible light, something that was not at all evident at the time.

Be that as it may, Arago's observations spread trouble among the believers of the Newtonian theory. Arago was still among them, but not for long. Humboldt wrote in his preface to Arago's *Œuvres complètes*[11]:

Applying a prism in front of the objective of a telescope, [Arago] proved not only that the same refraction tables can be used for the light emanating from the Sun and for that of the stars [this is at least a practical application of Arago's observations], but also that the rays from the stars towards which the Earth moves, and those of the stars from which the Earth moves off, are refracted exactly in the same way. This threw already many doubts on the emission theory.

Arago mentioned other experiments linked to aberration whose purpose was to search for differences in the velocity of light. They consisted in observing the position of stars with a telescope filled with water. They were initiated by Rudjer Boskowich (1711–1787), then by Alexander Wilson (1714–1786), and resumed during the second half of the nineteenth century by Airy and others. They did not yield more results than Arago's own observations.

The Intervention of Fresnel

Perturbed by his result, Arago, who was more and more persuaded of the validity of the undulatory theory, asked his friend Fresnel to try to account for his observations in the frame of this theory. Fresnel got to work, and produced a strange explanation: On 5 September 1818 he wrote to his brother Léonor:

I have done recently a little work that I find of some importance. I proved that, assuming that the Earth is porous enough to communicate to the ether that penetrates and

[10] Fourth edition (1813), p. 326. Laplace concludes: "*Would not it be possible to suppose that luminous bodies send an infinity of rays with different velocities, and that the only ones whose velocity is comprised between certain limits have the property to excite the sensation of light, while the others only produced an obscure heat.*"

[11] *Arago, F. *Œuvres complètes*, t. 1, p. I–XXXII, see p. VI.

surrounds it a very small part of its velocity, which does not exceed one hundredth for example, one could explain in a satisfactory way not only the aberration of stars, but also all the complicated optical phenomena of the terrestrial motion.

This work was published in 1818 in *Annales de chimie et de physique* as a letter to Arago.[12] The demonstration is subtle, and the figure of the paper is wrong. Fortunately, this demonstration has been correctly analyzed by Pierre Costabel (1912–1989)[13] (see Box 4.1). Fresnel understood that the velocity of light in the prism moving with the Earth did not add simply with its velocity u, prefiguring in some way the theory of relativity. He expressed this by assuming that the ether inside the moving prism did not take the velocity u, but only the velocity $u(1-1/n^2)$, n being the refraction index of the glass.

Arago's interest in the problem waned, because he was now working with Ampère on electromagnetism. He did not mention it while presenting in 1853 his paper of 1810. Fresnel's hypothesis of the "partial drag of ether" would have been completely forgotten if Fizeau had not mentioned it at the beginning of the paper reporting an experiment that we will now examine.

Box 4.1 Fresnel's Interpretation of Arago's Observation

Let us recall first that during the nineteenth century light was supposed to propagate in an ether that filled the whole Universe and was the absolute reference frame proposed by Newton, with respect to which the Earth was moving. Fresnel assumed that the light from a star penetrated from the left the prism of Arago, which is moving to the right with the Earth with the velocity u (Fig. 4.4). The light penetrated the prism perpendicularly to its face with the velocity v. Inside the prism, it would have a velocity v/n, n being the refraction index of the prism, if the prism was fixed with respect to ether. But this velocity might have been different if the prism moved. When leaving the prism, the light ray was deviated toward the direction H, and propagated anew in the fixed ether. But the telescope with which this ray is observed moved with the Earth and the prism with velocity u. Due to the time of propagation of the light in the fixed ether between the prism and the telescope, the ray was not seen as coming from the direction H, but from a nearby direction H′: this is the aberration phenomenon, which did not depend on the distance prism-telescope. But Arago always saw the same deviation by the prism whatever

(continued)

[12] **Annales de chimie et de physique* (1818) 9, p. 57–66, also Fresnel (1866, 1868, 1870) t. 1, p. 627–636.

[13] **La vie des sciences* (1989) 6, p. 327–334. Mascart (1889–1893) t. 3, chap. 15, criticizes the demonstration by Fresnel and gives another one that is not more satisfying.

Box 4.1 (continued)

the velocity *u*, thus the light did not propagate inside the prism with the velocity *v*/*n*, but with a slightly modified velocity, which changed the deviation by the prism such as to compensate aberration. Fresnel's interpretation was that the ether was only partly dragged inside the prism, with a velocity much smaller than *u*. His calculation yielded the formula given in the text. The velocity of light *v* was eliminated in this calculation, so that Arago would not have been able to see a change of the velocity of light anyhow. In the modern interpretation, whatever the velocity of the prism, the light always enters it with the velocity c, and the velocity of light inside the prism is always c/*n*.

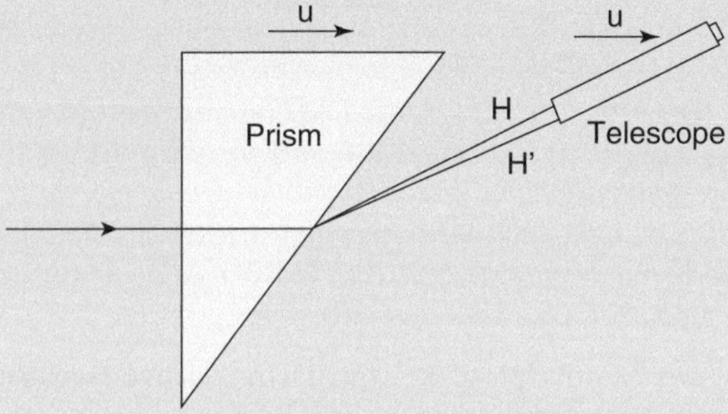

Fig. 4.4 Fresnel's interpretation

Fizeau's Experiment

Box 4.2 Hippolyte Fizeau[14]

Fizeau was born into a wealthy family. He studied medicine while doing various experiments in physics, and finally gave up medicine for physics. He worked by himself or with Léon Foucault, his contemporary, to perfect the daguerreotype and used it for astronomy and photometry. Together they performed beautiful interference experiments in visible and infrared light. Fizeau discovered, independently of Christian Doppler (1803–1853), the frequency shift of waves emitted by a source in motion. He directly measured for the first time the velocity of light in 1849. He also carried out several remarkable experiments in optics. He had a strong influence on the physics of his time.

[14] For a detailed scientific biography of Fizeau, see Lequeux (2014).

Fizeau (Box 4.2, Fig. 4.5) had probably attended Arago's courses of "popular astronomy," and was encouraged by him for his research on the velocity of light. He directly measured for the first time this velocity in 1849 (see later). In 1851, prompted by Arago, he undertook to check Fresnel's ideas on the partial drag of ether by a new, and very ingenious, experiment. In the paper reporting this experiment,[15] he first recalled Fresnel's hypothesis:

If only a part of the ether is drag [by a moving body], the velocity of light is increased, but only by a fraction of the velocity of the body... This consequence is not evident, but Fresnel has shown that it is confirmed by very probable mechanical considerations.

Then he explained what he wanted to do:

One owed to M. Arago a method based on the interference phenomenon, which is able to show the smallest variations in the refraction indexes of bodies [we will see that in Chap. 10]. The observations of MM. Arago and Fresnel of the difference in the refraction index of dry and wet air showed the extraordinary sensitivity of this method.

I adopted the same principle, and adapted the double tube of M. Arago to the apparatus with two conjugated telescopes that I used to determine the absolute velocity of light, so that I could study directly in two media, air and water, the effects of the motion of a body on the light that crosses it.

Figure 4.6 shows the principle of the experiment, which re-used the two telescopes with which Fizeau measured the velocity of light 2 years before. The light coming from the Sun through an heliostat, either that of Gambey belonging to the Observatory or more probably a Silbermann heliostat belonging to Fizeau (Fig. 4.7) was concentrated by a cylindrical lens upon a slit located at the focus of one of the telescopes. A parallel beam was produced. Two slits were placed in front of the objective. The light that crossed each of these slits was sent along a 1.5-m tube, then reached another telescope with a mirror at its focus. This telescope reflected the light coming from one of the tubes into the other one. Air or water circulated in the two tubes so that, with the direction indicated in the figure, the effects added to each other. The two reflected beams crossed the slits, entered the first telescope and interfered at its focus. One saw with the eyepiece the interference fringes formed in this way.

With water circulating in series in the tubes with a velocity of 7 m/s, Fizeau observed a displacement of 0.23 fringe with respect to quiet water. The calculation of partial drag adapted from Fresnel predicted 0.20 fringe, instead of 0.40 fringe if the drag was total. Fizeau concluded that Fresnel was right.[16] If he filled the tubes with air circulating with a velocity of 25 m/s, he saw no displacement.

[15] **Comptes rendus* (1851) 33 p. 349–355.

[16] There is a fundamental difference between the observation of Arago and of Fizeau's experiment. In the latter, the refringent medium moves with respect to the observer, while this is not the case in the former.

Fig. 4.5 Hippolyte Fizeau (1819–1896). Private collection

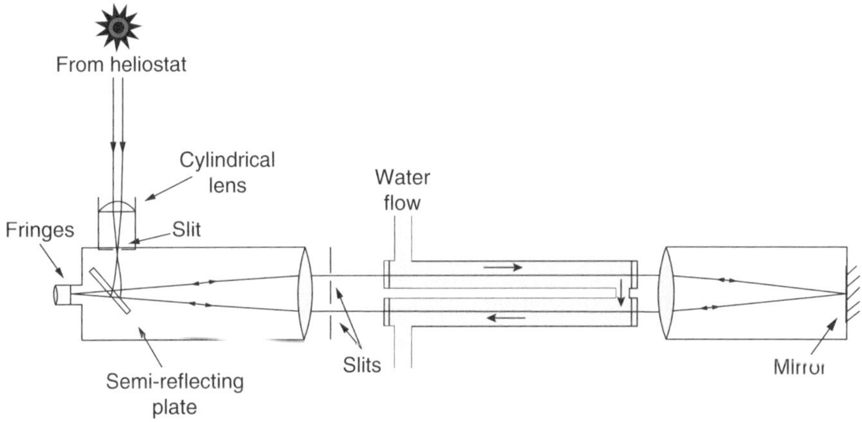

Fig. 4.6 Fizeau's experiment. Explanations in the text

The calculation predicted a negligible displacement of 0.00023 fringe with partial drag, instead of 0.41 fringe with total drag. Once again, Fresnel was right, or at least his formula was exact. However, Fizeau was not fully convinced by his explanation. He wrote as a conclusion of his paper:

The success of this experiment seems to show me that Fresnel's hypothesis should be adopted, or at least his formula for expressing the change of the velocity of light under the effect of the motion of bodies. Indeed, although this law is verified, and is

Fig. 4.7 A Silbermann heliostat that belonged to Fizeau, mid-nineteenth century. Heliostats were largely used as light sources by physicists, chemists and also for photographic enlargements. Musée d'histoire urbaine et sociale de Suresnes

a very strong confirmation of the hypothesis of which it is only a consequence, perhaps the conception of Fresnel would look so extraordinary, and difficult to admit in some points, that other proofs are needed as well as a deep analysis by theoreticians before it can be adopted as the expression of reality.

How right was Fizeau to be skeptical!

The Modern Interpretation of Arago's and Fizeau's Experiments

Fizeau's experiment was carried out again by several scientists, in particular by Albert A. Michelson (1852–1931) and Edward W. Morley (1838–1923), who had better accuracy.[17] Their results confirmed Fizeau's. Michelson and Morley also realized in 1881, and again in 1887, a "gigantic apparatus" with which they showed that it was impossible to detect the motion of the Earth with respect to ether. As to Fizeau,

[17] Michelson, A.A. & Morley, E.W. (1886) *American journal of science* 31, p. 377–86, accessible by http://en.wikisource.org/wiki/Influence_of_Motion_of_the_Medium_on_the_Velocity_of_Light

he spent a large part of his life trying to detect this motion by various means, but in vain. No one had yet abandoned the idea of ether. But things were going to change progressively. For example, the French physicist Eleuthère Mascart (1837–1908), wrote as early as 1874, hence before the Michelson–Morley experiments[18]:

The general conclusion of my Memoir would be (with the exception of the experiment of M. Fizeau on the rotation of the plane of polarization by a pile of glass plates[19]) that the translation motion of the Earth has no observable influence on the optical phenomena produced with a terrestrial or solar light, that these phenomena do not allow to determine the absolute motion of a body, and that only relative motions can be determined.

In 1893, in his remarkable *Traité d'Optique,* where one can find a lot of interesting information on nineteenth century physics, Mascart was much more affirmative.[20] The theory of relativity was germinating. Its construction was a long and gradual process to which Hendrik Anton Lorentz (1853–1928), Henri Poincaré (1854–1912) and others contributed. However, Albert Einstein (1879–1955) was the one who made the decisive step, by accepting fully the implied overthrow in the concepts of space and time. His fundamental postulate was the constancy of the velocity of light in vacuum, designated by the letter c. It is only then that one understood that the ether did not exist, that there was no absolute reference frame, and that the light could propagate in vacuum. No physicist before could consider this, even Maxwell, who established the electromagnetic nature of light.

It is not difficult to interpret Arago's and Fizeau's experiments in the frame of the theory of relativity.

First, it is clear that Arago could not see any change in the deviation of his prisms, because light arrived on them always with the same velocity c whatever the nature of the source and its motion.[21] Only the wavelength of the light is modified by the Doppler–Fizeau effect. The same reason explains the failure of the experiment of Michelson and Morley.

Fizeau's experiment is more subtle. In this case, light in water has a velocity lower than c, and this velocity can compose itself with that of the water, but not by direct addition. The relativistic calculation shows that, if v is the velocity of light in water at rest with respect to the observer ($v = c/n$, n being the refraction index), and if now the water moves with respect to the observer in the same direction as light with the velocity u, the resulting velocity V of the light with respect in the frame of the observer is:

$$V = (v + u) / (1 + v^2 / c^2)$$

[18] Mascart, E. (1874) *Annales scientifiques de l'École Normale Supérieure.* 2e série, 3, p. 363–420, see p. 420. Accessible by http://archive.numdam.org

[19] This experiment seemed initially to yield a positive result, but the hopes failed after some time.

[20] Mascart (1889–1893) see t. 3, p. 114.

[21] Actually a slightly lower velocity than c because the prisms were in air. But this velocity still did not depend on the source and its motion.

In vacuum, $v=c$, $v=(c+u)/(1+u/c)=c$ is as expected from the theory of relativity. As v^2/c^2 is much smaller than 1, one can perform a limited development of the denominator of the above formula to get:

$$V=(v+u)\left(1-v^2/c^2+\ldots\right)=v+u\left(1-1/n^2\right)+\ldots,$$

To the first order, this is Fresnel's formula.[22]

If the velocities were added directly, one would have the Newtonian formula $V=v+u$, which is quite different. In the version of the experiment where water is replaced by air, the difference between the relativistic and the Newtonian formula is even larger, because v is barely smaller than c, then V in relativistic mechanics.

These results look quite simple to us today, but we have seen how long and tortuous the way was to obtain them.

Arago's "Crucial Experiment"

In spite of evidence, the followers of the corpuscular theory of light had a hard time accepting the wave theory. Arago was well placed to observe this, and probably suffered much from it. For example, Biot still believed in 1839 in the initial explanation that Arago and Laplace gave of the 1809–1810 observations.[23] But Arago was not someone to disarm, and although he was increasingly busy with his political and administrative activities, he decided to carry out what he called a "crucial experiment" with the purpose of baffling the adepts of the Newtonian theory. This was the comparison of the velocity of light in air and in water (remember that in the emission theory the velocity of light was supposed to be greater in the denser medium, while it was the reverse in the undulatory theory). The history of this experiment is well known,[24] so I will give here only a brief summary.

The Rotating Mirror

Because light has a very high velocity, Arago needed a way to see and if possible to measure extremely fast phenomena. In 1834, during a trip to England, he heard about an apparatus just built by the physicist Charles Wheatstone (1802–1875, Fig. 4.8),

[22] This demonstration was first published by Max von Laue (1879–1960) 2 years after Einstein's fundamental paper: von Laue, M. (1907) *Annalen der Physik* 23, p. 989–90. English translation in http://en.wikisource.org/wiki/Translation:The_Entrainment_of_Light_by_Moving_Bodies_in_Accordance_with_the_Principle_of_Relativity

[23] See **Comptes rendus* (1839) 8, p. 259–272 and 315–327.

[24] See Tobin (2003) and Lequeux (2014).

Fig. 4.8 Charles Wheatstone (*left*) and Louis Breguet (*right*). Private collection

whose purpose was to measure the "speed of electricity".[25] Wheatstone placed above each other two sparks produced by an electrostatic machine at the two ends of a long wire. He looked to them by reflection on a small mirror rotating very rapidly. If the spark in the distant end were delayed with respect to the other, he would have seen a small shift between the two images because the mirror would have slightly turned in the meantime. Indeed, he saw this shift and concluded that electricity propagated with a very high velocity, that he estimated as about 460,000 km/s.[26] The method was clever, and Arago proposed to use it for his own experiment in 1838, when he presented Wheatstone as a foreign member of the Academy of Sciences.

Arago imagined producing a long single spark (Fig. 4.9).[27] Wheatstone showed that such sparks last for a very short time. The light from the lower part of the spark would cross a tube full of water, while that from the upper part would propagate in air. Both rays would fall of a mirror rotating around a vertical axis; the lower ray would be delayed with respect to the upper one and would be reflected later by the mirror. One would have looked at the reflected rays with a telescope, and seen which one had taken more time; Arago calculated that with a tube of 14 m and a mirror rotating at 1000 rps the difference between the two images would be half an arc minute.

---

[25] Wheatstone, C. (1834) *Philosophical transactions* 124, p. 583–591, accessible as http://www.jstor.org/stable/108080

[26] This velocity is that of propagation of a disturbance in an electric circuit, not that of electrons in the conductor. It depends on the properties in the environment of the conductor. Fizeau and Gounelle (1850, *Comptes rendus* 30, p. 437–440) measured more reasonable velocities of 100,000 and 180,000 km/s in long telegraphic lines.

[27] **Comptes rendus* (1850) 30, p. 489–495; also *Arago, F., *Œuvres complètes* t. 7, p. 569–599.

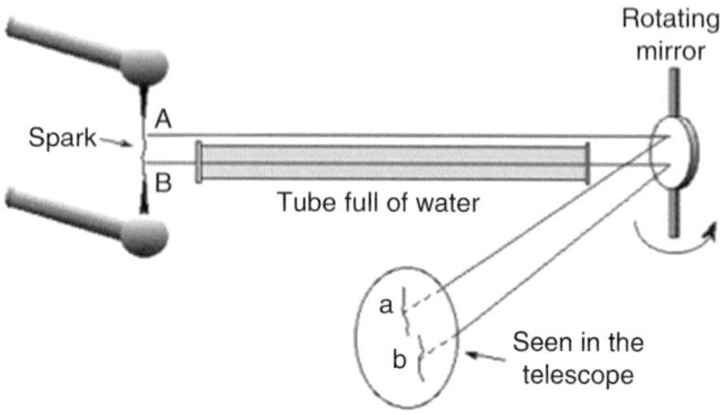

Fig. 4.9 Arago's project to compare the velocity of light in water and in air

To increase the chances of seeing the spark, Arago proposed having several observers, each one with a telescope, all around the rotating mirror, and to replace the simple rotating mirror by 8–10 mirrors on the same axis. Instead of water, he also thought to use carbon disulfide CS_2, which has a very large refraction index and disperses light strongly: one would have seen a spectrum and it would have been sufficient to see in which order the colors were disposed to solve the problem. Arago, at this time named "director of observations," with probably an increased budget, had Louis Breguet (1884–1883, Fig. 4.8) build several rotating mirrors with their clockworks. One of them is in the collections of the Paris Observatory (see the title-page of this chapter) and could turn at 2000 rps and even 3000 rps. In spite of numerous trials, Arago could not obtain any result and abandoned the project. He was overwhelmed by other work and already weakened by diabetes. A reason for his failure is that he did not follow Wheatstone, who managed to synchronize the spark ignition with the rotation of the mirror, insuring in this way that the direction of the reflected spark would remain fixed and could be seen by a single observer.

Fizeau and Foucault Pick Up the Torch

Fizeau and Foucault visited the Observatory frequently and knew Arago well. It is not surprising that they would take over the abandoned experiment. The light from the spark being weak, they thought of replacing it by the light from the Sun reflected by a heliostat. Fizeau said that Foucault had the idea to send back towards the rotating mirror the ray it had reflected, thanks to a plane mirror. The direction of the reflected ray would be fixed, avoiding the need of many observers as proposed by Arago. Finally the two men acknowledged that the idea was due to the German astronomer Friedrich Wilhelm Bessel (1784–1846). Another advantage of this idea was that the rotating mirror would have slightly turned during the travel to the flat mirror and back, so that the return beam would be slightly deviated: the measurement of this deviation would allow to obtain the velocity of light. In any case, the idea of a flat mirror was not a really good one. The flat mirror would only have sent back the ray to the rotating mirror for the very short time it would have fallen perpendicularly to its surface, the light being lost during the rest of the time. Fizeau and Foucault, who collaborated in the trial, did not succeed and abandoned the experiment.

Fig. 4.10 Léon Foucault (1819–1868). Wikimedia Commons

This was only temporary. Probably after his measurement of the velocity of light in 1849, Fizeau had an excellent idea: to replace the flat mirror by a concave one, which would reflect the light beam during all the time it would reach it, increasing enormously in this way the luminosity of the experiment. It is then that Fizeau and Foucault ceased their collaboration and tried the experiment separately. Fizeau, who was well acquainted with Arago, installed his experiment in the very large meridian room of the Observatory (now called Salle Cassini), using one of the rotating mirrors of Breguet. Foucault (Fig. 4.10) worked in his home on Rue d'Assas in Paris with a rotating mirror driven by a small steam-driven turbine built by Paul-Gustave Froment (1815–1865). Apart from this, the two set-ups were very similar (Fig. 4.11). Then a steeplechase began, depending very much on the weather because both scientist used a heliostat. Foucault won the competition on 27 April 1850,[28] while Fizeau and Breguet succeeded only on 17 June.[29] The results agreed: the velocity of light was slower in water than in air, and the corpuscular theory was definitely dead.

Problems of Deontology

It is probable that Arago was somewhat upset by the way things occurred. If at present it is generally considered as normal for a scientist to work from a published idea, this was certainly not the case in his time. Arago was careful to give his point of view in a note describing his project for the comparison of the velocity of light in water and

[28] *Foucault, L. (1850) *Comptes rendus* 30, p. 551–60.

[29] *Fizeau, H. & Breguet, L. (1850) *Comptes rendus* 30, p. 771–4.

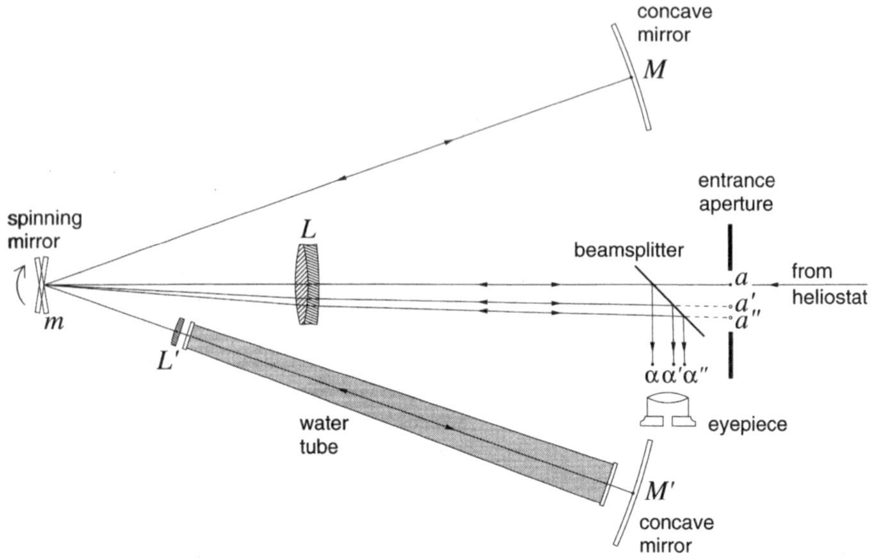

Fig. 4.11 Foucault's set up to compare the velocity of light in water and in air. The tube filled with water was long of 3 m. A thread was placed in *a*. The concave mirrors *M* or *M'* and the lens *L* produced an image of it in the same plane when the rotating mirror *m* had the appropriate angle. When this mirror turned slowly, the beams were reflected upon each other and the image was in *a*. But when its rotation was very rapid, it turned slightly during the time that light takes to go from *m* to *M* and return, or from *m* to *M'* and return, so that the images were deviated to *a*' (through air) or *a''* (through water). These images were reflected to the side by the semi-transparent glass plate to *α*, *α'* and α'', which could be observed through the eyepiece. From Tobin (2003), with thanks

in air.[30] He mentioned that Wheatstone had thought in 1835 to make this comparison and also to measure the velocity of light, but did not published anything, so that:

Because this idea did not result in any publication, it was for everybody as non-existent: publication is the only means to establish a priority in such matters.

As Arago had published the principle of the measurement, he considered himself as the owner of the idea, adding:

The publicity given to my system of observation has probably been considered by M. Wheatstone himself as a reason to abstain: he did not carry out the experiment, although he had the necessary equipment.

He also mentioned that

M. Bessel, after my publication in the Comptes rendus, told me that he had considered a modification of my apparatus [the use of a plane mirror]... While sending me this information, the illustrious astronomer of Koenigsberg added: "Although my

[30] See note 27.

process looks simpler, because it is only a modification of yours, I will refrain to try it: the idea of the experiment is yours; you have demonstrated the possibility for its realization; whatever the result, it will belong to you."

Then Arago added:

The things were such, when M. Fizeau determined, by a very ingenious experiment, the velocity of light in the atmosphere. This experiment not being indicated in my Memoir, the author had the right to carry it without exposing himself to the slightest reproach for indelicacy.

As to the experiment for the compared velocity of light in a liquid and in air, the author wrote to me: "I have done yet no attempt in this direction, and I am waiting for your explicit permission to work on it." This loyal reservation could only add to my regard for the character and the works of M. Fizeau, and I hastened to authorize M. Breguet to lend him one or several of my rotating mirrors.

M. Foucault, whose inventive mind is known by the Academy, came by himself to inform me of his desire to submit to the test of an experiment a modification that he wanted to make to my set up.

Given the present status of my sight, I can only accompany with my wishes the experimentalists who want to follow my ideas.

Thus Arago gave to Fizeau the permission to carry out the experiment. Although Foucault had, according to his friend the physicist Jules Lissajous (1822–1880), "the extreme delicacy to ask Arago for an authorization that he could barely decline,"[31] Arago was probably upset, although Lissajous had added: "The permission was given so readily that one does not know what one should admire most, of the modest respect of the young scientist or of the noble condescension of the old man." Foucault probably realized that it must have pained Arago, who he thanks at the end of his communication to the Academy with uncommon flattery.

The Direct Measurement of the Velocity of Light

The experiments we just described did not aim at measuring the absolute velocity of light. However, in the middle of the nineteenth century, this value had a great interest by itself, as we will see now.

The Precursors

From antiquity to the middle of the seventeenth century, a kind of "unique thought" assumed the instantaneous propagation of light. Galileo was one of the first to have doubts on the matter. He was the first to attempt to measure the velocity of light by

[31] Lissajous, *Notice historique …*, in Foucault (1878), vol. Plates, p. 1–18, see p. 5.

a famous nocturnal experiment.[32] He had a lantern that he uncovered. Another person located at a large distance had another lantern that he uncovered when he saw the light of the first lantern. Galileo tried to evaluate the time between the uncovering of his lantern and the instant when he saw the light from the second lantern. If the velocity of the light was, for example, similar to that of sound (340 m/s), he would have been able to appreciate this delay. But clearly the light was too fast for the success of the experiment, and Galileo could not decide whether light propagated instantaneously or was simple extremely rapid.

The next step was the demonstration that the velocity of light was finite. This was due to Jean-Dominique Cassini (1625–1712, Fig. 4.15), the first responsible for the Paris Observatory, and Ole (or Olaus) Rømer (1644–1710, Fig. 4.16), a Danish astronomer who was working at the Observatory.[33] Before he came to Paris in 1669, Cassini set up a program of systematic observations to forecast the eclipses of Jupiter's satellites of Jupiter in the shadow of the planet, whose observations would allow the synchronization of clocks in different places, and then the determination of longitudes. At the Observatory, these observations were made by several astronomers, in particular by Rømer. Cassini reported first, on 22 August 1676, that the eclipses of the first satellite, Io, came late with respect to the ephemerides when the Earth moved off Jupiter, and earlier when it came closer. He understood that this could be due to the fact that light had a finite velocity and took more and more time when the Earth moved away from Jupiter, and less and less time in the other case (Fig. 4.12). Cassini soon had doubts about this hypothesis, because the other satellites of Jupiter did not seem to share this property. Actually, their motion is complicated due to their mutual gravitational perturbations, but Cassini could not know that. However, Rømer was confident in the explanation and published it in spite of Cassini's reservations,[34] so that Rømer was the only one who is remembered by history. This explanation was accepted without problem by Huygens, then by Newton after Edmond Halley (1656–1742) had succeeded in 1694 in finding the effect for all satellites, by studying carefully their motion from a large number of observations.

Cassini estimated the transit time of light from the Sun to the Earth as 10 or 11 min; but neither he nor Rømer tried to calculate the numerical value of the velocity of light.[35] They simply were satisfied to see how large it was. Huygens was apparently the first to calculate it, in his *Traité de la lumière* of 1690. He used the distance

---

[32] Galileo Galilei (1638) *Discorsi e dimostrazioni matematiche intorno à due nuove scienze*, Leiden, Elsevier, English translation in http://oll.libertyfund.org/titles/753, see p. 48 of the Ebook PDF version.

[33] This is a complex story, reported in detail by Bobis & Lequeux (2008) *Journal of astronomical history and heritage* 11, p. 97–105, available freely via http://cdsads.u-strasbg.fr

[34] **Journal des sçavans,* (1676) p. 233–236. Poor and arranged translation in English in *Philosophical transactions* (1677) 12, p. 893–894, accessible via http://www.jstor.org/stable/101779

[35] Many physics treatises, even famous ones, give values of the velocity of light that they attribute to Rømer but are different from each other, and they often give wrong dates for the discovery of the finite nature of this velocity; see Wroblewski (1985) *American Journal of Physics* 53, p. 620–630.

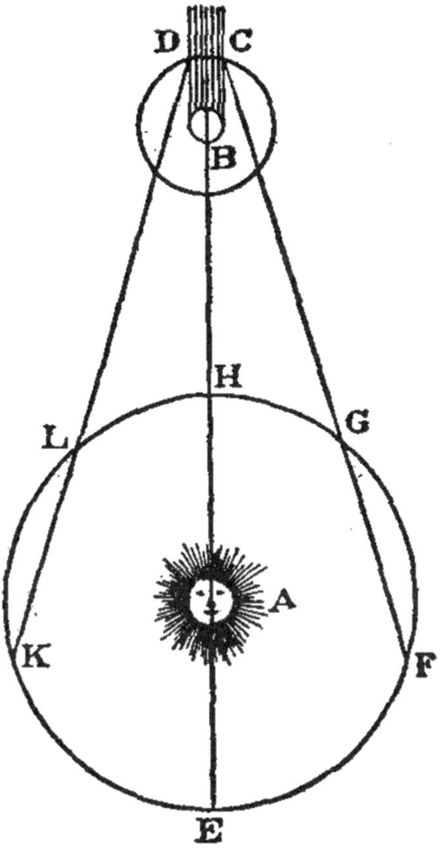

Fig. 4.12 The explanation by Rømer of the delays in the eclipses of the first satellite of Jupiter, Io. The Sun is in A, Jupiter in B with its shadow cone, and the drawing is in the reference system Sun-Jupiter. Two positions of the Earth, L and K, are represented at the times of two emersions of Io out of Jupiter's shadow in D. The Earth moved away from Jupiter between these two emersions, and the second one seems late because of the extra time required for the light to propagate. Conversely, the immersions of the satellite in the shadow, in C, seems increasingly early when the Earth moves from F to G. Wikimedia Commons, Gij1952

Sun–Earth, then estimated as 12,000 times the diameter of the Earth (the true value is 11,750 times), divided by the transit time, to obtain 230,000 km/s (in modern units). Of course, the accuracy was poor and the result is too small by 20 %.

The discovery of aberration by Bradley in 1728 allowed new progress. Bradley estimated that the light is 10,210 times faster than the Earth on its orbit, and consequently that the time taken by light to cross the radius of this orbit is $3.16\times10^7/(10{,}210\times2\pi)=492$ s; or 8 min 12 s (3.16×10^7 is the number of seconds in a year). This is a very good result, close to the present value of 8 min 19.005 s. However, Bradley considered that the radius of the Earth's orbit was still poorly known, and he refrained from giving a numerical value for the velocity of light.

Later, there was other progress in the determination of the crossing time of light from the Sun to the Earth. Delambre gave 8 min 13.20 s in 1817, from more than a thousand observations of eclipses of the satellites of Jupiter.[36] In 1843, this value

[36] Delambre (1817) *Tables écliptiques des Satellites de Jupiter d'après la Théorie de M le Marquis de Laplace, et la totalité des observations faites depuis 1662 jusqu'à l'an 1802*, Paris, Courcier.

was superseded by that obtained by Wilhelm Struve (1793–1864) in Russia from a very accurate determination of the constant of aberration, i.e. 8 min 17.8 s. The knowledge of the distance from the Earth to the Sun was also improved thanks to observations of the passage of Venus in front of the Sun in 1761 and 1769, discussed by the German astronomer Johann Franz Encke (1791–1865). Putting everything together, Arago gave in his *Astronomie populaire* a value of 77,076 *lieues* per second, i.e. 308,300 km/s, for the velocity of light in vacuum. Because this book was based on lectures given from 1813 to 1846, it is difficult to know when he estimated this value.

Fizeau's Toothed Wheel

We do not know why Fizeau was sufficiently interested by the velocity of light to envisage a direct measurement. Perhaps he was pushed by Arago. He asked Froment to build at his own expense the apparatus depicted on Fig. 4.13, with which he made this measurement in July 1849.[37]

Figure 4.14 shows the principle of the instrument, and Fig. 4.15 shows what Fizeau saw in the eyepiece. The basic principle is similar to that of Galileo's experiment, but the great novelty is the use of a toothed wheel, based on that used by Felix Savart (1791–1841) to produce sounds with known frequency. The distance between the house of Fizeau's parents in Suresnes, where the main station was located in a belvedere, and Montmartre where the returning telescope was placed, was 8633 m. The wheel had 720 teeth and an occultation occurred when it turned at 12.6 rps. At twice this velocity, light was seen again and at a still higher velocity there was another occultation. From this, Fizeau derived a velocity of 70,948 *lieues* per second, i.e. 315,300 km/s.[38] Fizeau did not attach much importance to this value, because he knew that the rotation speed of his wheel was not known precisely and that it was difficult to appreciate the extinction and reappearance of light in the eyepiece; but he had shown that it was possible to directly measure the velocity of light on the Earth. His experiment was considered as a major event, and the Academy of Sciences, prompted by Arago, supported the construction by Froment of a larger instrument that should have allowed a more precise determination. However, what happened to this apparatus is not known.

[37] **Comptes rendus* (1849) 29, p. 90–92. The original apparatus has disappeared because Fizeau used some of its elements for other experiments. A replica can be seen in the museum of the École Polytechnique. For a detailed analysis of the experiment, see Frercks J. (2000) *Centaurus* 42, p. 249–287, accessible freely via http://onlinelibrary.wiley.com/doi/10.1034/j.1600-0498.2000.420401.x/abstract

[38] The metric system was not yet fully used in practice, even by scientists, half a century after its promulgation, although its use was declared compulsory in 1840.

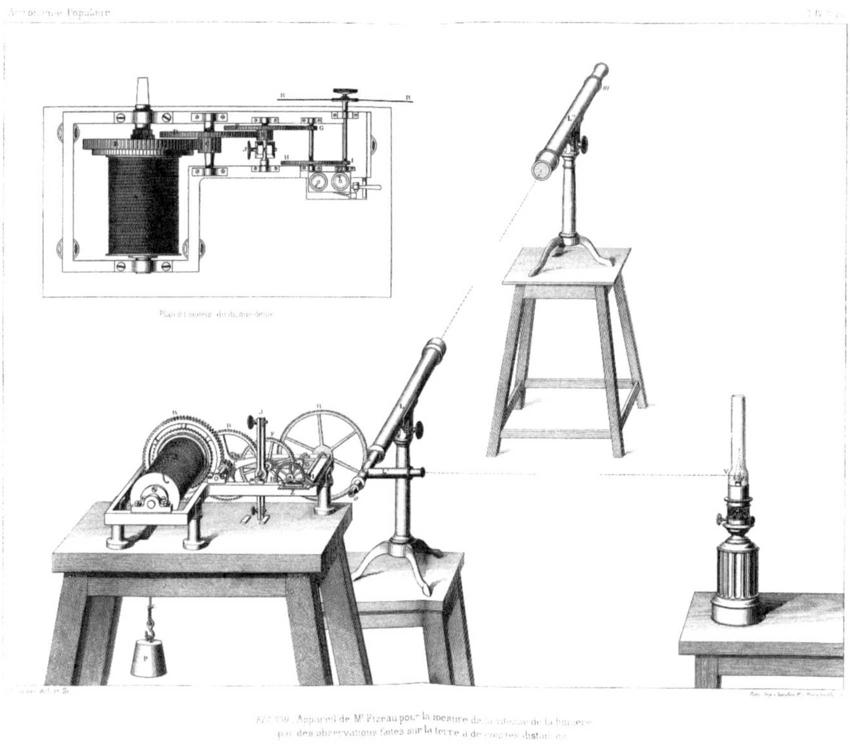

Fig. 4.13 Fizeau's apparatus to measure the velocity of light. The light beam from the lamp (actually a heliostat or a Drummond lamp made of a piece of chalk heated by an oxyhydrogen burner) was chopped up by a toothed wheel driven by a clockwork (also seen from the *top* in the *box*). The telescope sent the beam from Suresnes to Montmartre, where another telescope with a flat mirror at its focus sent it back to the first one. The beam again crossed the toothed wheel and was observed through the eyepiece. A counter and a chronometer allowed the determination of the rotation speed of the wheel. From Arago's *Astronomie populaire*, author's collection

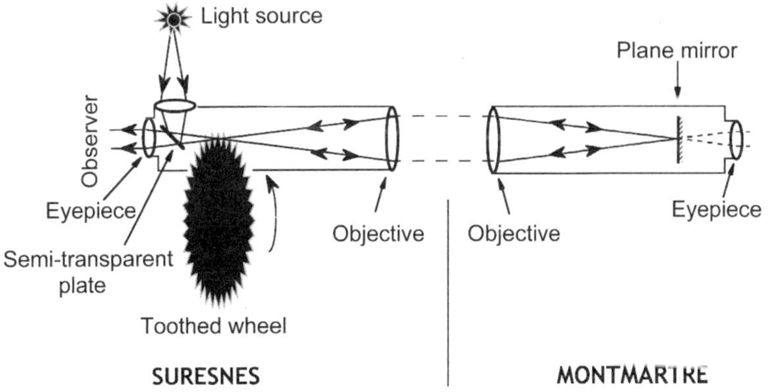

Fig. 4.14 Principle of Fizeau's apparatus. (*Right*) A modified telescope had to send back the light beam in its direction of arrival. For this, Fizeau put a flat mirror exactly at the focus: this device has the property of sending back light exactly in the arrival direction, even if the alignment is not perfect. To focus the mirror, Fizeau looked with the eyepiece of this telescope at distant objects through a non-silvered, ruled part of the mirror, and superimposed their image to that of this part

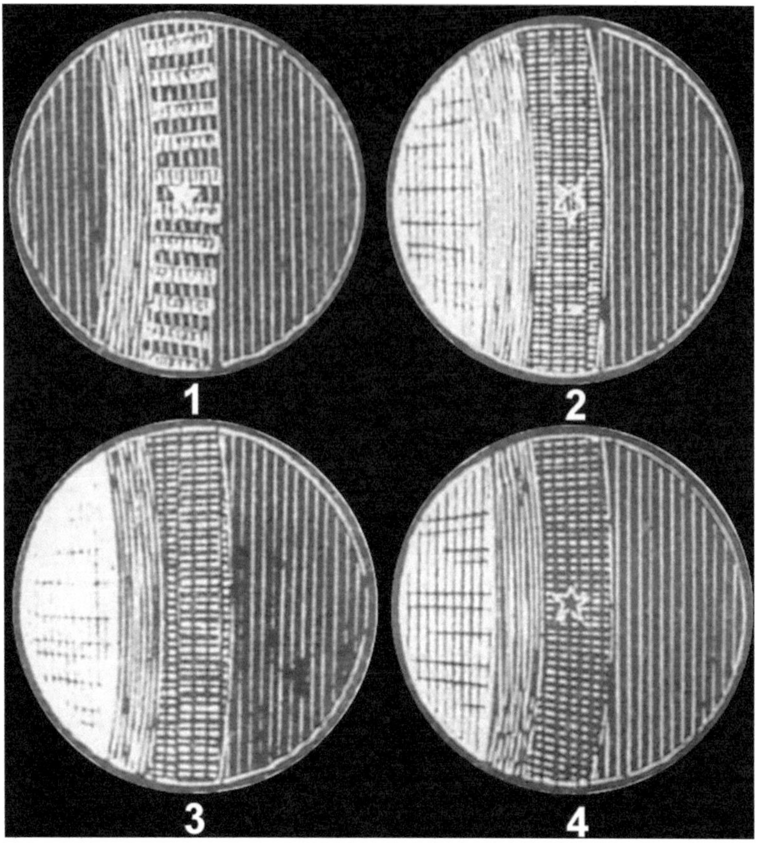

Fig. 4.15 What Fizeau saw in the eyepiece. At low rotation velocities of the toothed wheel, the return light beam passes in the interval between two teeth (*1*). When the wheel rotates slowly, the return beam is still visible because the time taken by light to cross the distance Suresnes–Montmartre and back is smaller than the time of rotation between two teeth (*2*). At some higher speed, the return beam strikes a tooth and no light is seen (*3*). At a still higher speed, the return beam comes to the next gap between teeth and is visible again (*4*). From Arago's *Astronomie populaire,* private collection

The Velocity of Light After Fizeau and Arago

Although Arago died in 1853, soon after the experiments I have described, I found it of interest to describe succinctly what followed.[39]

Le Verrier, Arago's rival, was appointed director of the Paris Observatory in 1854. His scientific goal was to apply Newtonian mechanics to the whole solar system, including of course the mutual perturbations between planets. He discovered Neptune

[39] Mainly from Tobin, W. (1993) *Vistas in Astronomy* 36, p. 253–294 and Tobin (2003).

in this way in 1846. In 1858, he concluded from his calculations that the then-adopted value for the mean distance from the Earth to the Sun, 153.5 million kilometers, was overestimated.[40] He proposed instead 147 million km. At that time, the transit time of light through that distance was rather well known: 497.8 s, the value obtained by Struve (see above). If one could precisely measure the velocity of light, one would obtain a precise value for the distance Earth–Sun, and Le Verrier hoped that it would confirm his theoretical determination. He then asked Foucault, who was now the *physicien de l'Observatoire,* to precisely measure the velocity of light (Fig. 4.16).

Foucault decided to use a rotating mirror as in his previous experiments and asked Froment to build an apparatus, the principle of which he described as early as 1850,[41] and which took into account the experience gained in the comparison of the velocity of light in water and in air (Fig. 4.11). The original rotating mirror is preserved at the Paris Observatory, and there are replicas in various places. The mirror (Fig. 4.17) was driven by a compressed air turbine. The compressed air was produced by a very stable bellow built by the famous organ builder Aristide Cavaillé-Coll (1811–1899), which is probably the one preserved in the *Musée des Arts et Métiers/CNAM* in Paris. Its speed—a critical point for the measurement—was set as 400 rps by stroboscopy. The distance crossed by light was 40.5 m, thanks to five intermediate mirrors. The displacement of the image was measured accurately with a micrometer. In 1862, Foucault measured in this way a velocity of

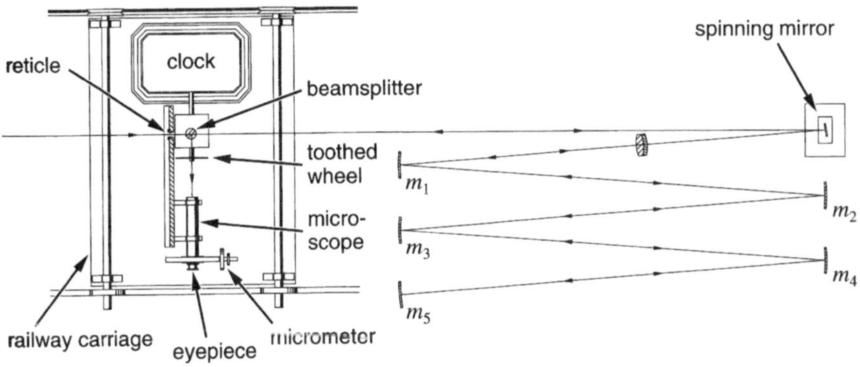

Fig. 4.16 Foucault's apparatus to measure the velocity of light in 1862. The solar light reflected by a heliostat illuminated a micrometer and falled on the rotating mirror. The latter sent the light via the mirrors m_1 to m_4 on the concave mirror m_5, which sent back the beam on the rotating mirror along the same path. This mirror having turned slightly during the course, it reflected the light in a direction different from that of arrival. A semi-transparent glass plate sent the beam sideways, and the displacement of the image of the micrometer could be measured with respect to crossed wires. One also saw in the eyepiece the edge of a toothed wheel driven by a clockwork at 1 rps. From Tobin (2003), with thanks

[40] *Le Verrier (1858) *Annales de l'Observatoire de Paris—Mémoires* 4, p. 1–262.

[41] *Foucault, L. (1850) *Comptes rendus* 30, p. 551–60.

Fig. 4.17 Foucault's rotating mirror (1862). The mirror itself is located in the dark circular barrel. The compressed air turbine driving the mirror is above it. Photo René Cavaroz

298,000 km/s, with an uncertainty estimated somewhat optimistically as 500 km/s.[42] Combined with the time taken by light to cross the distance from the Sun to the Earth, this measurement decreased the distance to 148.3 million km, close to Le Verrier's prediction.

If astronomers were satisfied with Foucault's result, some physicists had their doubts. One of them was Alfred Cornu (1841–1902, Fig. 4.18), professor of physics at the École Polytechnique, who was a pupil of Fizeau. In 1871, he tried to measure the velocity of light by Fizeau's method and Foucault's, and was disappointed by the rotating mirror. He decided then to use Fizeau's method but without keeping the speed of the toothed wheel fixed. Instead, he had it vary continuously, and an electric contact on one of the wheels of the driving clockwork wrote a mark at each turn on a recording cylinder, while other marks were produced by a chronometer.

[42] It was one of the first times a physicist gave the uncertainty on his measurement.

Fig. 4.18 Alfred Cornu. Wikimedia Commons, Mu

Cornu also wrote on the cylinder the times of appearance and disappearance of the light, and could later calculate the rotation speed of the toothed wheel at these different times.

The first attempts were between the École Polytechnique, which was at this time on the Montagne Sainte-Geneviève on Paris's Left Bank, and an old telegraph tower 2.5 km away, then in 1872 between the school and Mont Valérien, 10 km away. Cornu measured the distance by triangulation. He determined that the speed of light was 298,500 km/s, a figure close to Foucault's. Thus Cornu no longer had reservations with Foucault's method and his skepticism turned into admiration. Encouraged by Fizeau and Le Verrier, Cornu made new measurements in 1874, this time between the Observatory and the Montlhéry tower (no longer visible from the Observatory) with new apparatuses built by Breguet. The distance was already well known, but Cornu measured it again, with geodetic points on the Panthéon and at the suburb of Villejuif. He measured the speed at 300,400 km/s with an uncertainty estimated as 1/1000.[43]

This was the last manifestation of French supremacy in measuring the velocity of light. In 1878, Simon Newcomb (1835–1909) and Michelson in the United States started a series of increasingly precise measurements using a modified Foucault method. These measurements culminated in the 1927 and gave a velocity of 299,796 km/s, ±4 km/s. However, Cornu's apparatus was re-installed in 1902 at the

[43] Cornu's results are detailed in a very long Memoir: *Annales de l'Observatoire de Paris-Mémoires* (1876) 13, p. 1–317, accessible via http://adsabs.harvard.edu/historical.html

Nice observatory by its director Henri Perrotin (1845–1904), who made several measurements.[44] The best one, in 1902, gave 299,880 km/s, ±50 km/s, acceptably close to Newcomb's and Michelson's results.

After Perrotin and Michelson, neither the rotating mirror nor the toothed wheel was used, because better means were available, such as magneto- or electro-optical shutters. Then came other methods that were very accurate. In the 1970s, several laboratories succeeded in simultaneously measuring the wavelength and the frequency of an infrared laser, whose ratio gives the velocity of light, with an accuracy better than 0.2 km/s. Because time can also be measured with extreme accuracy, the *Conference générale des poids et mesures* (General Conference of Weights and Measures) decided in 1983 to fix the velocity of light as 299,792,458 m per second exactly, and to derive from it the definition of the meter, which is no more a fundamental unit.

[44] For a history of these measurements, which caused a controversy between Cornu and Perrotin, see Bogaert, G. & Blanc, W. (2011) *Reflets de la Physique* n 26, p. 20–22 accessible as http://www.refletsdelaphysique.fr/articles/refdp/pdf/2011/04/refdp201126p20.pdf

Chapter 5
The Birth of Electromagnetism

The Electric Battery

At the end of the eighteenth century, electricity was limited to electrostatics. However, the invention in 1799 of the electric battery by the Italian physicist Alessandro Volta (1745–1827, Fig. 5.1) opened new horizons.[1] Soon, important ensembles of batteries were constructed: assemblies of many "trough batteries" (Fig. 5.2) invented in 1802 by William Cruickshank (1745–1810 or 1811); these batteries had zinc and copper electrodes immersed in a diluted solution of sulphuric acid. In 1807, Wollaston built a large battery for the Royal Institution of London, with 2000 pairs of 4 × 4-in. plates, generating about 2200 V. It was used by Humphry Davy, who discovered sodium and potassium through the electrolysis of melted salts and experimented with the electric arc. Soon after, Napoléon financed a battery of 600 elements at the École Polytechnique (Fig. 5.3). Another one, similar to the London battery, existed in Geneva, and the list is certainly not complete.

These very expensive instruments were the equivalents of our modern particle accelerators. Of course, there were criticisms: some commentators claimed that the École Polytechnique battery was a bad investment:

A rather considerable sum was given a few years ago in France to construct a voltaic apparatus that would be used by the most able chemists. One expected marvels from this beautiful dotation; its effect was only some sort of moral galvanism, one of the jerks of opinion that the head of the government always attempted to trigger…

Arago answered to this[2]:

For a good history of electromagnetism, see http://users.wsg.net/bedrosian/em history.htm

[1] Volta, A. (1800) *Philosophical transactions* 90, p. 403–431, accessible via http://www.jstor.org/stable/107060. This paper, written in French, contains the figures reproduced here.

[2] **Œuvres complètes*, t. 4, p. 405–408.

J. Lequeux, *François Arago*, Astrophysics and Space Science Library 421,
DOI 10.1007/978-3-319-20723-0_5

A primitive electric motor built by Froment, mid-nineteenth century. Three pairs of coils (a vertical one is barely visible under the axis) successively attract bars of iron located around the rotor. A commutator in front distributes the current to the coils so that the rotor and the attached flywheel turn in the desired direction. Bibliothèque de l'Observatoire de Paris

Fig. 5.1 Volta's "artificial electric organ." Initially, it was formed by a pile of zinc and silver disks separated by cloth disks dampened by salt water or some alkaline solution. Wikimedia Commons, Luigi Chiesa

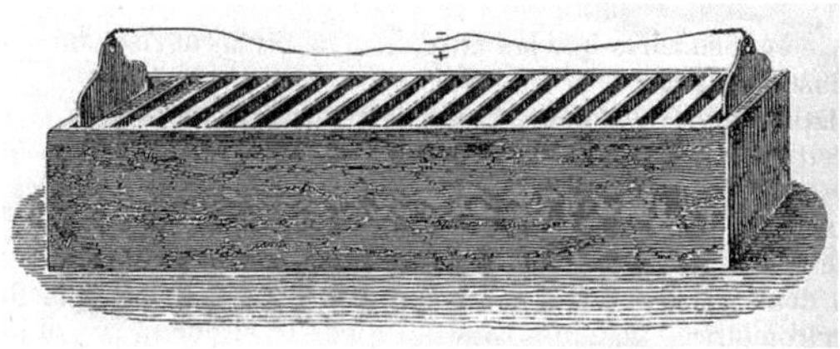

Fig. 5.2 Cruickshank's trough battery rapidly replaced Volta's original battery. Wikimedia Commons, Kurzon

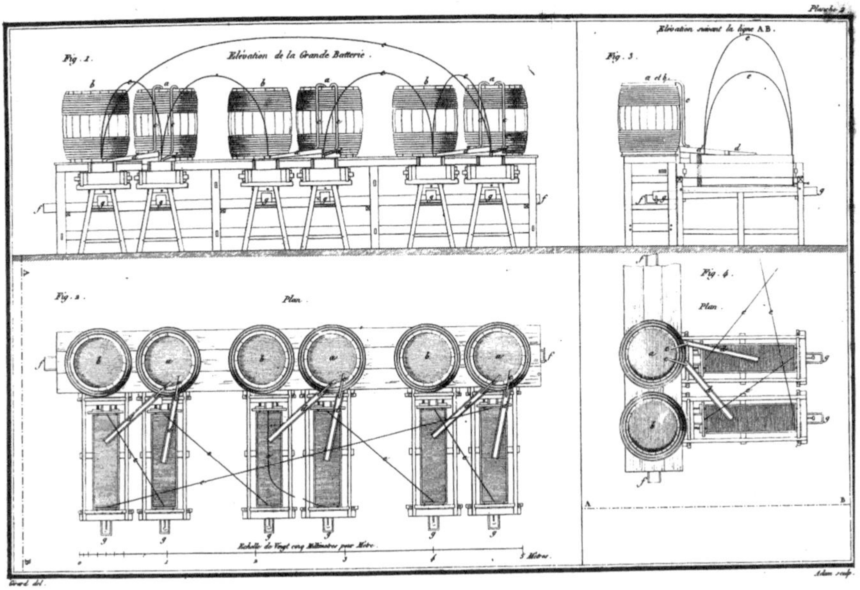

Fig. 5.3 The large battery installed by Gay-Lussac and Thenard at the École Polytechnique. It consisted of six wooden containers with 100 pairs of plates each. The surface of the plates was 9 dm^2. Each pair was made of a copper plate weighting 1 kg and a zinc plate of 3 kg. The conductors connecting each container to the next one are visible. The barrels contained an acid solution to be poured into the containers, or water to rinse them after draining. The battery could supply about 10 A under 660 V. From Gay-Lussac and Thenard (1811). Cnum—Conservatoire numérique des Arts et Métiers—http://cnum.cnam.fr

Since the writers never heard that science benefited from this equipment, I am happy to tell them that there exists a two-volume book by MM. Gay-Lussac and Thenard; that this book, entitled "Physico-chemical researches made with the battery, etc.," contains a very extended chapter on the causes of variations in the energy of a galvanic battery…

The batteries were mainly used by chemists for electrolyses. However, they would soon be used for research on electricity.

Fig. 5.4 Christian Œrsted. Private collection

Œrsted's Experiment

We do not know what type of battery was used by the Danish physicist Christian Œrsted (1777–1851, Fig. 5.4). He is often pictured in his laboratory with a Volta pile like that in Fig. 5.1, but these representations are much later than the experiments. Anyway, he discovered in 1820 a new property of electricity: A magnetized needle is deviated when placed close to an electric current (Figs. 5.5 and 5.6).

This experiment was a great surprise, because it showed for the first time that a force was not necessarily directed along the line joining the two interacting bodies, but was in this case in a perpendicular direction. It became immediately celebrated, and was the starting point for many experiments in Europe and the United States. Arago saw it in August 1820 in Geneva, where Œrsted's Memoir, which was written in Latin, had just been translated into French. He promoted it in France, and wrote[3]:

The readers of the Annales [de Chimie et de Physique] should have remarked that we only publish with care the announcements of extraordinary discoveries, and until now we can be satisfied with these reservations. However, as far as the Memoir of M. Œrsted is concerned, the results it contains, how strange they may look, are accompanied by too many details to be suspected of any error. I would add that M. the professor de la Rive [Charles Gaspard, 1770–1834], of Geneva, who himself discovered extremely curious phenomena with the powerful voltaic

[3]Arago himself published in 1820 a French translation of Œrsted's Memoir, with a note, in **Annales de Chimie et de Physique* 14, p. 417–425. The cited text is from this note.

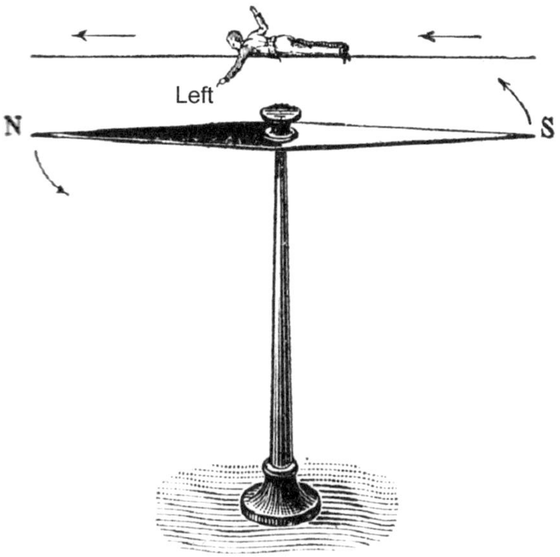

Fig. 5.5 Œrsted's experiment. A magnetized needle with a blackened north pole, placed above a horizontal electric current, is deviated as indicated in the direction of the left arm of the "Ampère's fellow". Author's collection

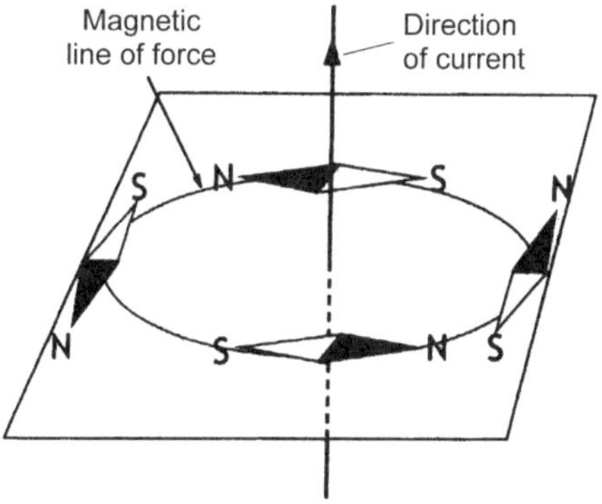

Fig. 5.6 A variant of Œrsted's experiment. Magnetized needles are placed around a vertical electric current and are deviated as indicated (at least if the Earth's magnetic field can be neglected). The magnetic line of force that passes through the needles is represented

batteries he owns, allowed me to see the verification he made of Œrsted's experiments together with MM. Prévost [Pierre, 1731–1839], Pictet [Marc-Auguste (1752–1825)], de Saussure [Nicolas Théodore (1767–1845)], Marcet [Alexandre (1770–1845)], de Candolle [Augustin Pyrame (1778–1879)], etc. I could thus be convinced of the exactness of the main results obtained by the learned Danish scientist.

Œrsted mentioned an "electric conflict" surrounding the conducting wire, noting that "this conflict makes circles." He probably wanted to say that the magnetized needles around the wire as in Fig. 5.6 are tangent to a circle centered on this conductor. Michael Faraday would understand later that this is a line of force of the magnetic field.

Fig. 5.7 André-Marie Ampère, engraving by Ambroise Tardieu, 1825. Wikimedia Commons, Astrochemist

Ampère's First Experiments

Back from Geneva, Arago again carried out Œrsted's experiment at the Academy of Sciences, on 11 September 1820. He spoke of it with enthusiasm to his friends, and looked for batteries for his own experiments. He could use several of them, probably including that of the École Polytechnique, where Ampère (Fig. 5.7, Box 5.1) was a professor. Ampère had seen Arago's demonstration of Œrsted's experiment at the Academy, and was very excited. The two scientists first worked separately, then together. Later, Ampère specified their respective contributions and that of Faraday:

One knows that M. Œrsted was first to see the action of voltaic conductors on magnets; M. Faraday, the spinning action, always in the same direction, between a conductor and a magnet [see later]; M. Arago, the property of these conductors to magnetize iron and steel; and M. Ampère, everything relative to the mutual action between conductors and to the action exerted on them by the terrestrial globe, as well as the rotation of a magnet or a conducting wire around its axis.

Ampère made a series of important experiments, described in numerous publications.[4] He very clearly distinguished the electric tension (what we now call potential difference, or voltage) from the electric current:

---

[4]The first Ampère's publications on electromagnetism were published by Arago in **Annales de chimie et de physique*; (1820) 15, p. 59–76 and 170–218; (1821) 18, p. 88–106 and 313–333; (1822) 20, p. 60–74 and 398–421. Then they were published by the Academy of sciences: **Mémoires de l'Académie des sciences* (1823) 6, p. 175–388. See also Ampère and Babinet (1822). More information on http://www.ampere.cnrs.fr

Box 5.1 André-Marie Ampère

Born in Lyons in 1775, Ampère showed at a very young age a strong interest for science, and especially for mathematics. At 13, he sent a Memoir of geometry to the Academy of Lyons. But the Revolution, to which he adhered, was for him a cause of disarray because his father was guillotined (he was a judge and had protested against revolutionary excesses). He married in 1799, but his wife died only 4 years later. His second marriage, in 1806, was unhappy and he divorced 2 years later. Named professor of physics and chemistry at the *École centrale* of Bourg-en-Bresse, he sent to Delambre in 1802 a Memoir entitled "Considerations on the mathematical theory of games," hoping to be transferred to Lyons where his wife and son were living. His value being appreciated by Delambre and by Laplace, he was first transferred to Lyons in 1803, then the next year to the École Polytechnique in Paris as a *répétiteur* in mathematics. Two years later, he was promoted to professor of analysis, a post he held until 1822. However, he was very interested in chemistry. He showed that "fluoric acid" was made of hydrogen and fluorine, at that time an unknown element. In 1814, he stated independently of Amedeo Avogadro (1776–1856) the law that bears their two names: "Equal volumes of gas in the same conditions of temperature and pressure contain the same number of molecules." He also participated with Fresnel in various experiments in optics. He was elected to the Academy of Sciences in 1814. In 1819, he began to teach philosophy at the Faculty of Humanities in Paris, giving a course on "the classification of intellectual facts." But Œrsted's discovery was to change his interests: Through an enormous effort that affected his health, he created electromagnetism in less than 4 months, in a fierce competition with Biot. He was elected professor at the prestigious Collège de France in 1824. The paper he published in 1826 on electromagnetism was his last great scientific work. He spent the rest of his life working on the classification of sciences. Named in 1828 general inspector of public instruction, he died from pneumonia in 1836, during a round of visits in Marseilles.

The electromotive force manifests itself by two sorts of effects that I must first distinguish by a precise definition:

I will call the first electric tension, the second electric current. The first is observed when the two bodies between which the electromotive action occurs are separated from each other. The second occurs when, conversely, they are part of a circuit of conducting bodies.

He admitted, as did everybody at this time, that the electric current was double, one of positive electricity and the other of negative one. But he said that he would only speak of a single sort of electric current, for simplification. He mentioned that the tension was measured with an electrometer, and the current with a magnetized needle, what he called a *galvanometer* (from the name of Luigi Galvani [1737–1798], the predecessor of Volta).

He described his fundamental discovery as follows:

[I discovered remarkable effects] in setting in parallel directions two rectilinear parts of two conductors connecting the poles of two voltaic batteries... I observed that in sending a current in each of them they attracted mutually if the currents were in the same direction, and that they repelled each other if their directions were opposite.

This discovery came only 7 days after he had seen Œrsted's experiment. For his experiments, Ampère carefully constructed several apparatuses.[5] In one of these experiments, a conducting wire was wound around an isolating rod, and he saw that this coil behaved like a magnet when the current was on. This suggested to him that magnets should contain close-loop currents. He attempted in this way—and this was his ultimate purpose—to unify magnetism and electricity.

Ampère established the formula that gives the force between two elements of electric currents, but without a numerical coefficient. How could he have done, since there was no means to quantitatively measure intensity? Later, the unit of intensity, which bears his name, was defined from this force. Its official definition is:

The ampere is that constant current which, if maintained in two straight parallel conductors of infinite length, of negligible circular cross-section, and placed 1 m apart in a vacuum, would produce between these conductors a force equal to 2×10^{-7} *N per meter of length.*

Ampère was not the only one to set the basic formulae of electromagnetism. There was in 1820 a race among Biot, Savart and Ampère to give a mathematical expression able to represent the effects of electric currents. In October of that year, Biot and Savart presented to the Academy "a Memoir whose object is to determine, through precise measurements, the physical laws according to which metallic wires put in communication with the two poles of the voltaic apparatus act on magnetized bodies".[6] They suspended a magnetized needle in various positions close to a linear, very long conductor, and appreciated "the action felt by a molecule of austral or boreal magnetism," observing the orientation of the needle at equilibrium and the period of its oscillations around this position. They inaugurated in this way a method to estimate the strength of a magnetic field that was used for a long time, for example in the "intensity compass" described later in Chap. 9. They found that the equilibrium

[5] Very clumsily, Ampère most often missed his experiments when he was facing an audience. Arago and Fresnel, who were more skillful, helped him when possible.

[6] °*Annales de chimie et de physique* (1820) 15, p. 223–223.

direction was perpendicular to the direction of the current and that the "magnetic force" was inversely proportional to the distance to the wire.[7]

A few days later, Laplace showed that the "magnetic force" created by a small element of current was inversely proportional to the square of the distance to this element. It was a satisfactory result for Newtonians, for whom the actions at distance should indeed have been proportional to the inverse square of this distance, as for gravity, electrostatics and even magnetism as shown by Charles-Augustin Coulomb (1736–1806). After much effort, Ampère gave the differential formula that gives the force exerted by one element of current on another one.[8] For the first time a formula contained the essential parameter of electric current: its intensity. Ampère could then calculate by integration the effect of a rectilinear or circular current, and that of a solenoid.[9]

Arago Intervenes

In the meantime, Arago carried out other experiments[10]:

I adapted a rather thin, cylindrical wire to one of the poles of the voltaic battery, and I noticed that when this wire was in contact with the opposite pole, it attracted iron filings as a real magnet… As soon as the conjunctive wire [e.g. the wire that connected the two poles] ceased to be in contact with one of the poles, the filings detached from it and fell.

Arago realized in the same way a permanent magnetization of small pieces of steel and of a steel needle. He added:

Ampère, to whom I showed the experiments I just described… had just made his important discovery… He derived from it, by analogy, the consequence that the attractive or repulsive properties of magnets depend on electric currents that circulate around iron or steel molecules… These theoretical considerations suggested to him immediately that one would obtain a stronger magnetization by replacing the straight wire I used by a wire wound in helix, along which the needle would be placed.

---

[7] One generally calls the *law of Biot and Savart* the formula that gives the magnetic field produced by an element of current, whose intensity is inversely proportional to the square of the distance to this element. This denomination is inappropriate. Actually, this formula was demonstrated not by them, but by Laplace and by Ampère; Biot and Savart only observed the magnetic field produced by a very long current, which is inversely proportional to the distance to the wire.

[8] **Annales de chimie et de physique* (1822) 20, p. 398–421. This formula was completed by Demontferrand and by Savary: **Annales de chimie et de physique* (1823) 22, p. 91–100.

[9] **Annales de chimie et de physique* (1824) 26, p. 134–162 and 246–258.

[10] **Annales de chimie et de physique* (1820) 15, p. 85 and 93–102.

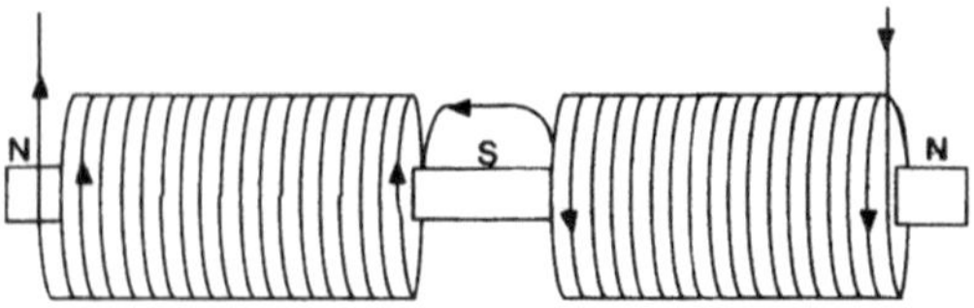

Fig. 5.8 Arago, placing a long iron bar inside two coils wound in opposite senses, obtained the magnetic poles illustrated in the figure

No sooner said than done. Ampère and Babinet wrote about this in 1822[11]:

[Arago and Ampère] thought that if one were to wind a conducting wire around a bar of steel, this would produce inside this bar currents in the same sense that those in the surrounding helix: in this way, a south magnetic pole would appear to the left of the currents in the helix for an observer located in these currents and looking at the bar... This was entirely confirmed by experiment...

With a single steel wire inserted into two different helices wound in opposite senses around the same tube, one obtains in the part of the wire between the two helices one of these intermediate poles named "consequent points," so that if, for example, the two extremities of the bar have each a north pole, the middle has the properties of a south pole [Fig. 5.8].

M. Arago also noticed that a straight conducting wire does not magnetize a needle placed parallel to it.... M. Arago completed the results he obtained with the battery by producing the same effects with the ordinary electricity. It suffices for this to send in the turns of a coil a series of sparks produced by an electrical machine.

This text is interesting because it describes the invention of the electromagnet, due to Ampère and Arago together, and not only by Arago as often stated. Also, one sees that Arago showed the identity of the electricity produced by a battery and by an electrostatic machine, something that was already clear in Ampère's mind but deserved an experimental confirmation.

The ways Ampère and Arago conducted their research were very different, but this did not affect their harmonious collaboration. Arago never tried to crush, through his considerable authority, a less well-known scientist, whose genius he acknowledged. This attitude is similar to that he had before with Fresnel. Arago was basically an experimenter who did not try to explain the causes of what he observed. Ampère had a deeper mind: He was not only an imaginative experimenter, but he set out to understand the laws of the phenomena he observed, and even to some extent to understand their physical causes. However, on the last point, he would be overshadowed by Faraday.

---

[11] °Ampère and Babinet (1822) p. 68–69.

Arago and Faraday: Transformation of Energy

The First Electric Motors

Ampère and Arago had both understood that the action of electric currents between them or with a magnet resulted in forces, and Ampère had carried out an experiment in which a conductor was given "a continuous motion of revolution under the effect of a magnet, of the Earth or of another conductor".[12] However, Faraday (Fig. 5.9, Box 5.2) had observed such a motion before Ampère, by carrying out the two experiments represented Fig. 5.10.[13]

Box 5.2 Michael Faraday
Faraday was born near London in 1791 into a poor family. His education was rudimentary, and he earned his living for 7 years as an apprentice bookbinder, a specialty in which he could develop his remarkable manual skills. His interest in science was piqued by reading an unorthodox paper on electricity in the *Encyclopaedia Britannica*. This also developed his critical faculties. In 1810 he entered a scientific society, the City Philosophical Society, and attended the courses of Humphry Davy at the Royal Institution. Davy took him on as an assistant, and made a long trip with him through France and Italy, where they met many scientists. Back in 1815, Faraday launched into chemistry, and soon became an acknowledged specialist. In 1820, he synthesized the C_2Cl_6 and C_2Cl_4 molecules, and discovered benzene in 1825. This did not prevent him from developing an interest in the nascent electromagnetism, contributing actively to it. This culminated by his discovery of induction in 1831. In 1834, he set up the laws of electrolysis. Four years later, he gathered the scattered existing knowledge of electricity with the purpose of producing a coherent theory, discovering in passing diamagnetism and paramagnetism, as well as the rotation of the plane of polarization of light crossing some materials submitted to a magnetic field. Then, he began to introduce the notion of field that would be developed by his pupil James Clerk Maxwell (1831–1879). Unfortunately the deterioration of his mental faculties and his insufficient knowledge of mathematics did not allow him to follow Maxwell's progress. He retired in 1862 and died 5 years later. Given the variety and novelty of his fundamental and applied research, Faraday was certainly one of the greatest experimentalists of all times.

---

[12] **Annales de chimie et de physique* (1822) 20, p. 60–74.

[13] Faraday's discoveries have been analyzed very well by Tyndall (1868).

Fig. 5.9 Michael Faraday, painting by Thomas Phillips, 1842. Wikimedia Commons, Succu

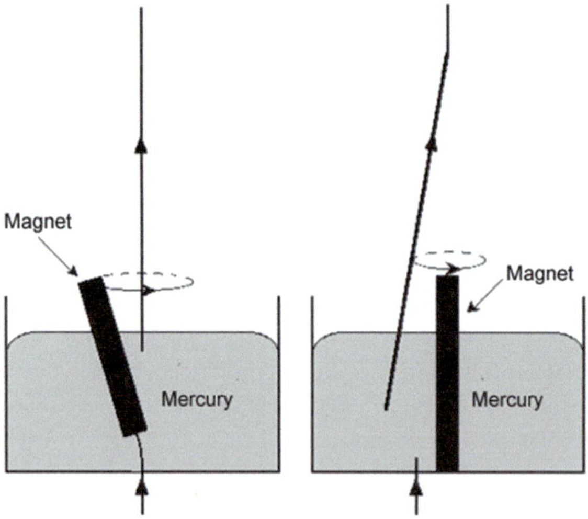

Fig. 5.10 Faraday's first motors. (*Left*) A magnet supported by a flexible conducting wire was in equilibrium in a mercury bath thanks to the Archimedes force; when the current circulated as indicated, the magnet experienced a force that caused its rotation. (*Right*) The magnet is fixed and a suspended conducting rod plunged in mercury. When the current was on, the force exerted by the magnet caused the rotation of the rod

Ampère and Faraday soon had followers who built various motors. The best known is that of Peter Barlow (1776–1862) depicted in Fig. 5.11.

Ampère, meanwhile, described a universal "rather complex electro-dynamical apparatus"[14] that also produces motion.

[14] Anonymous (Ampère) (1824); see also **Annales de Chimie et de Physique* (1822) 20, p. 60–74.

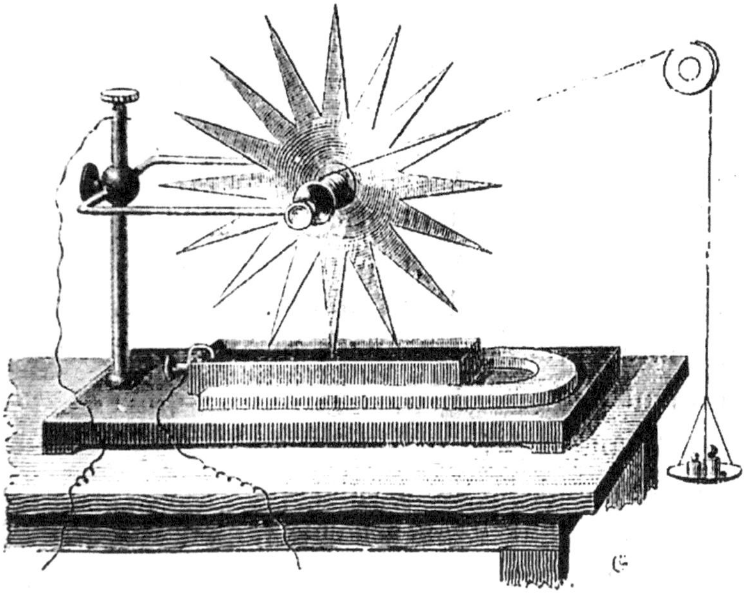

Fig. 5.11 Barlow's wheel. This is the best known primitive electric motor, because its functioning looks easier to understand than that of Faraday's motors (Fig. 5.10), the topography of the magnetic field being straightforward. The current enters by the mercury bath in which the edge of the copper wheel plunges and exits by its axis. A horseshoe magnet placed on the wooden support creates a magnetic field perpendicular to the wheel, and produces its rotation. Wikimedia Commons, Yodah06

These instruments were the first electric motors, where a new form of transformation of energy was produced: the transformation of "electrical strength" into "mechanical strength" (read: energy). With his famous experiment, Œrsted had already shown the transformation of this "electrical force" into "magnetic force." Already, the transformation of heat into mechanical energy, and vice versa, was known thanks to the steam engine. One also knew the transformation of electrical energy into chemical energy, and the reverse transformation in batteries.

Faraday had an original view of what occurred in his motors.[15] Instead of considering, like Ampère and the Newtonians, the forces exerted in a straight line on a current by a magnet, or by the internal closed currents that were supposed to give the magnetism, he thought of a "circular magnetic field" that acts on the magnet or the conducting rod in his experiments (Fig. 5.10); this came directly from Œrsted (see Fig. 5.6). Here, we see the birth of the concept of line of force, which was to be so fruitful.

[15] See the very interesting article of L. Pearce on Faraday in Gillespie (1970–1990) vol. 4, p. 527–540.

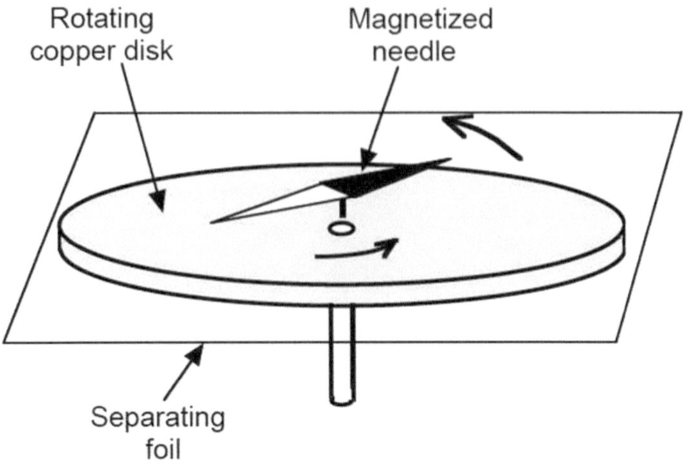

Fig. 5.12 The magnetism of rotation. A magnetic needle is placed above a rotating copper disk, and is driven at a lower velocity by the magnetic field produced in the disk by the currents induced by its own magnetic field. The disk and needle are separated by a thin foil in order to avoid the perturbations of the wind created by the rotating disk

The "Magnetism of Rotation"

After his short but intense collaboration with Ampère, Arago turned to other topics: a collaboration with Fresnel on lamps for lighthouses, then the geodesic connection of England and France. However, during a trip to England with Humboldt, he made an observation that was to have important consequences. He presented it at the Academy of Sciences on 22 November 1824. He wrote about this toward the end of his life[16]:

While measuring with my friend Alexandre de Humboldt the intensity of the magnetic field on the slope of the hill in Greenwich, I noticed that the declination needle, if put in motion, came to rest faster when it was placed in its box[17] *than when isolated from any surrounding body.*

In 1825, Arago carried out the simple experiment shown on Fig. 5.12. He had a conducting plate rotate under a magnetized needle "placed in a glass box and separated from the moving plate by a membrane in order to keep it away from the motions of air." He saw that the needle was driven in the same direction of rotation as the plate. As "the clockwork which gives the motion was entirely made of copper," this was necessarily a new phenomenon, what Arago called "magnetism of

[16] **Œuvres complètes* t. 4, p. 424–448.

[17] This box was probably made of brass, but Arago says nothing about this.

rotation"; indeed, the rotation of the disk had produced a magnetic field able to drive the needle. On 7 March 1825, he presented his experiment at the Academy.[18] Arago claimed that the effect existed whatever the material of the rotating disk.

This experiment is easy to reproduce, and indeed Arago wrote in 1850 that it could be seen in all physics laboratories. He also noticed that the action on the needle was much less visible when radial slits are cut in the rotating disk. He also showed that the experiment succeeds with an electric current instead of a magnetized needle.

Ampère's electromagnetic theory was unable to explain the phenomenon, and indeed no one could understand the magnetism of rotation, considered as "the darkest physical phenomenon of the day".[19] Arago's discovery was considered so important that the prestigious Copley Medal of the Royal Society was awarded to him in 1827.

Controversies

Several scientists reproduced Arago's experiment, among them John Herschel and Charles Babbage (1792–1871), the famous pioneer of informatics. Like Arago, they substituted the rotating metallic disk by disks of non-conducting materials, but found no effect while Arago said he had found a weak effect. They noticed that the effect was directly dependent on the conductivity of the material. Another physicist, Brewster, even pretended to be the first to have performed the experiment. Arago's response was crushing!

An experiment that recalled Arago and Humboldt's initial observations was performed by two Italian scientists. Arago, who considered that their results were wrong, commented[20]:

MM. Léopold Nobili and Bacelli, from Modena, produced oscillations of magnetized needles above non-metallic substances... without finding appreciable differences in the oscillations of the needles above the disks or without them. If the Modena physicists had given the distance between the needle and the non-metallic plate, I could perhaps have demonstrated their error; but all I can do is to oppose exact measurements to their denials.

We know now that Nobili and Bacelli were right, as were John Herschel and Babbage. Arago was somewhat insincere in the controversy. He gave results on the damping of the oscillations of the magnetic needle close to various substances, with a lot of details but little precision on the circumstances of his experiments, and his

[18] *Annales de Chimie et de Physique (1825) 28, p. 325–326.

[19] From °Tyndall (1868), p. 25. Amongst the many attempts to explain the phenomenon, one can cite those of Poisson: **Annales de Chimie et de Physique* (1826) 32, p. 225–240 and 306–307.

[20] **Annales de Chimie et de Physique* (1826) 32, p. 213–223.

work is not a model of scientific rigor.[21] When he placed the needle above a plane surface, it was so close (sometimes as close as 0.7 mm) that it was certainly damped by the viscosity of air.

The Discovery of Induction

Faraday is the one who succeeded in explaining the magnetism of rotation. Indeed, this phenomenon was one of those that led him to induction, which he discovered in 1831.[22] Faraday had long been obsessed by the idea that an electric current could excite a current in another conductor. First, he superimposed two coils, sent a current in one of them and hoped to obtain a current in the other, but with no success. However, he noticed that there was a weak, temporary current in the second coil at the time he opened or cut the circuit in the first one. He then built a closed magnetic circuit excited by a primary coil and comprising a secondary coil. Now, the effect was very strong: clearly, an electric current was "induced" by the variation of another current, via the intermediary of a magnet (we recognize here the principle of the transformer). After this, Faraday carried out all the classical induction experiments. For example, he observed the creation of a temporary current when he plunged a magnet into a hollow coil.

He was now in a position to explain the magnetism of rotation. He first carried out an experiment in which a copper disk rotated between the poles of a permanent magnet, and collected the currents induced in the disk (Fig. 5.13). He understood, thanks to this experiment and others, that the currents acted on the magnet in order to oppose the relative motions. He could then account for Arago's experiment itself: the magnetic field[23] of the magnetized needle induced currents (that were called later Foucault's currents) in the rotating disk. These currents themselves created a magnetic field that acted on the needle to oppose the relative motion between it and the disk.

When carrying out the preliminary experiment of Fig. 5.13, Faraday invented the first machine able to produce a continuous electric current: this is a *magneto*. The magneto is the reverse of Barlow's motor of Fig. 5.13. We will see later the developments of this principle.

---

[21] Indeed Babbage and Herschel wrote as a note to p. 485 in the *Philosophical Transactions* (1825) 115, p. 467–496 (accessible via http://www.jstor.org/stable/107753): "The meager statements and imperfect reports which have hitherto reached us of Mr. ARAGO's researches, had prepared us to expect a much more appreciable amount of magnetic force in non-metallic bodies than we have observed… This naturally renders us desirous of seeing that eminent philosopher's own account of the means employed by him to render sensible such very minute forces, which must have been unusually delicate."

[22] *Philosophical transactions* (1832) 122, p. 125–162 (accessible via http://www.jstor.org/stable/107956): Faraday devotes p. 146–162 to his "explication of Arago's magnetic phenomena".

[23] Here we anticipate. Faraday did not speak of magnetic field, a notion that he would develop much later. But he already used the term of lines of force.

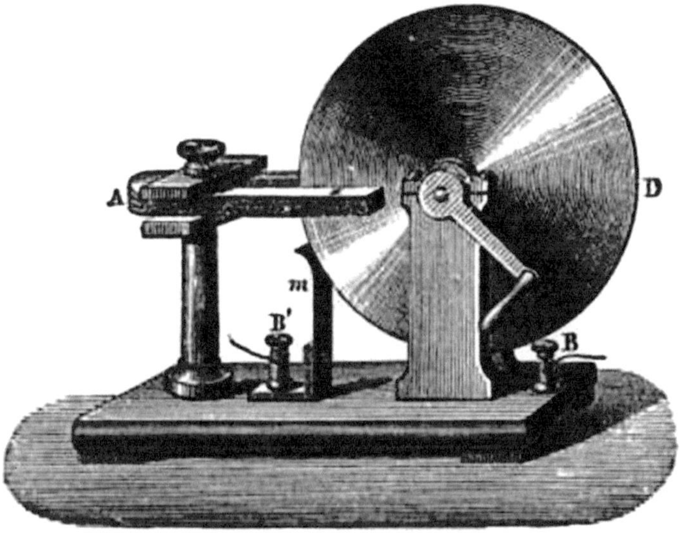

Fig. 5.13 An induction experiment of Faraday. A copper disk D was rotated between the poles of a permanent magnet A. The induced current was collected by the sliding spring contact m, the other pole being the axis f the wheel. The current was detected by a galvanometer. This is not the original device, but probably a later one owned by Léon Foucault. Wikimedia Commons, Chetvorno

Arago's Stubbornness

Arago would never be really convinced by Faraday's explication, nor by his very complete experiments and those of the other scientists I have mentioned. He wrote[24]:

Faraday in 1832 was the first to show, using a galvanometer whose wires were placed on different parts of metallic disks above which there was a fixed magnet, that there are in these metallic disks currents induced by the magnetic needle, and one thought that there would yield a complete explanation of all the phenomena I discovered. I do not share this opinion. On 25 September 1844 I informed the Board of longitudes of my doubts. Here is what the minutes of the meeting contain on this topic:

M. Arago recalls the experiments he carried out, a long time ago, on the diminutions of amplitude of the oscillations of a magnetized needle when these oscillations take place at a short distance from a glass plate, from ice (frozen water) or from the surface of a liquid. M. Arago describes the circumstances of his experiments, from which it results that in the special cases of glass, ice of liquids, the phenomenon has nothing to do with induction.

---

[24] *Arago, F., *Œuvres complètes* t. 4, p. 424–448, see p. 437–438.

At the end of his life, when he was almost blind, Arago asked "his friends Laugier and Barral" to perform new experiments. These experiments, of which he gave a detailed account at the Academy on 7 March 1853, 7 months before his death, were no more rigorous than the previous ones.[25] He concluded nevertheless:

One will acknowledge, I hope, that my experiments are different from those imagined by Faraday, and that they cannot be completely accounted for by the mere existence of temporary induction currents.

Too bad he had been so stubborn!

Faraday and the Magnetic Field

In the meantime, Faraday progressed through new experiments, and above all through deep reflections. He set up the laws of electrochemistry, and showed that the current in an electrolyte was carried by "molecules" that acquired a positive or negative "exaltation": our modern ions. He understood that this "exaltation" required energy. At this time, the nomenclature that we still use in electrochemistry appeared: Faraday invented the term of *electrode* himself, while those of *cathode, anode, electrolysis, electrolyte, anion* and *cation* are mainly due to William Whewell.

However, the most important intellectual advance of Faraday is the notion of *field*, in order to account for actions at a distance. This idea arose during his work on magnetism. Faraday ignored the cause of magnetism, but stated a fundamental point, the origin of which can be found in his ideas on the lines of force that were mentioned earlier: magnetism shows itself in the medium that surrounds the magnet or the current, and the energy is there, rather than in the magnet or the current. He showed this by tracing with iron filings the lines of force around a magnet (Fig. 5.14), and also by observing that these lines of force concentrate in a ferromagnetic or paramagnetic medium, and conversely diverge in a diamagnetic medium. He wrote,[26] acknowledging that his paper "contains much of a speculative and hypothetical nature":

All the phenomena... seem to me to show the physical existence of an atmosphere of power about a magnet... I conceive that when a magnet is in free space, there is such a medium (magnetically speaking) around it. That a vacuum has its own magnetic relations of attraction and repulsion is manifest from former experimental results... What that outer magnetic medium, deprived of all material substance, may be, I cannot tell, perhaps the æther.

Faraday proposed that the same "atmosphere of power" existed around electric charges and around masses, introducing in this way the electric and the gravitational fields. He stopped there, being conscious that he had touched something fundamental[27]:

---

[25] *Arago, F., *Œuvres complètes* t. 4, p. 424–448, see p. 448.

[26] Faraday (1855) p. 407–437. See p. 422 and 425.

[27] ibid. See p. 437.

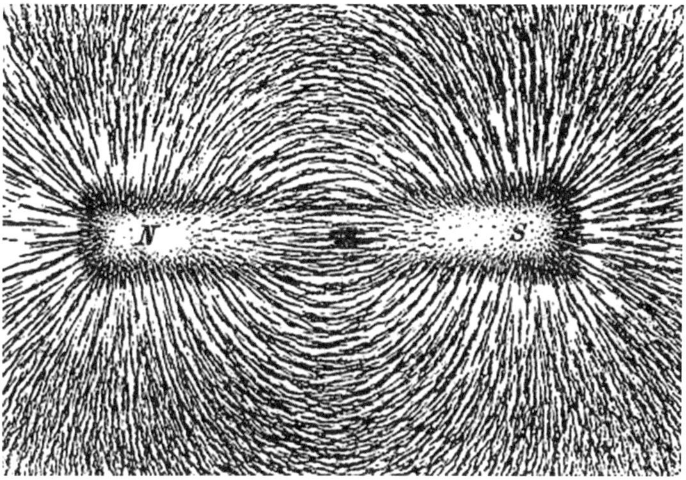

Fig. 5.14 Lines of force of the magnetic field of a magnet, traced by Faraday with iron filings. Wikimedia Commons

I still hold the opinion [on the lines of force] with some hesitation, with as much, indeed, as accompanies any conclusion I endeavour to draw respecting points in the very depths of science, as, for instance, regarding one, two or no electric fluids; or the real nature of a ray of light, or the nature of attraction, even that of gravity itself, or the general nature of matter.

This revolutionary concept is at the basis of the works of his pupil Maxwell in the years 1850–1860. Maxwell said of Faraday's concept[28]:

[Faraday] conceives all space as a field of force, the lines of force being generally curved, and those due to any body extending from it on all sides, their directions being modified by the presence of other bodies.

We should not be surprised that this concept, so opposed to the ideas of Newtonians from Laplace to Ampère, had a hard time being accepted.

Wild Imaginings and Discoveries on Electricity

The reader might think from the preceding pages that the history of the beginnings of electromagnetism was smooth and linear. This was not the case. If the geniuses that were Œrsted, Ampère and Faraday are responsible for the main progresses, the literature of the time is full of papers describing the supposed properties of ether, and accounting for experiments sometimes interesting but wrongly interpreted.

[28] *Maxwell (1873) t. 2, p. 164.

The magnetism of rotation was an example, which had the merit of having put Faraday on the way to induction. It would be too long to summarize all these attempts and imaginings, due among others to de la Rive and Becquerel and their physicist sons. Of course, these erring ways are not unexpected at a time when all the classical physics was created at once.

However, two important discoveries deserve attention: that of the fundamental laws of electric circuits, due to the German Georg Simon Ohm (1789–1854), completed by the French physicist Claude Pouillet,[29] and the discovery of thermoelectricity. The latter is due to the Estonian Thomas Johann Seebeck (1770–1831), who observed in 1821 that a difference of electrical potential appeared between the ends of a metallic rod submitted to a temperature gradient. Closing the circuit with another metal, he noticed that a weak current could be detected with a galvanometer, proportional to the temperature difference between the two junctions.[30] He made in this way a pair of thermocouples that would soon be used to measure temperature differences. Thirteen years later, Jean-Charles Peltier (1785–1845), a French clock maker, discovered the effect to which his name was given, which is the reverse of the one discovered by Seebeck: if one sends a current in a circuit made of two different metals, one of the junctions cools itself while the other becomes hotter. This property was used later to refrigerate small objects. Finally, we should not forget the already mentioned discovery of diamagnetism and paramagnetism by Faraday himself (a discovery of which Becquerel had an inkling in 1821), and that of rotational magnetic polarization (the Faraday effect).[31]

Some of the First Applications of Electricity

The enormous interest provoked by the discovery of the voltaic battery, then of electrodynamics, did not fail to foster many experiments; most of them are now forgotten, except for those I cited before. There was also an increasing number of applications. I will concentrate on physical applications, because they were the ones that most interested Arago.

The Electric Telegraph

As soon as research on electricity started, it was thought that it could be used to transmit information at a distance[32]; for example, Samuel Thomas von Sœmmering (1755–1830) of Poland imagined in 1811 a telegraph in which one would have

[29] **Comptes rendus de l'Académie des sciences* (1837) 4, p. 267–279; for a detailed history, see **Comptes rendus de l'Académie des sciences* (1845) 20, p. 199–212.

[30] See **Annales de chimie et de physique* (1823) 22, p. 199–201.

[31] *Philosophical transactions* (1846) 136, p. 1–20, accessible via http://www.jstor.org/stable/108303

[32] For a history of the electric telegraph, see Highton (1852).

Fig. 5.15 Chappe's telegraph. The three arms were oriented by pulleys in different positions corresponding to letters. Private collection

observed an electrolysis in containers corresponding to the letters of the alphabet, produced when the distant operator would have closed the corresponding circuit. There was no risk that this clumsy apparatus would overcome the optical telegraph of Claude Chappe (1753–1805), which was then in full development, at least in France (Fig. 5.15). However, the Chappe telegraph required many stations with one or two operators each, could only work in daytime with good visibility, and was slow. Scientists were thus looking for something more practical.

This possibility arose in 1820. Three weeks after Arago announced Œrsted's discovery, Ampère wrote[33]:

One could, with as many conducting wires as letters, and placing each letter on a different needle, establish with a battery located far from these needles and which would communicate successively with its two extremities to those of each conductor, form a kind of telegraph able to write all the details one would like to transmit, whatever the obstacles, to the person in charge of observing the letters placed on the needles. Putting on the battery a keyboard whose keys would bear the same letters, and establishing the communication by lowering them, this means of transmission

[33] **Annales de chimie et de physique* (1820) 15, p. 59–76, see p. 73. Ampère imagined that a mobile conductor at the reception would be moved by the attraction by the same current in a nearby fixed conductor, and show which letter was transmitted.

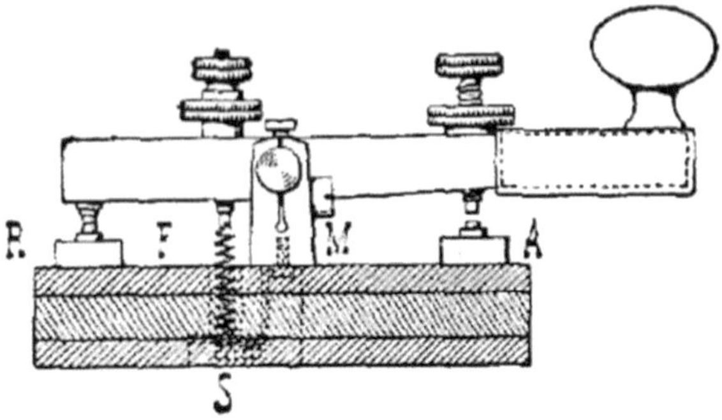

Fig. 5.16 A manipulator for the Morse telegraph. Wikimedia Commons, Kolossos

could be rather easy, and would only require the time necessary to touch the letters on one side and to read them on the other.

Recalling that the "magnetic virtue" of a current is stronger if the wire is wound in a spiral, Arago suggested that this "increase of force" could allow to build a recording apparatus using a weak current acting on an electromagnet, and in this way to construct a true electric telegraph. The idea was launched, but it took time to come to a practical realization.

Indeed, it is only in 1834 that Carl Friedrich Gauss (1777–1855) and Wilhelm Weber (1804–1891), respectively specialists of magnetism and of electricity, built in Göttingen, Germany, the first working electric telegraph, which could send a signal over the 1 km that separated Gauss's observatory and Weber's laboratory.

Later, although the electromagnet was known, ironless galvanometers were used for some time to detect the current. Their response was slow and no recording was possible. In 1837, a telegraph with 30 double copper wires corresponding to as many letters and signs was tried in Edinburgh. More astute, the telegraph of Pavel L'vovitch Schilling (1786–1837) in Saint Petersburg used only five pairs of platinum wires corresponding to five numbers, each letter being coded by a combination of these numbers. That of Wheatstone in 1837 was based on the same principle. The first telegraph working on a distance of many kilometers was that of Karl August Steinheil (1801–1870) in Munich, constructed in the same year. It could record the signals on a band of paper. Steinheil had the merit of discovering that the current could be returned via the ground, dividing by two in this way the number of necessary wires.

Samuel Morse (1791–1872) built the first practical telegraph, in the USA. His first attempts, which used the electromagnet from the start, date from 1832. In the final version, the transmission was done with a very simple manipulator (Fig. 5.16) and the received signal was recorded on paper tape (Fig. 5.17). Thanks to a coding of the letters by short and long signals (the Morse alphabet, still in use by amateur radio enthusiasts), a single conducting wire was required. The first regular telegraphic line was installed in 1844 between Washington, DC, and Baltimore, and extended to Boston in 1845. In 1852, there were already 19,000 km of telegraphic

Fig. 5.17 A Morse telegraph receiver, around 1837. Wikimedia Commons, Zubro

lines in the United States. The electric telegraph was also rapidly spreading in England.

Despite the efforts of Arago, who in 1842 promoted the electric telegraph at the Chamber of Deputies and described its success in England, its development was rather slow in France. The reason was that it competed with Chappe's telegraph, which was then in full development. Also, one feared that the electric lines could be destroyed by malevolence, and the influential physicist Pouillet had reservations about the system. Eventually, the electric telegraph developed in parallel with the railways, because there was a clear need in this case, and the rail tracks were watched over with sufficient care for insuring the safety of the telegraphic lines that were running along side. King Louis-Philippe gave a credit of 240,000 francs on 23 November 1844 for building a telegraphic line along the railway from Paris to Rouen. Louis Breguet (see Fig. 4.8) was in charge of the work: it was finished on 18 May 1845, and the first messages were sent that same day. Breguet had invented for the purpose a system that sent the information letter by letter (Fig. 5.18) at a speed similar to that of the Morse telegraph. Although less robust than the latter, the Breguet system was used until 1880 at least along some of the railway lines. As to the Chappe telegraph, it was abandoned in 1855.

Initially reserved in France for the army and Government, the telegraph was opened to the public in 1851, with considerable success. At the end of that year, a telegraphic submarine cable connected France to England. It was used in 1854 to determine the longitude difference between Paris and Greenwich (see the next chapter). Soon, a telegraphic bureau appeared in all of the important cities of France.

Fig. 5.18 A Breguet dial telegraph. The transmitter (*left*) sent as many impulses as there were steps when the needle was turned from its rest position to the letter. Thanks to an electromagnet and a ratchet mechanism, the receiving needle (*right*) turned by the same amount, so that after an initial adjustment the two dials always indicated the same letter. Wikimedia Commons

Electric Motors

The first functional electric motors appeared soon after the experiments of Faraday, Barlow and others who were mentioned earlier. None of these motors, which did not use induction, had any future. Some of them were rotating engines in which pieces of iron attached to the rotor were attracted by fixed electromagnets placed around this rotor, and fed successively in electricity by a commutator driven by the axis of the rotor. Such a device is photographed on the first page of this chapter. The first motor of this type was probably made in 1834 by a German scientist working in Saint Petersburg, Moritz Hermann von Jacobi (1801–1874). Another type of motor used hollow coils containing a bar of iron on half of their length, and in which another iron bar plunged, attracted by the first one when the current was turned on. With two identical coils fed alternatively by a commutator one obtained a reciprocating motion that was transformed into a rotation by a crank-arm, as in a steam engine. In the United States, Joseph Henry built a motor of this type called a "rocker arm engine," followed by others.

In France, the builder Gustave Froment (1815–1865, Fig. 5.19) dominated the market of electric motors. He conceived motors of both types from 1844. He also superimposed several rotary motors to produce large vertical ones. However, all of these motors would soon be abandoned because of the high cost of batteries, and also because the commutators were rapidly destroyed by the sparks produced at

Fig. 5.19 Gustave Froment. Private collection

each opening of the electric circuit.[34] There were only four electric motors at the 1867 Universal Exhibition in Paris. But the situation was to change after the appearance of Gramme's dynamo.

Magnetos and Dynamos

Soon after Faraday had shown that moving a magnet could produce a current, some engineers and scientists hoped that it would be possible to build machines able to replace the very expensive batteries. The first such machine, which dates from 1832, is that of Antoine Hippolyte Pixii (1808–1835). It used the induction produced by moving permanent magnets in front of fixed coils, where a current was created (Fig. 5.20). This type of apparatus, called magneto-electric machine, or simply magneto, was still in use not long ago for bicycle lights. In the magneto of the Englishman Henry Hyde Clarke (1815–1895), the permanent magnet was fixed and the coils rotated (Fig. 5.21). Both machines produced an alternating current, which was not

[34] These sparks are an effect of the self-induction discovered by Joseph Henry and studied by Faraday. The energy stored in the coil produces the spark when the circuit is open. This problem was not yet solved in 1860. Probably the builders did not know that Fizeau had found the solution in 1853 by inserting a condenser in parallel (**Comptes rendus de l'Académie des sciences* 36, p. 418–421). But the primitive condensers were quite cumbersome.

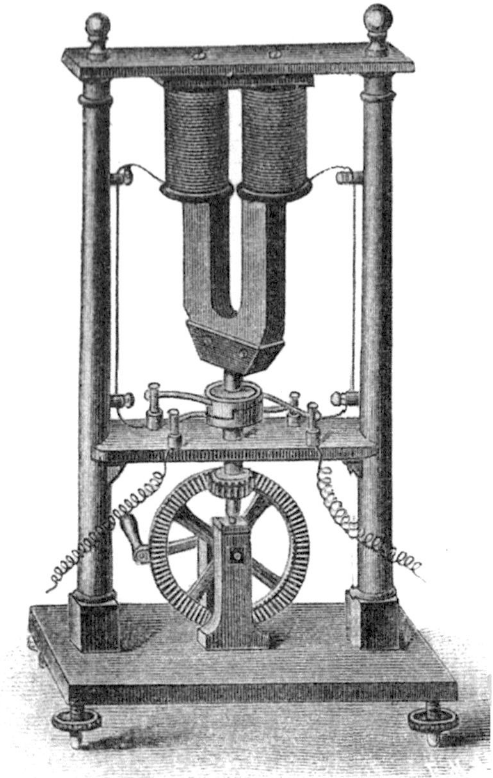

Fig. 5.20 Pixii's magneto. Two straight magnets were rotated by a crank in front of a piece of iron arranged as a horseshoe. An alternating current was created in the two coils mounted in series. This current was rectified by a commutator on the rotation axis. Wikimedia Commons, Niethammer

Fig. 5.21 Clarke's magneto. Two coils were driven by a crank in front of the large horseshoe magnet B. The induced current was rectified by a commutator visible on the axis to the right. Here, it served for electrolysis, the most common use of these small magnetos. Private collection

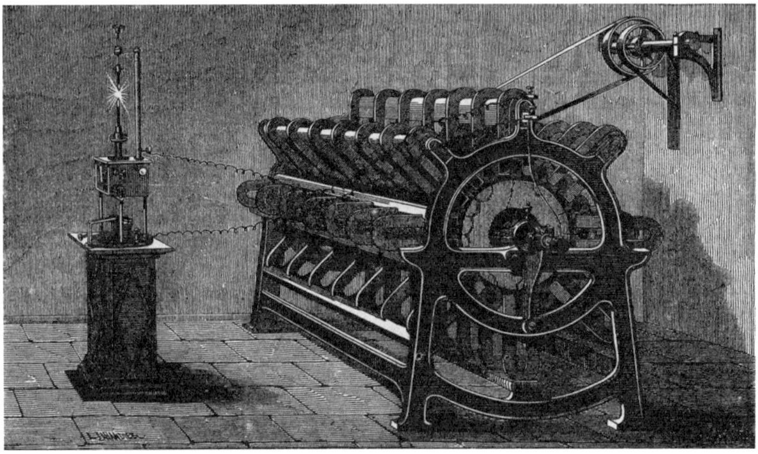

Fig. 5.22 The Alliance magneto, ca. 1860. It was a combination of many Clarke magnetos, the horseshoe magnets of which are visible. The rotor consisted of many coils in series or in parallel, according to the needs, which passed in front of the magnets (the north and south magnetic poles alternated along a series). It was driven by a belt. Here, the magneto fed an electric arc mounted on a regulator of Victor Serrin (1829–1905). No rectifier was needed in this case. For obtaining a continuous current, a Gramme collector was adapted to the axis. Private collection

considered as useful at the time: it was rectified with a commutator on the rotation axis, with the inconvenience that sparks rapidly destroyed the contacts. Who had the idea to replace this commutator by a collector with many contacts connected to several different circuits, diminishing in this way the intensity of the sparks? Some people say that it was Breguet, but no such collector was to be found among his instruments. More probably, the inventor was the Belgian engineer Zénobe Gramme (1826–1901), who built Clark-type magnetos with collectors around 1869.

Gramme was working in the Parisian company *L'Alliance*, directed by Auguste Berlioz (1819–1880), a distant relative of the composer. This company had built the famous "Alliance" magneto (Fig. 5.22), driven by a steam engine or by the gas engine invented by Jean-Joseph Étienne Lenoir (1822–1900) in 1860. It was used for a few years to feed the electric arc of some lighthouses. But it had to be abandoned because of the difficulty of maintaining such a machine in remote places.

The electromagnetic machine of the Englishman Henry Wilde (1833–1876) was a great improvement over the Clarke and Alliance magnetos (Fig. 5.23). The permanent magnet was replaced by an electromagnet fed by an auxiliary magneto, so that the magnetic field was considerably stronger. In 1867, Werner von Siemens (1816–1892), and, independently, Charles Wheatstone, remarked that the iron used in electromagnets keeps a residual magnetization when the current is turned off. Then, even in this case, a weak electric current was produced in the electric coils when the rotor was driven. If one used this current to feed in derivation the electromagnet, the magnetic field increased. Through a cascade process, the current grew

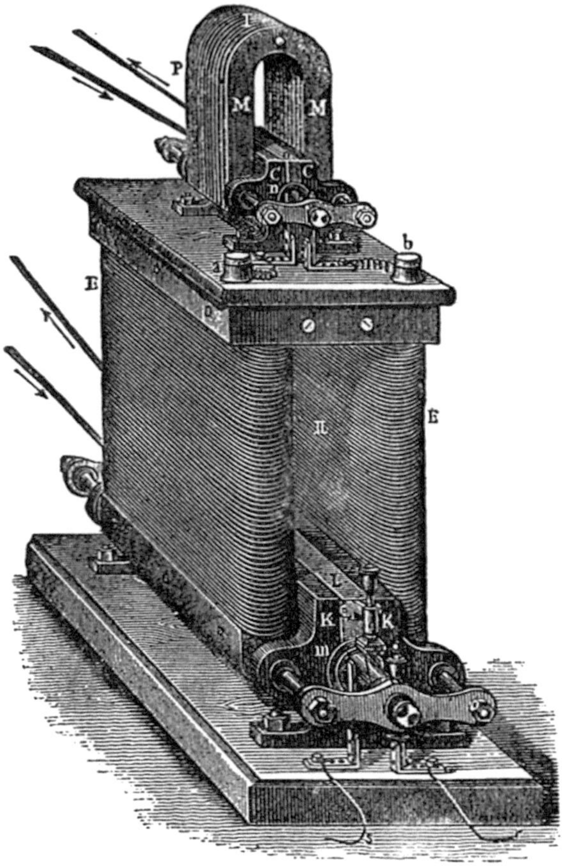

Fig. 5.23 The Wilde electromagnetic machine. An auxiliary magneto, on the *top*, fed a large electromagnet. The rotor, due to Siemens, was made of wires wound parallel to the axis on an iron core. It was driven by a belt; another belt drove the magneto. A collector is visible in front. Private collection

considerably when the coils rotated faster. The self-excited machine, christened electric dynamo, or simply dynamo, was born. The best ones were constructed by Gramme (Fig. 5.24).

In 1873, the French engineer Hippolyte Fontaine (1833–1910), while setting up an exhibition on electricity in Vienna, discovered by chance that a dynamo fed by an electric current acts as a motor: the dynamo is reversible. He then connected a dynamo to another one acting as a motor at a distance of 1 km, carrying for the first time a large amount of energy over a large distance. This was the beginning of a new era. A residual but important problem was the strong losses in the conductors at the relatively low DC voltages produced by the dynamos. This was solved by the use of

Fig. 5.24 A Gramme dynamo, "superior type." The two square coils of the horseshoe electromagnet were fed by a small part of the current produced by the machine. This current was collected and rectified by the large collector located on the rotation axis, to the left of the rotor. This dynamo weighted 4.4 metric tons and could give 187 A under 210 V. Author's collection

alternating current and transformers to carry it at high AC voltages, and back at low AC voltages through another transformer; but motors working with alternating current had to be developed. All this was carried out by the Croatian-American scientist Nikola Tesla (1856–1943).

Chapter 6
Measuring the Earth

Geodesy Before Arago

The Origins

Men have measured distances and surfaces on the Earth since antiquity. This was necessary in Egypt, because the yearly flood of the Nile covered all the arable land with silt, so that everyone had to recover his own field after the river subsided. Thus, the first land surveyors were Egyptians. Later, the Romans set up a corps of *agrimensores*, field measurers. But it was only in the twelfth century that new measurements leading to land registry took place, first in the Dauphiné (the region around Grenoble), then extended progressively to the whole of France.

Measuring long distances was a distinct problem. Until the seventeenth century, they were estimated from the duration of travels and the maps were very inexact. In 1518, Jean Fernel (1497–1558), using an idea of Eratosthenes (284–192 BC), measured the distance from Paris to Amiens by counting the numbers of turns of the wheels of a carriage. In this way he obtained a value of 56,747 toises[1] (about 110 km) for the length of one degree of latitude, a rather accurate value. A new technique called triangulation (Box 6.1) was invented by the Dutchman Gemma Frisius (1508–1555) and used by the great astronomer Tycho Brahe (1546–1601) to obtain the distance between the coast of Denmark and the island of Hven, where his observatory was located. In the Netherlands, Willebrord Snellius (ca. 1580–1626), who had found the law of refraction, measured the length of the meridian between Alkmaar and Bergen op Zoom. Giovanni Battista Riccioli (1598–1671) made a similar measurement in Italy. The results were not very accurate because the tech-

[1] In France, the lengths were measured in *toises* of approximately 1.95 m. The toise was divided into 6 *pieds* of 12 *pouces* each. The French pouce, of 27.07 mm, is larger than the inch of 25.4 mm.

J. Lequeux, *François Arago*, Astrophysics and Space Science Library 421,
DOI 10.1007/978-3-319-20723-0_6

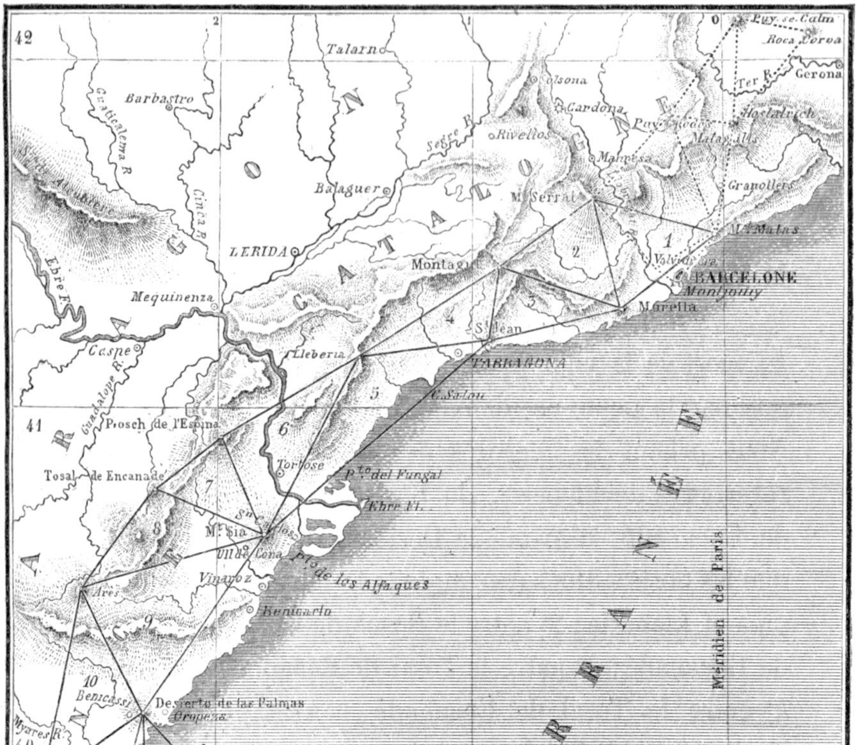

A. Vuillemin, del.

Biot and Arago's triangulation in Spain. Bibliothèque de l'Observatoire de Paris

Box 6.1 The Principle of Triangulation

First, the length of a base AB is measured with rulers or chains (Fig. 6.1); in general the length of a base is of the order of 10 km. Then, considering a target C, for example a steeple, one measures the angles *a* and *b*, which require that a reciprocal sighting is possible between the vertices of the triangle ABC. This triangle is now completely defined, and the length of its side BC can be calculated. BC can now be used as a base for a new triangle BCD, and so on. Some triangles are redundant, allowing checks. The measurement errors accumulate when there are many triangles, but fortunately the angles can be measured very accurately. One has to take into account the altitude difference between the vertices of the triangles using spirit or plumb levels, and this complicates the calculations. Astronomical observations allow to determine the azimuth (orientation with respect to the north), of the base or of any side of a triangle, setting the orientation of the whole set of triangles. This is necessary, for example, when one wants to measure the length of a portion of a meridian.

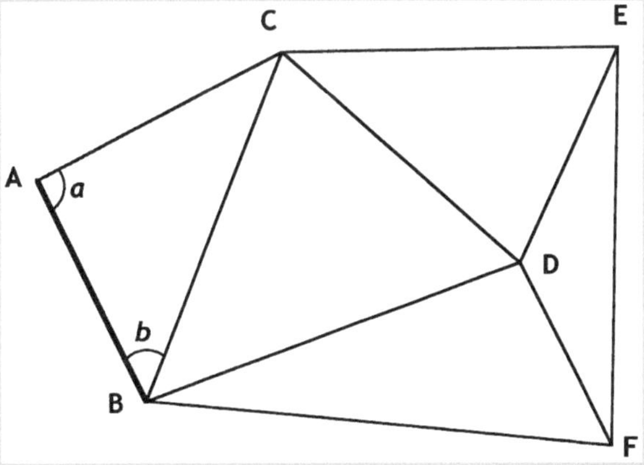

Fig. 6.1 Principle of triangulation

nique was imperfect. The French astronomer Jean Picard (1620–1682) developed the practical bases of the method that was to be used without important changes for three centuries. The principle of triangulation is actually very simple, although its implementation can be complicated.

The Length of the Degree and the Shape of the Earth

Soon after the Academy of Sciences was founded in 1666, Picard was in charge of measuring the length of a degree of the meridian of Paris. The basis of his triangulation, about 11 km, was measured in joining end-to-end wooden rulers of two toises (about 4 m). The triangulation covered 153.5 km. The latitude was measured at both ends by measuring the height of the pole star above the horizon with a new instrument, the astronomical sector, and all of the angles of the triangulation were measured with a horizontal quadrant. It was then possible to calculate the length corresponding to a difference of latitude of one degree. Picard obtained 57,060 toises, corresponding to 111.3 km. Then, multiplying this number by 360, he could obtain the circumference of the Earth, which was then assumed to be spherical. It was the first time that the Earth was measured accurately.

Sometime later came the question of the shape of the Earth, an important question for map making. Was it flattened like a pumpkin or elongated like a rugby ball? Newton predicted that the Earth should be flattened due to the centrifugal force created by its rotation. One expected in this case that the length of a degree of meridian would be a function of latitude (Fig. 6.2). Jean-Dominique Cassini (Cassini I) decided to extend Picard's measurements from Dunkirk to Perpignan, at the extreme northern and southern points of France. This was achieved in 1718 by his son Jacques Cassini (Cassini II, 1677–1756). Comparing the northern and the southern parts of the meridian, he decided that the Earth was elongated. A dispute between his followers and those of Newton followed, so that the Academy decided to send two geodesic operations, one to Peru (in what is now Ecuador) near the terrestrial equator and the other one to Lapland (in northern Sweden), as close as possible to the pole. The result confirmed Newton's idea: the Earth was flattened at the poles. To be sure, the Academy decided to again measure the meridian in France. This was done by Cesar-François Cassini de Thury (Cassini III, 1714–1784), the grandson of

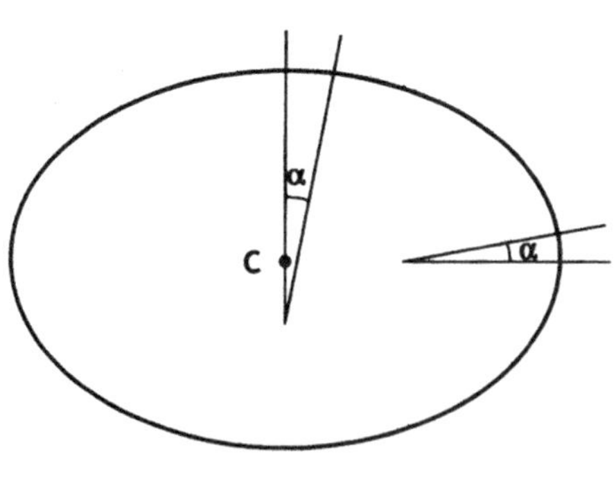

Fig. 6.2 The length of the arc corresponding to the same difference of latitude depends on the latitude because the Earth is not spherical. The difference of latitude α is the angle between the verticals at both ends of the arc. For the same value of α, this arc is longer near the poles for a flattened Earth

Cassini I and son of Cassini II, together with Nicolas-Louis de la Caille (1713–1762); they found that the initial measurement was slightly erroneous and that the Earth was indeed flattened.[2]

The flattening of the Earth obtained in combining the results of Peru and Lapland was 1/200; this number is the difference between the polar and equatorial radii, divided by the latter. This was close to the value of 1/230 predicted by Newton who assumed that the Earth had a uniform density and was (or had been) fluid. However, the measurements in Lapland were not of very good quality. A new triangulation in the same region made from 1801 to 1803 by Swedish scientists gave somewhat different results, hence a flattening of 1/329. This was harder to reconcile with Newton's value. But was the Earth of uniform density? Surely not. Alexis-Claude Clairaut (1713–1765) in 1743 solved the problem of the equilibrium shape of an Earth made of layers with variable density parallel to the surface. Somewhat later, Jean le Rond d'Alembert (1717–1783) gave a more general solution, and showed how precession and nutation (the two movements of the axis of the Earth) depend on the shape of the Earth.

The Beginnings of Gravimetry

Clairaut had also shown that it is possible to determine the flattening of the Earth from the measurements of gravity made at different latitudes. Gravity can be obtained either by measuring the oscillation period of a pendulum with a known length, or by determining the length of a "seconds pendulum", so named because it was set so that each swing took exactly 1 s. It was in this way that Jean Richer (1630–1696) found in 1672 that the seconds pendulum was slightly shorter in Cayenne (French Guyana) than in Paris. Thus the gravity was slightly lower in Cayenne. This was explained later as due to the flattening of the Earth. At the end of the eighteenth century, geodesists used to make such gravity measurements together with triangulation. The French used a pendulum (Fig. 6.3) designed by the famous engineer and sailor Jean-Charles de Borda (1733–1799). With it, very accurate values of the gravity could be obtained.

The two types of geodesy—the direct one via triangulation and the dynamic one via the measurements of gravity—gave initially different values for the flattening of the Earth. However Laplace, putting together all available measurements, found two very close values: 1/308 from triangulation and 1/310 from pendulums. The effect of this flattening on the motion of the Moon yielded a very similar value: 1/305. This value was taken as the reference for a long time. A recent determination (1980) based on the motion of artificial satellites gives 1/298.26 (GRS80 data).

[2]*Cassini de Thury (1740) *Suite des Mémoires de l'Académie royale des sciences*, published in 1744.

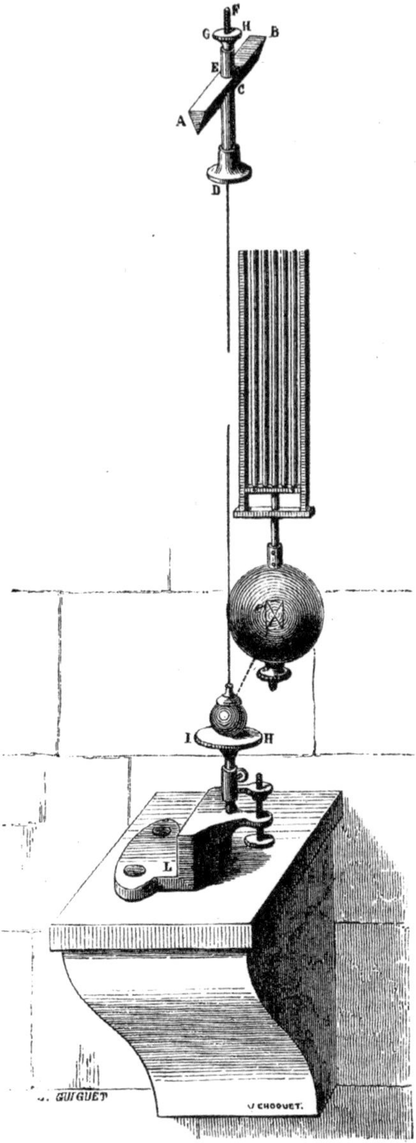

Fig. 6.3 A Borda pendulum for gravimetry. This pendulum, the upper and lower parts of which are represented, was supported by the knives AB sitting on a plane surface. The lower ball was made of platinum, and sat on a plate that was lowered for operation. One compared the oscillations of the pendulum with the almost synchronous ones of the pendulum of a clock, seen on the back. The period of the clock was determined by astronomical observations. Then the period of the pendulum could be determined very accurately. One could also derive from it by a simple law the length of a seconds pendulum. The gravity was obtained with a high accuracy from one or the other number. From Arago's *Astronomie populaire*, author's collection

Geodesy and Cartography: The Cassini Map of France

During the seventeenth and eighteenth centuries, geodesy and gravimetry were made by astronomers, but cartography—the construction of maps—was the domain of specialized engineers. To build them, one needed determinations of longitude as well as latitude. Measuring latitude was not difficult: it sufficed to measure the angular height of the pole star above the horizon, and to make a small correction because this star is not exactly at the pole. Thus, one obtained the latitude directly. The determination of longitude was much more problematic. It required good clocks at the two points between which the difference of longitude had to be measured. These clocks had to be adjusted by astronomical observations to the local time at each place. The difference between these times was equal to the difference of longitude. But the problem was to obtain this difference. Until the end of the eighteenth century, it was not possible to carry a working clock on long distances, so that the only way was to observe the same signal from the two places. This signal could consist of some astronomical event visible from a large fraction of the Earth. The first attempt at using such an event was organized in 1635 by Nicolas-Claude Fabri de Peiresc (1580–1637). He coordinated observations of an eclipse of the Moon from different harbors around the Mediterranean Sea. From the longitudes determined in this way, he found that the Mediterranean sea was 1000 km shorter in the east-west direction than on the maps of the time. However, lunar eclipses are rare and slow, so that their use lacks accuracy. It was much more appropriate to observe the eclipses of Jupiter's satellites in the shadow of the planet; they are frequent and better defined (see also Chap. 4). Their observation became a major program for astronomers. Picard and Philippe de la Hire (1640–1718) used this method between 1676 and 1681 to determine the position of a number of places in France, with the result that France shrank by 10 % in the east-west direction! Unfortunately, this method is not practicable at sea because of the motions of the ship, and one had to wait for the second half of the eighteenth century to properly determine longitudes at sea using either the position of the Moon with respect to reference stars or good transportable chronometers.

There is however another possibility to obtain distances in the east-west direction: to measure them directly by triangulation along parallels. The first such measurements were made by Jacques Cassini (Cassini II) from Brest to Strasbourg in 1733–1734. Their purpose was not cartography, but the desire to know the length of an arc corresponding to a given longitude difference. For a given latitude, this quantity depends on the shape of the Earth. The result was inconclusive because of the uncertainty on longitudes. More successful was a similar measurement made in 1739–1740 by La Caille and Cassini III from Mont Saint-Clair near Sète to Montagne Sainte-Victoire, near Aix-en-Provence, in the south of France. In this case, they were able to accurately measure not only the length of the arc but also the difference of longitude between these places. They used a new method, apparently for the first time: the fire signal. They burned 10 lb of gunpowder on the roof of the church of Saintes-Maries-de-la-Mer. The resulting flash was seen from the two mountains and was used to determine the difference of the local times between them, hence the difference of their longitudes. By combining their results with that of north-south surveys, La Caille and Cassini III confirmed that the Earth was flattened; the costly expeditions to Peru and Lapland could have been avoided!

Other measurements of parallels and meridian were also made, providing the frame for the construction of a detailed and accurate map of France. This map, the *Carte de Cassini*, was realized under the supervision of Cassini III and completed by his son Jean-Dominique Cassini (Cassini IV, 1748–1845). The measurements took 34 years, from 1750 to 1784, requiring about 20 men in the field and an equal number to engrave the maps. The result was a major progress in cartography. However there were not yet altitudes in the map.

Triangulations at the End of the Eighteenth Century

The end of the eighteenth century saw many more triangulations, in particular several geodesic linkages between France and England. They gave the opportunity to try new instruments, in particular in England the large theodolite of Jesse Ramsden (1735–1800) and in France Borda's repeating circle (Fig. 6.4, Box 6.2).[3] The bases were first measured by glass tubes, then by steel chains. It is remarkable that the

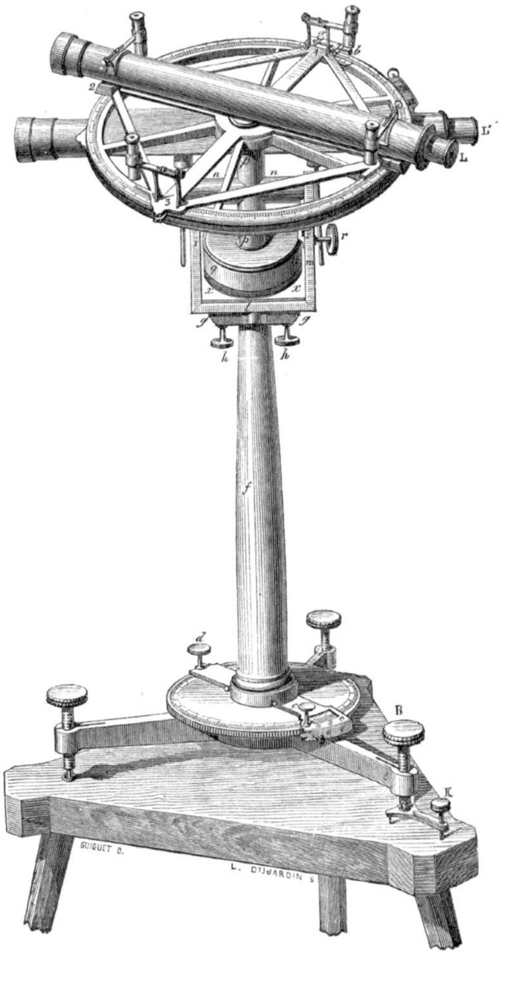

Fig. 6.4 A repeating circle of Borda, built around 1805 by J. Fortin (1750–1831), set in horizontal position for triangulation. From Arago's *Astronomie populaire,* author's collection

[3]The theodolite was well developed in England, but appeared only slowly in France. The first French one was built by Étienne Lenoir around 1800, with a single, horizontal graduated circle. Theodolites with two perpendicular circles, which allowed the measurement of azimuths and elevations, were constructed later by Gambey. They very progressively replaced the repeating circles, which wore down rapidly because of the large number of rotations they experienced.

Box 6.2 Borda's Repeating Circle

The repeating circle is an instrument to measure angles, made of two telescopes whose orientation is measured by a single divided circle. The telescopes can be fixed to the circle or dissociated from it. This ensemble can be oriented either more or less horizontally for triangulation, or vertically for measuring the height of a star at meridian passage, in order to determine the latitude. Figure 6.5 explains how it worked. The principle was due to the

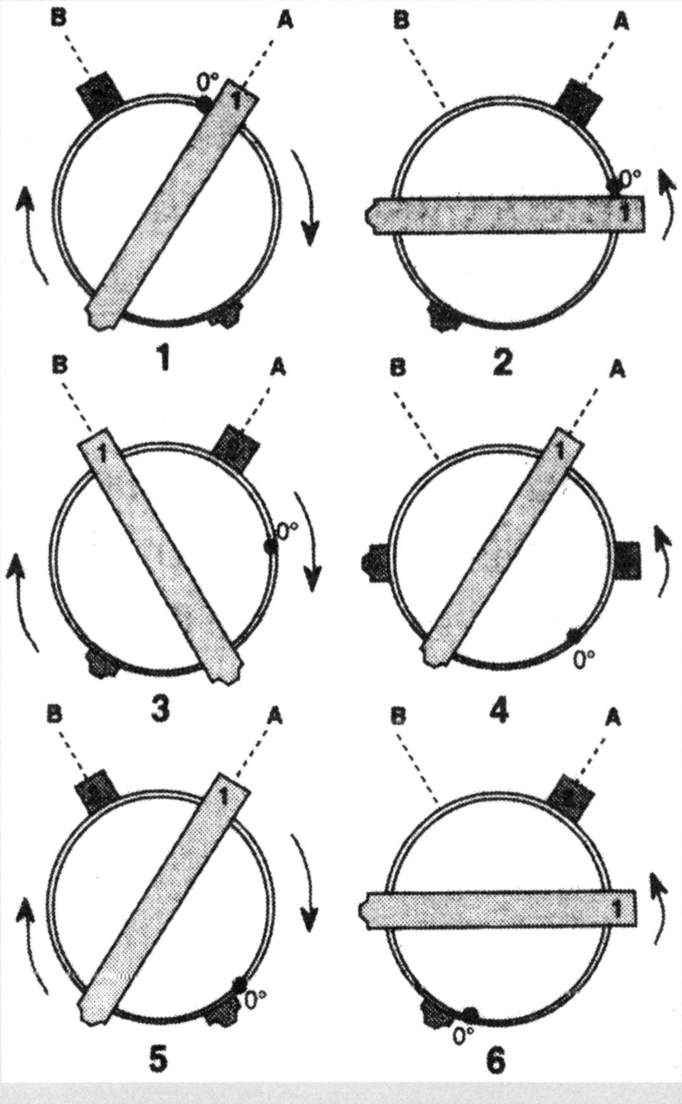

Fig. 6.5 Running the repeating circle for triangulation. Explanations in Box 6.2

(continued)

Box 6.2 (continued)

German astronomer Tobias Mayer (1723–1762), and the first realization to Borda and Lenoir. To measure the angular distance between directions A and B, one first points the telescope 1 to A and the telescope 2 to B, and clamps them to the circle (configuration 1). The point labeled 0° is the origin of the graduations of the circle. One turns it until telescope 2 points to A (2). Then, one unlocks telescope 1 and points A with it (3). It is then clamped to the circle and the whole is turned until telescope 1 looks at B again (4). Then telescope 2 is unclamped and turned to point A (5). The configuration is now the same as the initial one, but the circle has turned by twice the angle between A and B. Then one resumes the process (6) as many times as possible. At the end, one divides the total angle by which the circle has turned by twice the number of cycles to obtain the angle between A and B. The accuracy was improved because the errors in reading the circle and in pointing the targets were reduced. The errors due to the eccentricity of the circle and the imperfections in its graduation were also reduced. With a relatively small instrument, one obtained the same accuracy as that with the big Ramsden's theodolite. However, certain systematic errors were not eliminated in the process.

In order to measure the height of a star at its meridian passage, or rather its distance to the zenith, which is the complement to 90°, one sets the circle in the vertical position (Figs. 6.4 and 6.6). Telescope 2 is set to the horizontal position thanks to a spirit level, and telescope 1 is pointed to the star (configuration 1). Then the circle is turned by 180° around the vertical axis (2), and one looks again to the star with telescope 1, unclamped from the circle (3). Then the whole instrument is rotated again by 180° (4), telescope 1 is clamped to the circle and is turned to point to the star (5). The circle is turned by twice the zenith distance of the star from position (1), and one resumes the successive steps until a sufficient number of measurements has been obtained. For all the steps, an assistant adjusts the horizontality of telescope 2 by acting on the small auxiliary screws at the foot of the instrument (see Fig. 6.4).

(continued)

Box 6.2 (continued)

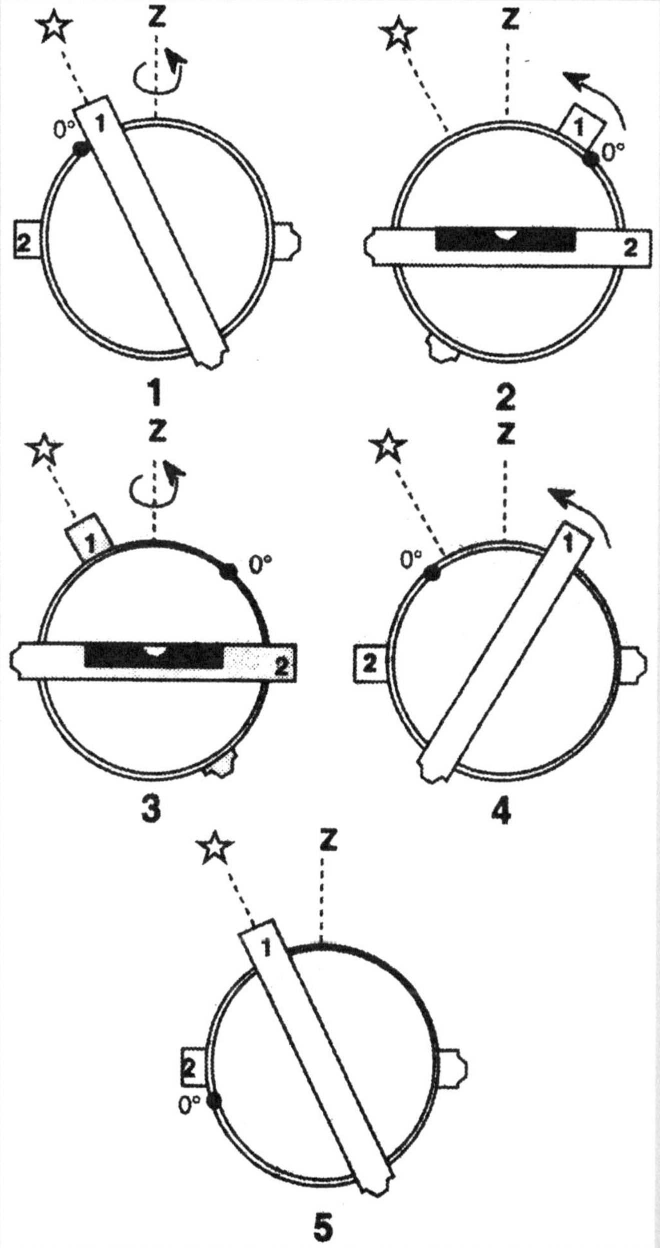

Fig. 6.6 Running the repeating circle for measuring a zenith distance. Explanations in Box 6.2

France–England linkage ran smoothly, in spite of the bad relations between the two countries. It is true, however, that the result was of interest to the two sides because it allowed combining astronomical observations made in Paris and in Greenwich.

Measuring the Paris Meridian During the French Revolution

In 1791, the new Assembly (*Assemblée constituante*) elected during the French Revolution decided to create a universal decimal system of units, which would replace the incredible diversity of measures used in France and elsewhere. The new unit of length, the meter, was to be linked to a fixed quantity, common to the whole world: the length of a meridian of the Earth. There would have been ten million meters in the quarter of this meridian.[4] It was considered necessary to measure again a part of the Paris meridian, because the previous measurements had been imperfect or not corrected for the altitude differences between the geodetic points. Delambre and Méchain (Fig. 6.7) were put in charge of a new triangulation between Dunkirk and Barcelona (Fig. 6.8). In the meantime, provisional meters were constructed from the determination by La Caille and Cassini III, and distributed throughout France so that people would become familiar with them.

The history of the new triangulation (1792–1798) is well known and described in many books,[5] so that we will only discuss some relatively poorly known points. It suffices to say that it was conducted with remarkable skill and rigor despite many difficulties due to the agitated period of the Revolution.

To prepare for the triangulation, Borda asked Lenoir to build four repeating circles, four rulers of two *toises* each (a bit less than 4 m) for measuring the bases,[6] and *réverbères* (reflectors) with parabolic mirrors and oil lamps for observation of geodetic points at night. The rulers (Fig. 6.9) were especially well designed, but the measurement of a base was as usual a major operation: that of the main 11-km base from Lieusaint to Melun, south-east of Paris, took 36 days. Another base was measured near Perpignan. Its measured length differed only by 30 cm from the value calculated after all the triangulations from Paris to Perpignan, a remarkable agreement.

Delambre and Méchain carefully measured not only the horizontal angles, but also the elevation of their targets over the horizon. It was the first time that this was done systematically. Progressing from the north for Delambre and from the south for Méchain, the two parties met in Rodez. Each one determined the altitude of the

[4] It was also foreseen that the meter would have been the length of a seconds pendulum at 45° latitude; but this idea was not retained because of the difficulties of practical realization.

[5] For specialists : Méchain and Delambre (1806–1810), in French, no English translation. Popular: Guedj (2001), Adler (2002), Murdin (2009).

[6] The rulers are preserved at the Paris Observatory. Two of the circles are at the Marseilles observatory and at the National Geographic Institute at Saint-Mandé, near Paris.

Fig. 6.7 Pierre-André Méchain. Wikimedia Commons, Grebenkov

steeple of the cathedral in Rodez, with respect to the Atlantic Ocean and the Mediterranean Sea respectively. Delambre found 706.98 m, and Méchain 706.06 m. The agreement is once again remarkable, but probably partly fortuitous.[7]

They also measured the latitude of the ends of the triangulation in Dunkirk and Barcelona with the repeating circles in vertical position. To calculate the total length of the Paris meridian from pole to equator, they had to assume a value for the flattening of the Earth: they took 1/335 from the gravimetric measurement with the Borda pendulum. This is a bit too small, but the difference is rather unimportant at latitudes around 45°. Following all these operations and painful reductions made principally by Delambre, the Commission of Weights and Measures fixed in 1799 the length of the meter as 443.296 *lignes* of the *toise du Pérou.*[8] This toise, an iron bar built by Langlois in 1735, was the one used by the astronomers who measured the length of a degree near the equator, and which was the standard for the Borda rulers. Twelve standard meters in iron and three in platinum were constructed by Lenoir to replace the provisional ones made before the triangulation. One of the platinum meters, deposited at the National Archives, was the primary standard; the two others were carried to the Paris Observatory and the Conservatory of Arts and Crafts. They are still extant, as well of the toise of Peru, which is preserved at the Observatory. It turned out to be difficult to substitute the final meter to the provisional

[7] Laplace, who made the first mathematical estimates of measurement errors, considered that the altitude of Paris obtained during the operations of Delambre had a probable error of 8 m.

[8] There are 864 lignes in a toise, so that the length of the toise was 1.94904 m.

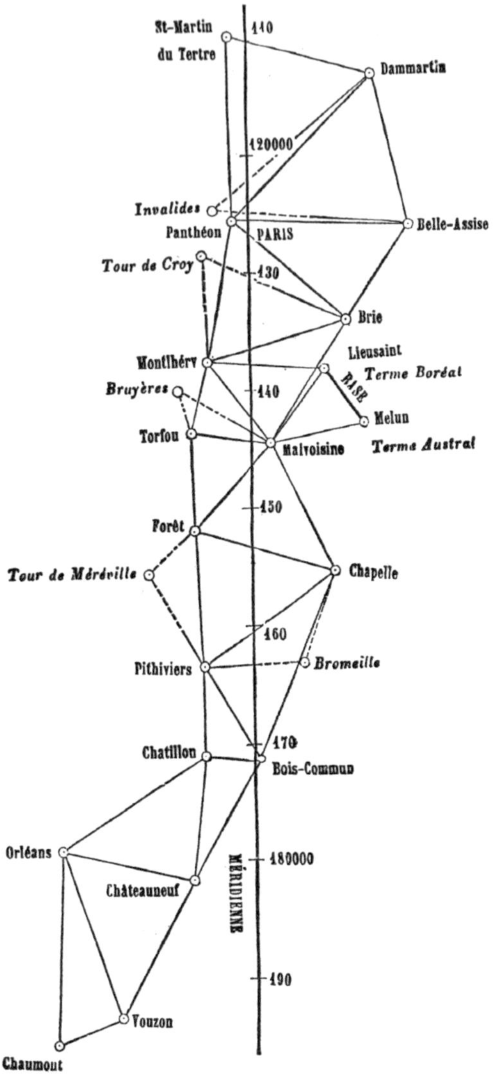

Fig. 6.8 A part of the triangulation of Delambre and Méchain. Note the base from Melun to Lieusaint. From Arago's *Astronomie populaire*, author's collection

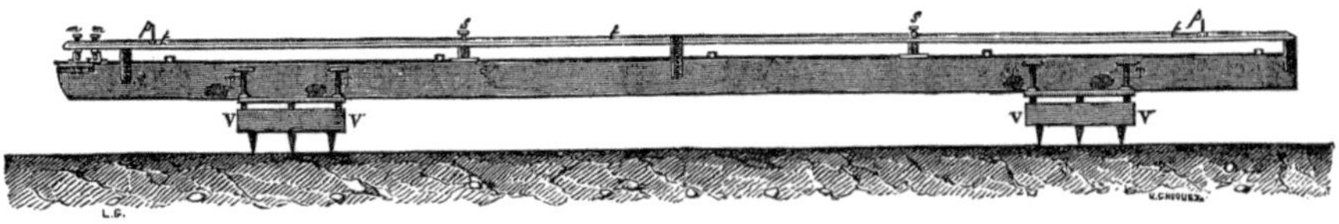

Fig. 6.9 Borda's ruler for measuring the bases of triangulation, lateral view. It was made of a brass ruler, superimposed on a platinum one, fixed to each other at one end in order to measure the relative expansion of the two rulers and to correct the measurements to a reference temperature. In order to preserve their ends, the two successive rulers were not in direct contact. A small graduated ruler slided in the platinum ruler and touched the end of the next ruler. The rulers were set horizontally and aligned through rods *p*. From Arago's *Astronomie populaire*, author's collection

one, because people were used to taking the latter as the standard; indeed, it was not very different: its length was 443.44 lignes.

However, it was initially foreseen to measure a longer arc of meridian: It would have been symmetrical with respect to the 45° parallel, so that the effect of the Earth flattening would have been minimal. When all the operations were finished, the Board of Longitudes gave attention to the problem. Méchain insisted that he be in charge of the extension south of Barcelona, to the Balearic Islands. This would have offered him the opportunity to correct an error that tormented him. He had measured the latitude of two places in Barcelona, and he had found that the difference was smaller by 3 arc sec than expected from the direct measurement of distance by triangulation. Thus, Méchain returned to Barcelona in 1803; but he died from malaria on 20 September 1804 in Castellón de la Plana. He did not have time to do much and to solve his problem. Later studies showed that he made in fact no error (Box 6.3): He undermined himself for nothing.

Box 6.3 Méchain's "Error"

Méchain's problem has been examined several times, in particular by Arago in 1812–1813. In a detailed discussion of the repeating circles,[9] Arago noticed that the focusing of the telescopes might produce differences of several arc seconds between different observers. The most interesting study is that published in 1828 by Jean-Nicolas Nicollet (1786–1843), an astronomer of the Paris Observatory who worked from Arago's ideas.[10] He wrote[11]:

One tried to explain [the difference observed by Méchain] by the lack of verticality of the plane of the instrument, or by the perturbation that the Mount Jouy could have exerted on the vertical during the observations in Barcelona. But the hypotheses that should have been exaggerated in the first case, and the examination of the places in the second one, did not confirm theses ideas.

One also indicted the use by Méchain of the star ζ Ursae Majoris, which was later found double. But this was not the main problem. Méchain made his observations in different conditions at the two locations. Now, the repeating circle he used had a systematic error that Méchain could not know, but that

(continued)

[9] **Connaissance des temps* for 1816, p. 345–358. See also *Biot (1843) *Comptes rendus* 16, 1019–1032.

[10] Nicollet was not well respected by Arago. He was ruined by hazardous speculations and emigrated in 1830 to the United States.

[11] *Bibliothèque universelle de Genève* (1828) 38, p. 165–179 and 250–263, accessible via Google books.

Box 6.3 (continued)

others discovered later: the latitudes measured from stars transiting north and south of the zenith differed by several arc seconds. Nicollet concluded:

As soon as this fact was known, this difference ceased to be a difficulty to the eyes of those scientists who had the opportunity to observe something similar. In the meeting of 11 November 1818 of the Board of Longitudes, M. Arago, reporting on the way he had determined with M. Biot the latitude of Dunkirk, noticed that the latitude of Barcelona as derived in the same way [as Méchain] from northern and southern stars agreed with the geodesic measurement.

Then Nicollet performed anew all the calculations, taking into account the systematic error of the circle, with better values of the positions of stars (including ζ Ursae Majoris) and more correct refraction tables; he found that the anomaly was only of 0.21 arc sec, which was within the margin of error. The southern end of the meridian arc had a latitude too small by 0.42 s with respect to that adopted by the Commission of Weights and Measures to define the meter, which is almost negligible. Méchain's measurements were in fact good, but incomplete.

Arago claimed that the defect of the instrument was due to a too small graduated circle.[12] This led Laplace to order to the Reichenbach's firm in Munich a larger repeating circle, not for geodesy but for astronomy. This instrument had a circle of one meter in diameter; the French makers did not have the means to build such a large circle. But Arago said that even this instrument had non-negligible defects.

Arago's Work

Extending the Measurements of the Meridian to the Balearic Islands: Should One Change the Length of the Meter?

On 2 May 1806, a year and a half after Méchain's death, the Board of Longitudes decided to resume the measurements south of Barcelona, and to extend them to the Balearic Islands. He charged Biot, assisted by the very young scientist that Arago was, with the operation. Arago spoke Catalan, and this was clearly the reason for his choice. Their instruments were two repeating circles and a portable meridian

[12] *Arago, F., *Œuvres complètes* t. 11, p. 115–137 and 138–139.

telescope. The work, described by Arago himself in one of his scientific Memoirs,[13] started in October 1806, with the help of two Spaniards, Chaix and Rodriguez. After a rapid beginning when they used Méchain's measurements, they were stuck for almost 6 months at the Desierto de las Palmas due to poor visibility and to the incorrect orientation of the lamp and reflector at Ibiza, they finally succeeded in reaching Ibiza.[14] After linking Ibiza to the continent, a very difficult operation because of the distance (up to 170 km), they continued to Formentera in April 1807. Biot came back to Paris in May, carrying with him one of the two repeating circles, which was damaged. He left Arago to continue the observations alone, with the other circle built by Lenoir. Biot returned to Spain at the end of November with another circle, built this time by Fortin (Figs. 4.3 and 6.4), but he returned to Paris on 18 January 1808 with the results of the observations. Arago was again alone with his two Spanish colleagues, and nevertheless succeeded in linking Mallorca to Ibiza and Formentera. This extension to the east was unnecessary for the measurement of the meridian, but Arago wanted to measure the length of an arc over a rather large difference of longitude (3° here), in order to study the flattening of the Earth.

Arago had just finished his measurements when war between France and Spain was declared. Then began the adventures he told in his *Histoire de ma jeunesse*,[15] which made him famous. Despite the turmoil, he managed to carry his instruments and notes back to France.

Biot, and the young Arago, must be commended for having succeeded in completing their mission in very difficult circumstances. The precision of their results is remarkable.[16] Laplace and Marie-Charles de Damoiseau (1768–1846) estimated the probable error of only 8 m on the distance from Perpignan to Formentera, 466 km in a straight line.[17] They also made many meteorological and gravimetric observations. The extension of the meridian arc to the Balearic Islands confirmed the value given previously to the meter; using the value of the flattening of the Earth of 1/305 derived by Laplace from the theory of the Moon, the Board of Longitudes gave a length of the meter equal to 443.2958 *lignes* of the toise of Peru, almost identical to the 1799 value. There was no need to build new standard meters. The Board wrote[18]:

One sees that the measurement of the meridian that was just done in Spain confirms the value of the meter, and gives to it a new certainty because it is almost

[13]* Arago, F., *Œuvres complètes*, t .11, p. 55–114.

[14]The measurements had to be done by night, with light signals produced by oil lamps at the focus of parabolic mirrors. Coordination must have been a problem, due to the slowness of communications.

[15]* Arago, F., *Œuvres complètes* t. 1, p. 1–102.

[16]Biot and Arago were virtuosos of the repeating circle. They succeeded in making more than 100 measurements of the zenith distance of a star in 1 h, while Cassini IV could make only 10 in 12 min.

[17]**Connaissance des temps* for 1822, p. 346–348.

[18]**Connaissance des temps* for 1810, p. 485–488; also **Annuaire du Bureau des longitudes* for 1809, p. 81–87.

independent from the flattening of the Earth ... For the extent, the location and the exactness of its means, [this measurement] is the most beautiful operation of this kind that has ever been made.

The measurement of the Paris meridian was not unique. From 1804 to 1818, William Lambton measured in India an arc of meridian of almost 10°, and found for the meter a value shorter by 1/25 ligne. The scientists began to realize that the Earth is not really an ellipsoid, and that, as a consequence, the length of the meter could depend on the meridian. On the other hand, the geodesian Louis Puissant (1769–1843) discovered in 1836 a subtle error made in calculating the distance between the parallels of Barcelona and Formentera, following Delambre's method.[19] This calculation assumed implicitly that the meridians of the different places are parallel, which is not true. Correcting this calculation and including some new measurements, the distance between the parallels of Majorca and Formentera was found too short by 70 toises. Should one again change the length of the meter? The answer was: No! The report of the new calculations says:

If one had ever the strange idea of changing the unit of length to follow the progresses of geodesy, one would have been forced to give up when looking at all these measurements of meridians and parallels that reveal very important local irregularities and prove that the terrestrial globe is not a solid of revolution.

Thus the 1799 unit was kept. The beautiful idea of the *Assemblée constituante* according to which the meter would have been linked to the dimensions of the Earth was over. We might consider that the saga of Delambre, Méchain, Biot and Arago was of no use whatsoever. But they stay engraved in the collective memory.

Arago kept an interest for the meter all along his life. He made himself many comparisons between the different meters and toises, and reported in particular on the differences in the lengths of the three platinum standard meters. Indeed, the difficulties in measuring the length of these standard meters were such that others were constructed later, along which was the famous international prototype preserved at the *Pavillon des poids et des mesures* of Sèvres, near Paris. It was selected for a new definition of the meter in 1889. In 1960, this definition was again changed for the following one:

The meter is the length equal to 1,650,763.73 wavelengths in vacuum of the radiation corresponding to the transition between the levels $2p_{10}$ *and* $5d_5$ *of the atom of krypton 86.*

Unfortunately, it was discovered that this spectral line of krypton was asymmetrical, so that this definition was still imperfect. Fortunately, several laboratories, including the Primary Laboratory of Time and Frequencies of the Paris Observatory, succeeded in 1970 in measuring simultaneously the wavelength in a vacuum and the

[19] **Comptes rendus* (1836) 3, p. 428–433, 450–453 and 483–486, and **Connaissance des temps* for 1844, supplément, p. 123–131.

frequency of a laser radiation. Their product is the velocity of light, which is obtained with very high accuracy. This yielded to another change in the definition of the meter, which is derived from the velocity of light, which is fixed as 299,792,458 m exactly. The new definition, adopted in 1983, is:

The meter is the length of the path travelled by light in vacuum during a time interval of 1/299,792,458 of a second.

All this might seem of minor interest to the layman. However, we must realize that the precise measurements of length on the Earth don't use rulers or chains anymore, but the GPS and laser telemeters, which measure the propagation time of electromagnetic waves (light, infrared or radio waves). If the GPS insures our localization on the Earth, it does so by using the velocity of light. The new definition of the meter is now part of our daily life, and corresponds to a practical logic that was not foreseen in 1983.

The New Geodesic Linkage of France and England

In spite of the novel instruments that were used, the 1787 linkage between the Paris and Greenwich observatories was not a full success, because the partial results did not always agree. What seemed the best value for the difference of longitude was 9 min 18.8 s. The relations between France and England being better after 1815, the two countries agreed that new measurements had to be made. They would benefit from Delambre's triangulation from Paris to Dunkirk, while Laplace had made, as we have seen, the first quantitative estimates of errors in geodesic measurements. Already, Arago and Humboldt had visited London with Biot in 1817 to compare the lengths of their seconds pendulums with those of Captain Henry Kater (1777–1837), with whom they had become acquainted. Biot and Arago also collaborated in 1818 with the English general William Mudge (1762–1821) to determine the latitude of Dunkirk, the common point of the French and British arcs.

In 1821, the Academy of Sciences and the Board of Longitudes designated Arago and Mathieu to resume the linkage, while the Royal Society chose Kater and Captain Thomas Colby (1784–1852). The measurements began on 24 September 1821 but stopped 3 days later because of bad weather. Then, "our much esteemed companion M. Arago" (according to Kater) returned to Paris. The operations resumed on 12 August 1822, to stop again in November, and then resuming in July 1823, this time without Arago. Figure 6.10 shows a map of the triangulation. The instruments were about the same as used in 1787. On the English side, Ramsden's original theodolite was replaced in 1823 by a similar one belonging to the Ordnance Survey, and on the French side a repeating circle was used, then a theodolite of Gambey[20] (Fig. 6.11). The most significant improvement was the replacement of the

[20] See *Arago, F., *Œuvres complètes* 3, p. 603.

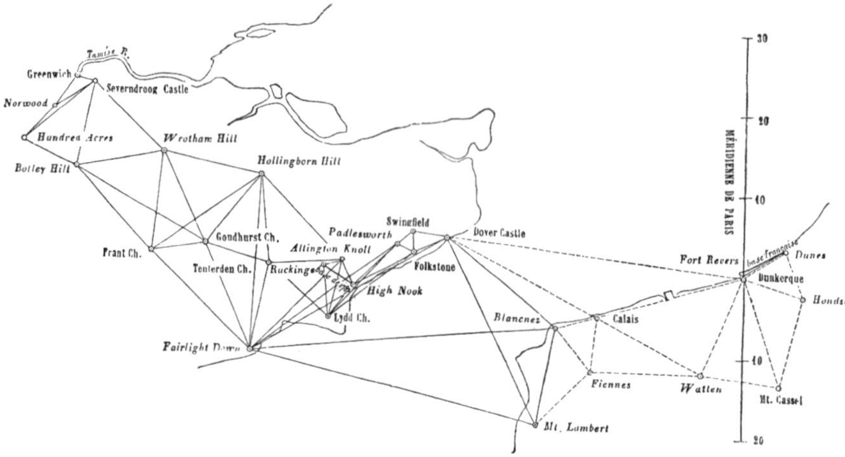

Fig. 6.10 The 1821–1823 triangulation from France to England. Note the French base near Dunkirk and the British one near Attington Knoll. From Arago's *Astronomie populaire,* author's collection

oil lamps with parabolic reflectors by "lamps with compound lenses," which were the first lighthouse multi-part lenses of Fresnel; they had a diameter of 3 ft and could be seen "at a distance of 48 miles like a star of the first magnitude."

The results of the French party were never given by Arago to the English one; Perhaps, they were never reduced, as suggested by Kater[21] 5 years after the end of the operations[22]:

It is to be regretted that our excellent associate M. ARAGO has not yet published the results of his operations in France; and I must therefore, in the absence of higher authority, take the longitude of Calais, as given in the Connaissance des Tems, to be 0° 28′ 59″ west of Paris. Adding this to 1° 51′ 18″.73 the east longitude of Calais from Greenwich, given by the present work, we obtain 2° 30′ 17″.73 for the difference of longitude between Paris and Greenwich. This converted into time is 9 m 21 s.18, differing only 0 s.28 in defect from the admirable results obtained by the operations with fire signals, reported in the Philosophical Transactions for 1826, by Mr. HERSCHEL.

What are these "admirable results"? John Herschel and Captain Edward Sabine (1788–1883) on the English side, and Colonel Bonne and Lieutenant Largeteau on

---

[21] *Philosophical transactions* (1828) 118, p. 153–239, accessible via http://www.jstor.org/stable/107842: see p. 192.

[22] Bigourdan (1928–1932, see 1932) p. A65–A66, and following him Levallois (1988) p. 131 write that the results of the operations, given to captain Ketter (in fact Kater) were lost after his death "soon after" the end of these operations. We know that this is not true.

Fig. 6.11 A theodolite with two *circles* and two telescopes, built by Froment, mid-nineteenth century. It is similar to the one that was used for the linkage between France and England in 1821–1823. From Arago's *Astronomie populaire*, author's collection

the French side had indeed measured the longitude difference between the two observatories, using fire signal observed from them and two intermediate stations.[23] They found a difference of longitude of 9 m 21.46 s. This result is rightly considered by Kater as more reliable than his own, and this is probably why he did not insist on obtaining the results of Arago, who had turned his attention to other topics as usual. However, the purely astronomical determination of longitudes was still sometimes in use, for example by the Paris astronomer Jean-Jacques-Émile Goujon (1823–1856), who in 1847 obtained a longitude difference of 9 m 21.6 s using the position of the Moon with respect to nearby stars. The best result was finally obtained in 1854 using the electric telegraph: 9 m 20.63 s, differing from the previous determinations by almost 1 s of time.

[23] *Philosophical Transactions* (1826) 116, p. 77–126, accessible via http://www.jstor.org/stable/107803

The France–England linkage by triangulation was done once more in 1861, rather badly because there were fewer and fewer competent people in France and the three repeating circles remaining at the *Dépôt de la guerre* (War Storehouse) were in poor condition. The means on the other side of the Channel were now superior.

Leveling and New Maps of France

Leveling

Until the nineteenth century, there was no elevation on the geographical maps. This does not mean that altitudes were not measured. After Blaise Pascal had shown by his celebrated experiment on the Puy-de-Dôme mountain near Clermont-Ferrand that the barometric pressure decreases with elevation, altitudes were often measured with barometers (barometric leveling). However these measurements were inaccurate, and for certain purposes elevations were measured by geometric means using spirit levels. For example, the altitude difference between the Loire river and Versailles was measured in 1674 by Picard and a few others in the hope that it would be possible to bring water from the Loire to the palace of Versailles.[24] The accuracy of these measurements is remarkable, as they differ from the modern ones by only 60 cm.

We mentioned earlier the measurements by Delambre and Méchain. They were not the only ones in France at the beginning of the nineteenth century. Such measurements became systematic after 1817, the date when it was decided to make a new map of France to replace Cassini's. To obtain the elevation of the top of the dome of the Panthéon in Paris, which was chosen as the reference point for the elevations of this map, a leveling measurement was made in 1820–1821 between Paris and Cancale on the Channel. The error was estimated as 9 cm using Laplace's method, a considerable improvement with respect to previous measurements. All the elevations of the geodesic points of Delambre and Méchain were then recalculated: this was one of the bases for the new map.

New Maps of France

A new geodetic survey of France was initiated by Laplace in 1811, and made under the direction of Colonel Bonne. It consisted of measuring a parallel from Bordeaux to Italy, and re-measuring the longitudes of many points among this parallel by

[24] The complex operations are described in detail in *Picard (1684), p. 140–175.

using fire signals. Together with other, similar measurements, this allowed a new determination of the flattening of the Earth, found to be 1/282. Sabine obtained a flattening of 1/289 from gravity measurements. These values are different from that determined previously by Laplace (1/305) because of measurement errors and of the departures of the Earth from axial symmetry.

All these operations appear to have been conducted in a somewhat anarchic way, but they were nevertheless useful for the realization of the new map of France, commissioned in 1817 and directed by Louis Puissant. Its scale was 1/80,000, slightly larger than that of Cassini's map, 1/86,400. All the new operations were done by military geographic engineers, including the longitude determinations, which were previously the prerogatives of astronomers. To complete the skeleton of the map, more triangulations were performed, yielding a total of four meridians and seven parallels across France. These measurements extended from 1818 to 1853, using repeating circles and, less often, theodolites. The new map showed altitudes of summits and important points, but no contour lines.

A new triangulation was started in 1891 by the *Service géographique des armées* (Military Geographical Service) and was completed only in 1991 by the *Institut géographique national* (National Geographical Institute), which replaced the military one in 1940. It mainly used theodolites, and the longitudes were obtained by telegraphic or radio signals. This resulted in the present maps with 1/50,000 or 1/25,000 scales. Today, the Global Positioning System (GPS) has replaced all the preceding methods of surveying.

Chapter 7
Arago and the Paris Observatory

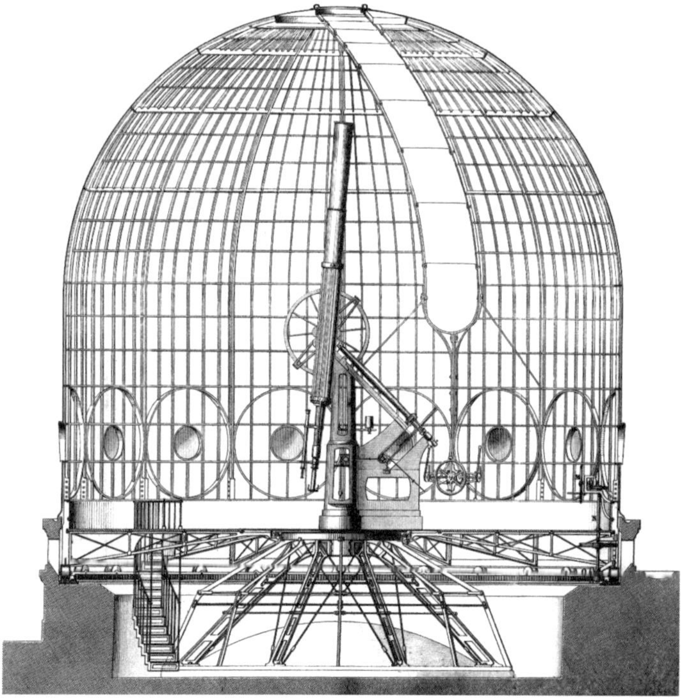

The large dome of the Paris Observatory, built by Henri Alphonse de Gisors (1796–1861) and the engineer Louis Travers in 1847. From Arago's *Astronomie populaire*, author's collection

J. Lequeux, *François Arago*, Astrophysics and Space Science Library 421,
DOI 10.1007/978-3-319-20723-0_7

Arago stayed all his adult life at the Paris Observatory.[1] He entered the Observatory in 1805 as "secretary-librarian," at age 19, became a titular member of the Board of Longitudes in 1822 and was promoted in 1834 to "Director of observations," making him *de facto* director. He kept this position until his death in 1853. We should not be surprised that his strong personality marked the institution, which we will discuss in this chapter. But let us first see in what condition he found the Observatory on his arrival.

The Paris Observatory Before Arago

The Beginnings

The history of the Observatory in the seventeenth and eighteenth centuries is well known thanks to Cassini IV's memoirs[2] and a book by Charles Wolf,[3] which is to some extent based on these memoirs. Founded in 1667 to serve not only as an observatory but also as a meeting place and depository of collections for the members of the Academy of Sciences, created a year earlier, the building was soon devoted entirely to astronomy, because the academicians considered that it was too far from the city center. However, as magnificent as it was (Fig. 7.1), the building, by Claude Perrault (1613–1688), was not well suited for astronomy, and is still not fit today. For Cassini I, who already asked for important modifications when he came to Paris in 1669,[4]

it sufficed to Perrault to give to the façade and to the mass of the Observatory this solemn and grand character corresponding to its purpose. This was the stamp that his genius wanted to impose. But it mattered little to him whether the astronomer could observe more or less conveniently.

The first person responsible for the Observatory, who did not have the title of director, was the Italian Giovanni Domenico Cassini (Cassini I), called to France by the prime minister, Jean-Baptiste Colbert (1619–1683). Cassini was already famous when he arrived. He adapted easily to the French way of life, changed his first names in Jean-Dominique and obtained French citizenship. An excellent observer and facilitator, he had already discovered the rotation of Jupiter and of Mars before coming to France and produced ephemerides (tables predicting the positions) of the four Galilean satellites of Jupiter. In Paris, he observed the division of Saturn's rings, to which his name was given later; discovered four satellites of the planet; and produced a magnificent map of the Moon. He also played a major role in the discovery of the finite nature of the velocity of light (see Chap. 4). Three other Cassinis

---

[1] For a general history of the Paris Observatory from its beginning to present, see Bobis, L. & Lequeux, J., dir., (2012) *L'Observatoire de Paris, 350 ans de science*, Paris, Gallimard & Observatoire de Paris.

[2] Cassini (1810).

[3] Wolf (1902).

[4] Cassini (1810), p. 45, note 1.

Fig. 7.1 The Paris Observatory at the time of its construction. For observing, the astronomers used long telescopes such as the one depicted on the left, or placed a lens on the edge of the upper terrace of the building or on the top of the wooden tower. The dome in the back is that of the church of the Val-de-Grâce, started by François Mansart (1598–1666) and finished by others in 1667. Wall painting at Perinaldo (Liguria, Italy), Jean-Dominique Cassini's birthplace. Wikimedia Commons, Tangopaso

followed at the Observatory, until the Revolution. They generally concentrated on geodesy rather than astronomy.

Meteorology and terrestrial magnetism were also among the assignments of the astronomers. Since its founding the Observatory kept a register of meteorological, and (later) magnetic observations.

The use of the astronomical instruments that one could find at this time is explained in Box 7.1.

Box 7.1 The Use of Astronomical Instruments in the Eighteenth and Nineteenth Centuries[5]

Most astronomical instruments during these centuries were devoted to the measurement of the position of stars, planets and comets. In some cases, they were also used for geodesy. One could find full circles, quadrants (1/4 of circle) or sextants (1/6 of circle) with graduations giving the orientation of a telescope, or sometimes of two perpendicular telescopes. Mural instruments were fixed on a wall in the plane of the meridian; they were used to measure the zenith distance of stars at their meridian crossing, from which one could derive the latitude of the place when their declination was known, or conversely their declination after the latitude was determined (Fig. 7.2).

Other instruments were transportable. One set them in a vertical plane and noted the graduation corresponding to the horizontal or vertical direction of the telescope thanks to a level or a plumb line; one could then measure the zenith distance of a celestial body or of a terrestrial target in any direction by pointing the telescope to it and reading the corresponding graduation. One could also measure the angular distance between two stars or two targets by orienting the instrument in the appropriate plane; this was the main use of Borda's repeating circle (see Chap. 6, Figs. 6.4 and 6.10). Theodolites had

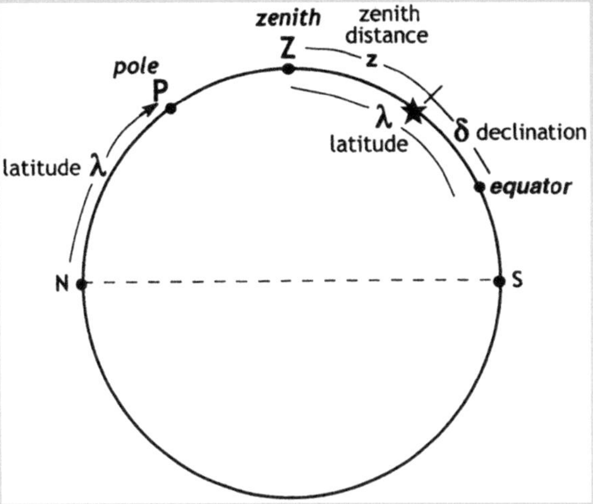

Fig. 7.2 Relation between the latitude l and the zenith distance *z* and the declination d of a star at its meridian passage. The figure is drawn in the plane of the local meridian. The latitude l is the angular height of the pole P of the Earth's axis over the horizon

(continued)

[5]One can find many illustrations representing astronomical instruments from the beginning to 1900, with interesting information on their builders and their use, in Repsold (1908, 1914).

Box 7.1 (continued)

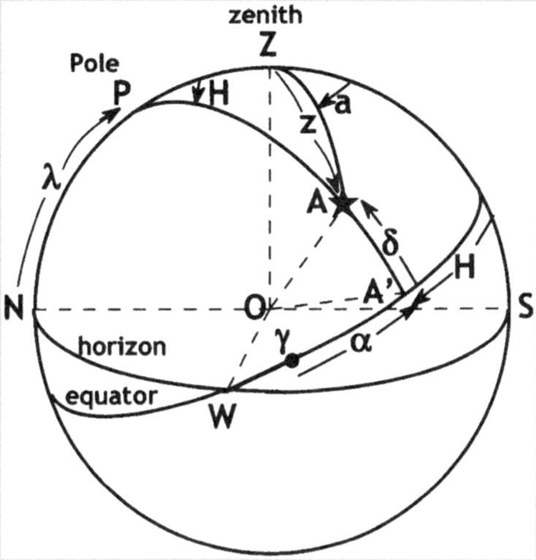

Fig. 7.3 The celestial sphere. The sky is represented by a sphere centered on the observer O. The horizon is represented with three cardinal points NWS. The diurnal motion is such that the celestial bodies seem to turn around the axis OP of the Earth, from east to west. The position of a star A is defined in the so-called horizontal coordinates by the azimuth a and the zenith distance z, and in the equatorial coordinates by the hour angle H and the declination δ. α is the right ascension, measured from the Gamma point γ, which is the intersection of the equator and the ecliptic (not represented), which is the annual trajectory of the Sun

initially only a horizontal circle, and were made later with another vertical circle and one or two telescopes (see Fig. 6.12), in order to measure azimuths and heights simultaneously.

The meridian telescopes could only be moved around a horizontal, east-west axis and could then observe only in the plane of the meridian. They were used with a clock to measure the time of the meridian crossing of stars, to determine time precisely or to measure the right ascension of the star if the time was known (Fig. 7.3). To insure the stability of the measurements, the meridian telescopes were adjusted on sighting-marks located south or north of the Observatory. Some meridian telescopes were transportable for use in the field; they had to be oriented, generally by observing the Pole star, and set horizontally thanks to a spirit level.

The ordinary telescopes were used for observing all kinds of celestial bodies. From Galileo to about 1758, the objective of theses telescopes consisted of a single lens and they were affected by chromatic aberration, minimized by a very large focal length. The invention by Chester Moor Hall (1703–1771) and John Dollond (1706–1761) of achromatic objectives made of two (sometimes three) lenses of different glasses, crown or flint, allowed to

(continued)

Box 7.1 (continued)

decrease considerably their focal length and made them handier. The reflecting telescopes invented by Newton and developed by William Herschel and others were achromatic by construction and could be made relatively short; but their mirror made of a special bronze, the *speculum*, was not very efficient and tarnished rapidly, so that a frequent resurfacing was necessary. It is only after Léon Foucault introduced silvered glass mirrors at the end of the 1850s that the reflecting telescopes (reflectors) overcame very progressively the refracting telescopes (refractors). The telescopes were also used to measure small angular distances (e.g. for double stars) and to determine the position of planets and comets with respect to nearby stars. One used for this micrometers inserted into the eyepieces, or other devices.

It was a great advantage to install the telescopes on *parallatic*[6] (now called equatorial) mounts, in which one of the axes was parallel to the rotation axis of the Earth: a slow motion around this axis was sufficient to follow the diurnal motion and to keep the object steady in the telescope. Some of the equatorial mounts were fitted with a clockwork for this motion. Graduated circles allowed finding the target when its position was known, something that was quite difficult with the alt-azimuthal mounts, in which the motion of the telescope was around a horizontal and a vertical axis.

Eighteenth Century Improvements

Because it was impossible to observe from the inside of the Perrault building, several observing rooms (*cabinets*) were constructed during the eighteenth century in a new wing on the east of the main building.[7] A large mural quadrant built in 1752 by the English maker John Bird (1709–1776) was set in one of the rooms to observe the stars passing south of the zenith, and another one by Jonathan Sisson (1690–1747) for observing the stars passing north. In 1803, a meridian telescope by another famous English maker, Jesse Ramsden (1735–1800), was installed in one of the observing rooms, replacing an older one by Étienne Le Noir (1744–1832), which was sent to the observatory of the École Militaire. It might seem strange to see those English instruments in the Paris Observatory. Cassini IV gives the reason[8]:

From the beginning to the middle of the reign of Louis XV, the Observatory housed [...] mural instruments and large mobile circles due to Langlois, Canivet and

[6] Citing Arago (*Œuvres complètes* t. 12, p. 32) "the parallactic or parallatic machine of the modern observer is so called because it serves to follow the celestial bodies along their diurnal parallels." The term *parallactique* is an unjustified attraction by the world parallax, which has no relation.

[7] For plans of these cabinets, see Cassini, *op.cit.*, Plates VII and VIII after p. 62.

[8] Cassini (1810), p. 4.

Lennel, who were the most famous builders of astronomy instruments of the time. But when I took the directorship of the Observatory [in 1784], these old talents were superseded by Bird and Ramsden. These English artists raised their art to perfection, leaving far behind the French, from whom they had taken off almost entirely the commerce of instruments of optics and mathematics.

Cassini IV had a small observatory built on the upper terrace, with a rain gauge and two observing turrets.[9] When Arago entered the Observatory, one of these turrets was occupied by a small equatorial telescope with a 65-mm diameter objective, installed by Bellet in 1804.[10]

There was also a small telescope made by Charité with a length of 3½ ft (1.13 m), a reflecting telescope of John Dollond with an aperture of 18 cm, and several astronomical clocks and meteorological instruments. There were many more, because the National Assembly had ordered that "all the astronomical instruments belonging to the Nation" be transferred to the Observatory. The list is in Cassini IV's Memoirs. It includes several clocks, a beautiful quadrant of 1.95 m radius built in 1742 by Langlois (active from 1730 to 1750), a meridian telescope by Charité, a dozen of telescopes including a Rochon telescope (see Chap. 4) and several reflecting telescopes, including one by Noël-Simon Caroché (ca. 1740–1813) of 6 ft (1.95 m) length with a 20-cm diameter platinum mirror mounted "according to Herschell's method" (sic).[11] The most impressive, if not the best, of all these instrument was "the grand telescope of Passy," with a 61-cm bronze mirror and a length of 6.3 m, built in 1772 by Dom Noël for the private observatory of king Louis XV in Passy, near Paris.[12] The mount of this telescope was modified in 1800 and a new mirror built by Caroché. It was housed inside the building, and was so big that the Observatory's architect, Antoine Vaudoyer (1756–1846), had the central door enlarged in 1801 so that the instrument could be rolled outside for observations. It was very cumbersome and did not produce any results. It was only preserved so it could be shown to the public, which had no access to the other instruments, until it was dismantled in 1841.

---

[9] Plan in Cassini, *op.cit.*, Plate IX.

[10] This instrument replaced probably an earlier one by Hautpois, purchased for 10,000 francs by the Observatory: see Lalande, in *Connaissance des tems pour l'année sextile VII*^{e} *de la République* (*22 Septembre 1798 au 22 Septembre 1799*), p. 282–318, accessible via http://gallica.bnf.fr. see p. 299.

[11] Caroché dreamt of an enormous reflecting telescope with a length of 13 m, with a platinum mirror for which 103 kg of platinum were already reserved. But it would have required a full metric ton, and the project collapsed.

[12] For details, see Lalande, in *Connaissance des tems pour l'an XIII* (*1805*), p. 253–255, accessible via http://gallica.bnf.fr. Images of this telescope can be seen in *Dom Noël (s.d.) *Instruments d'optique, Suitte de XXI planches gravées sous la direction de Dom Noël*.

The Observatory of the Board of Longitudes[13]

The Observatory upon Arago's Arrival

When Arago was named secretary-librarian of the Board of Longitudes in 1805 and was installed at the Observatory, then directed by Alexis Bouvard (1767–1843), he found the instruments I have just mentioned. Others were to arrive soon: an old reflecting telescope by James Short, purchased for 3000 francs (about $12,000) in 1807 to the heirs of Pierre Charles Le Monnier (1715–1799), a good Dollond telescope of 91 mm diameter, another one by Noël-Jean Lerebours (1761–1840), and a few others deposited at the Observatory and used occasionally. In 1810, the "telescope of the Emperor," actually a terrestrial telescope made by Lerebours and used by Napoléon, was deposited at the Observatory before being sent to the École Militaire in 1814. But there was no large telescope available.

A superb instrument was delivered in 1811 to Laplace by Georg Friedrich von Reichenbach (1771–1826) and collaborators, based in Munich. This was an astronomical repeating circle, costing 6750 francs (about $30,000). Laplace gave it to the Observatory immediately, and it was installed in the small observatory of the upper terrace. It is a sort of large theodolite with a 1-m diameter vertical circle. No instrument builder in France was yet able to produce such a large circle, and this is the reason why Laplace ordered it from a German instrument maker. It was supposed to measure the positions of the stars that could be occulted by the Moon to improve the ephemerides of Earth's satellite. However this instrument has never been used extensively, to the disappointment of Laplace.

The Instruments of Arago

Arago's influence at the Observatory rapidly increased after his return from Spain and his election at the Academy of Sciences. It was at its zenith when the Board of Longitudes named him director of the observations[14] in 1834. But the Board did not wait until this date to improve the instrumentation. The situation in 1823 is very well known through a report by Jean-Alfred Gautier (1793–1881), director of the Geneva observatory. He describes all the instruments already mentioned, and a new meridian circle with a diameter of 1.85 m built by Fortin, with an objective of Lerebours; this circle, installed in 1822 in the observing rooms, cost 12,000 francs (about $54,000) and was offered to the Board by the Duke of Angoulême (1775–1844), eldest son of King Charles X. Gautier also mentions sophisticated instruments for the study of the Earth's magnetic field.

---

[13] See the history of the Board of longitudes by Bigourdan (1928–1932).

[14] Arago was never named director, but this title made him the director *de facto*.

Immediately after this, in 1823, a 24.4-cm diameter refractor by Lerebours, costing 14,500 francs, was installed in the large octagonal room at the bottom of the east tower. As with several other instruments, it had to be rolled to the terrace for observations. In 1834, another telescope (or at least a telescope objective), with a 32-cm aperture, was installed. The following year, Arago observed Halley's Comet with the 24.4-cm telescope. Much later, in 1851, he mentioned still another telescope, with an 18- or 19-cm objective, which had been delivered by Lerebours in 1816; he planned to install it provisionally in the large dome of the east tower which was waiting for the 38-cm diameter equatorial telescope that will be described later. In 1848, the passage of Mercury in front of the Sun was observed with four instruments[15]: a telescope by "Cauche" (probably Robert Aglaé Cauchoix, 1776–1845) of 13.7 cm diameter, another one by Lerebours with a 14.7-cm aperture, perhaps purchased in 1831, the 18- or 19-cm telescope mentioned above, and the old Dollond telescope with a 9.1-cm diameter. The 24.4- and 32-cm telescopes are not mentioned, but at least the objective of the first one was still there in 1854 because Le Verrier had it mounted on a parallatic mount by Marc Secrétan (1804–1867).[16] We see that there is some confusion about the telescopes of the Observatory. Some of them were only deposited by their builders, probably with the hope that the Board of longitudes would purchase them.

Other instruments were under construction around 1823:

- a new meridian telescope by Gambey (Fig. 7.4) ordered by the Board to replace the Ramsden's instrument, whose optics were poor and which was transferred to the Marseilles observatory. It cost 12,000 francs; with its 15-cm diameter objective by Cauchoix, it was the largest such instrument in the world. It was installed in 1834;
- a 13,500-franc equatorial telescope, also built by Gambey, with a Lerebours objective of 10 cm in diameter that cost an extra 3000 francs (Fig. 7.5). It replaced the small "parallactic machine" by Bellet in the northwest turret of the small observatory on the terrace, which was also sent to Marseilles. Although completed in 1823, the new equatorial was only installed in 1826.

The Gambey's equatorial telescope, which was, according to Arago, "one of the most beautiful instruments ever built by man," was devoted to the measurement of the position of celestial bodies. For this, it had two large divided circles with a diameter of 3 ft (almost a meter), to measure respectively the hour angle and the declination of the target. A micrometer also due to Gambey allowed the determination of the position of comets with respect to the neighboring stars, and a powerful eyepiece was used for separating double stars. A clockwork motor was used to drive the rotation around the polar axis in order to track the target in its diurnal motion. This was a novelty at the time. The problem with this device was that the jerky movement of the escapement had to be transformed into a smooth, continuous motion.

[15] Anonyme (1848) *Comptes rendus de l'Académie des sciences* 27, p. 531.

[16] Yvon Villarceau, A. (1854) *Comptes rendus de l'Académie des sciences* 39, p. 949–961.

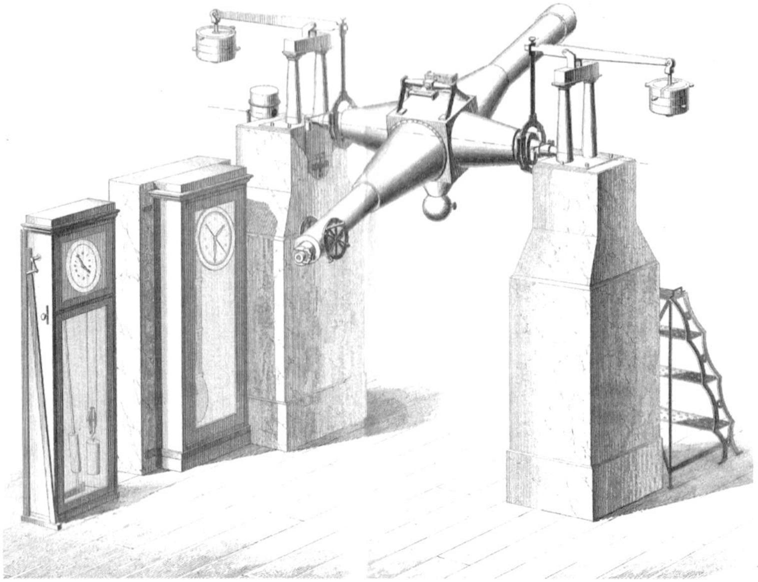

Fig. 7.4 Gambey's meridian telescope (1834). Note the two clocks on the *left*, one giving the mean solar time and the other one the sidereal time. From Arago's *Astronomie populaire*, author's collection

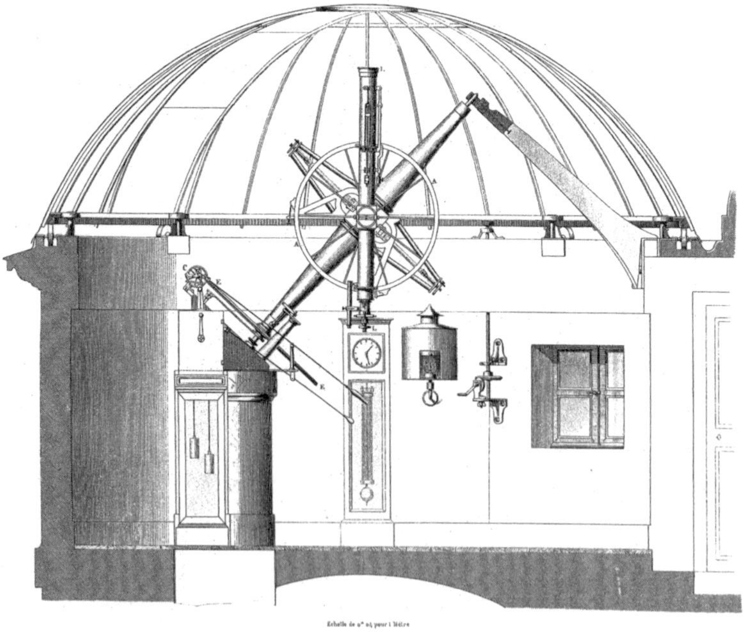

Fig. 7.5 Gambey's equatorial telescope (1826) in the southern turret of the small observatory of the terrace. Note the large graduated circles, and the driving clockwork C with its weights on the left. From Arago's *Astronomie populaire*, author's collection

According to Foucault,[17] "one was forced to introduce a spiral spring with a fly-wheel in order to suppress more or less completely the jerks of the clock... This caused a certain complication." Foucault made many efforts to solve the problem with conical pendulums and regulators, but without complete success. In 1832, as the turret in which the equatorial was placed was found to be too small, it was swapped with the Reichenbach circle and put in the southern turret, now covered by a costly rotating dome installed by an engineer named Picard.

The Board of Longitudes also considered rebuilding the observing rooms (*cabinets*), which were old and impractical. The Observatory's architect, Léon Marie Dieudonné Biet (1785–1856), prepared a project in 1828, but he was so indolent that the work only started in March 1832, to be completed in February 1834. As a consequence, there was no observations in these rooms from 1829 to 1837, the latter date when the instruments were re-installed and adjusted.

The last instrument installed during Arago's life was a mural circle by Gambey, placed in the reconstructed rooms in 1843. It had a 12.5-cm-diameter objective (Fig. 7.6). Figure 7.7 displays a view of the rooms with Gambey's meridian telescope and mural circle.

The deficiency of the French instrument builders at the end of the eighteenth century had by now been corrected. The highly praised English instruments were not without defects, and as the mid-century approached, the French builders were called

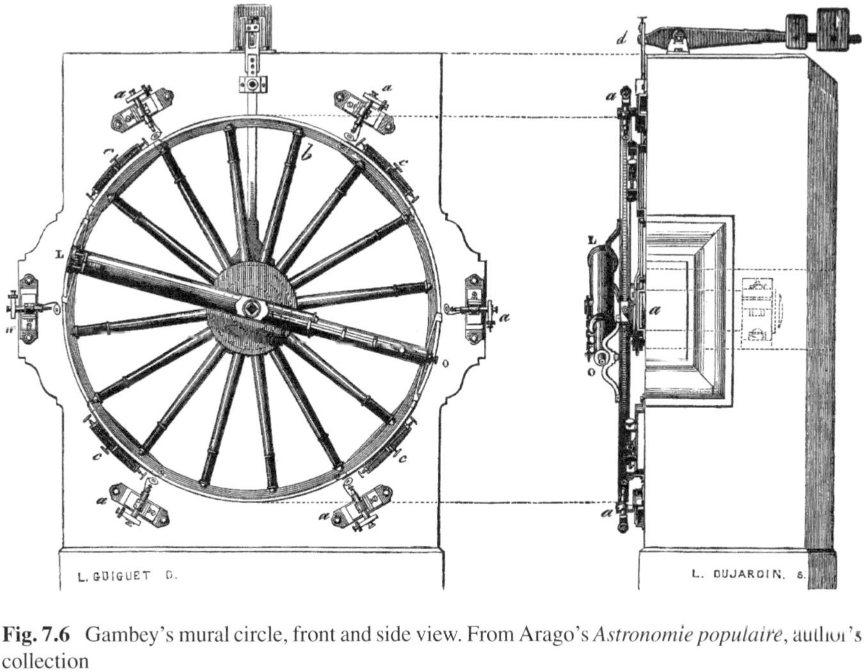

Fig. 7.6 Gambey's mural circle, front and side view. From Arago's *Astronomie populaire*, author's collection

[17] Foucault, L. (1847) *Comptes rendus de l'Académie des sciences* 25, p. 154–160, see p. 155.

Fig. 7.7 One of the meridian rooms of the Paris Observatory, with Gambey's meridian telescope in front and mural circle on the back. From *Le Monde Illustré*, 1862, Bibliothèque de l'Observatoire de Paris

systematically. Guillaume Bigourdan (1851–1932) notes with some chauvinism in his *Histoire du Bureau des longitudes,*[18] while listing the candidates (Bellet, Breguet, Cauchoix, Fortin, Le Noir, Henri Lepaute, Lerebours, Richer) for the position of "artist" of the Board, vacant due to the death of Caroché in 1813:

The presentation lists for the position… showed with what success, from the eighteenth century, the efforts of Cassini IV, continued by the Board of Longitudes, had set us free from foreign construction.

Indeed the members of the Board of Longitudes, in particular Arago, were quite effective in this progress, because they were constant promoters of French instrumentation.

The Great Equatorial Telescope of the East Tower[19]

However, once the English were dethroned, a new competition arose: that of German instrument builders. We have seen that Laplace had applied to a firm in Munich for his large circle in 1811. The most famous German builder was Joseph von Fraunhofer

[18] See note 13 of this chapter.

[19] For a detailed history, see Véron, P. (2003) *Revue d'histoire des sciences* 56-1, p. 191–220, accessible via http://www.persee.fr/web/revues/home/prescript/article/rhs_0151-4105_2003_num_56_1_2179.

(1787–1826). Associated with Joseph von Utzschneider (1763–1839), he had constructed in 1824 an equatorial telescope of 23 cm in diameter for the observatory of Wilhelm Struve (1793–1864) in Dorpat (now Tartu in Estonia). Astronomers around the world admired this instrument, the prototype of all the great nineteenth century equatorials.[20] Georg Merz (1793–1867) and Franz Joseph Mahler (1795–1845), successors of Fraunhofer and Utzschneider, built in 1839 a still larger equatorial telescope (38 cm in diameter) for the Pulkovo observatory in Russia, where Struve was named director in 1834. He and his son Otto (1819–1905) made many observations of double stars with it. Another 38-cm equatorial telescope was installed in Cambridge, England, in 1835, with an objective by Cauchoix and a mount designed by Airy. The Board of Longitudes wanted a similar instrument for the Paris Observatory, in order to observe double stars, a weak point of the Observatory.

There were Paris opticians who were able to build large achromatic objectives: Cauchoix, Noël-Jean Lerebours and his son Nicolas (1807–1873). They could use the flint glass produced by the Swiss engineer Pierre-Louis Guinand (1748–1824) and his son Henri (1771–1852). Cauchoix also knew how to make good parallatic mounts for equatorial telescopes (the Observatory purchased one from him in 1827). But he fell ill and died in 1845. Gambey was also able to build equatorial mounts, but he died in 1846. The mount for the large new telescope (Fig. 7.8) was finally ordered from the Austrian Johann Brunner (1804–1862), who worked in Paris since 1828.

Where to put this enormous instrument? The chosen solution was the top of the east tower. When Arago considered this possibility in 1839, this octagonal tower had no roof or terrace able to support the weight of the equatorial. The walls themselves did not seem strong enough to the architect of the Observatory, Henri-Alphonse de Gisors; they had to be reinforced. A cement dome was built to cover the tower, and the equatorial had to be mounted on a metallic spider resting on the walls. The engineer Louis Travers and his son, who had already built the iron roofing of the Chamber of Deputies in Paris, were put in charge of the project, which included a large rotating dome and its floor (see the picture in the title page of this chapter).

All this had a cost, indeed an enormous one. In total, 400,000 francs was spent on the project, or roughly $1.8 million. The work was completed in 1847, and the minister of the Public Instruction authorized the Board to start the construction of the equatorial mount, for which Brunner received 90,000 francs from 1851 to 1853. It was only finished in 1855, with a driving clockwork built by Breguet. But Arago had died 2 years before, without seeing the mount. Lerebours father and son had sold the 38-cm objective to the Observatory in 1849 for 40,000 francs. It had been tested at its completion in 1844 and found to be excellent.

In spite of this, the instrument was not a success. It had weak points in its conception, as did the dome. After a visit to the Observatory, Otto Struve, who had a considerable experience as an observer with the Pulkovo telescope, wrote in 1854 to

[20] It is preserved in Tartu: see pictures in http://www.obs.ee/obs/instrumendid/fr.htm.

Fig. 7.8 The parallatic (equatorial) mount of Brunner for the 38-cm telescope. From Arago's *Astronomie populaire*, author's collection

Le Verrier, who had just succeeded Arago, a rather critical but well-justified letter, of which there is an extract[21]:

1. The cast iron platform which will serve as the base for the equatorial rests on iron arcs. This base is not strong enough.[…] 2. This platform being supported by iron arcs will experience rapid changes in its position due to the sole effect of temperature.[…] 3. The diameter of the dome is 12 m for a height of 10 m. With respect to these dimensions the width of the trap-doors [through which one observes] is much too narrow. The equality between the internal and external temperatures is a vital condition to obtain good images.[…] 6. The motion of the

[21] Cited by Véron (op. cit.).

> *dome is not fast enough.*[22] *7. It is not at all useful that the floor on which the observer stands rotates with the dome. 8. It seems very inconvenient that the tower is raised above other observing rooms and housings.*

In spite of his doubts, Le Verrier continued the project. But when the Lerebours objective was to be installed on the telescope tube in November 1855, it was discovered that the surface of the crown glass was damaged by humidity; a large number of very small cracks were visible, preventing observation.[23] As for the equatorial mount, it was only delivered in 1859. The objective was not installed. After partial re-polishing, it was used by Cornu for his measurement of the velocity of light (see the end of Chap. 4), then after re-installation, for a few observations. The initial wooden tube was then replaced by an iron one (Fig. 7.9), and a new objective was built by the brothers Paul Henry (1848–1905) and Prosper Henry (1849–1903), but only in 1881. In the meantime, an equatorial with an iron tube, built by Friedrich Wilhelm Eichens (1820–1884), a German instrument builder based in Paris, was installed in 1858 on the east tower of the Observatory, under a more satisfactory dome due to the engineer Joseph Jean. It had an objective of 5.25 m of focal length and 31.6-cm diameter, perhaps that built in 1834 by Lerebours father (see above). This objective was only slightly smaller than the one in the east tower. It was in its turn replaced in 1884 by one made by Adolphe Martin (1824–1896). Then the giant telescope of Meudon, with its two objectives of 83 cm for visual observations and 62 cm for photography, was put into service in 1896.

The Brunner equatorial is still at the Paris Observatory, after some relatively minor modifications to the instrument and the dome. It has been used by several astronomers, including an assiduous amateur, Paul Baize (1901–1995), to observe double stars and planetary satellites. The dome of the 31.6-cm Eichens equatorial was transported in 1974 to an amateur observatory in Saint-Véran, in the Alps. It now contains a 60-cm reflecting telescope; the equatorial itself has disappeared.

A few remarks about the instrumental set I have described: Position astronomy was well served, which was normal at this time. The Board of Longitudes, probably perturbed by the failure of the great reflecting telescope of Passy, deliberately chose refractors. Still, everyone knew the superb results obtained by Herschel with his giant reflectors, and at its founding in 1795 the Board had considered building a 2-m-diameter reflector. Moreover, the refractors were not fit, or at least not used, to observe star clusters and nebulae, which were only occasional targets. This lack of interest is somewhat surprising on the part of Arago, but he was tied to some extent by the Board. French astronomy once again lagged behind after Foucault had built the first large reflector, the 80-cm-diameter one in Marseilles, soon followed by a similar one in Toulouse and a 1-m instrument placed in Meudon in 1893. They were among

[22] Indeed the observers have always complained about the difficulty in rotating the dome and the slowness of its rotation: one full turn would have taken 45 min at the beginning, and 10 min after a motor was installed.

[23] Later, Georges Bontemps corrected this problem in his glass factory of Choisy-le-Roi by changing the composition of the crown.

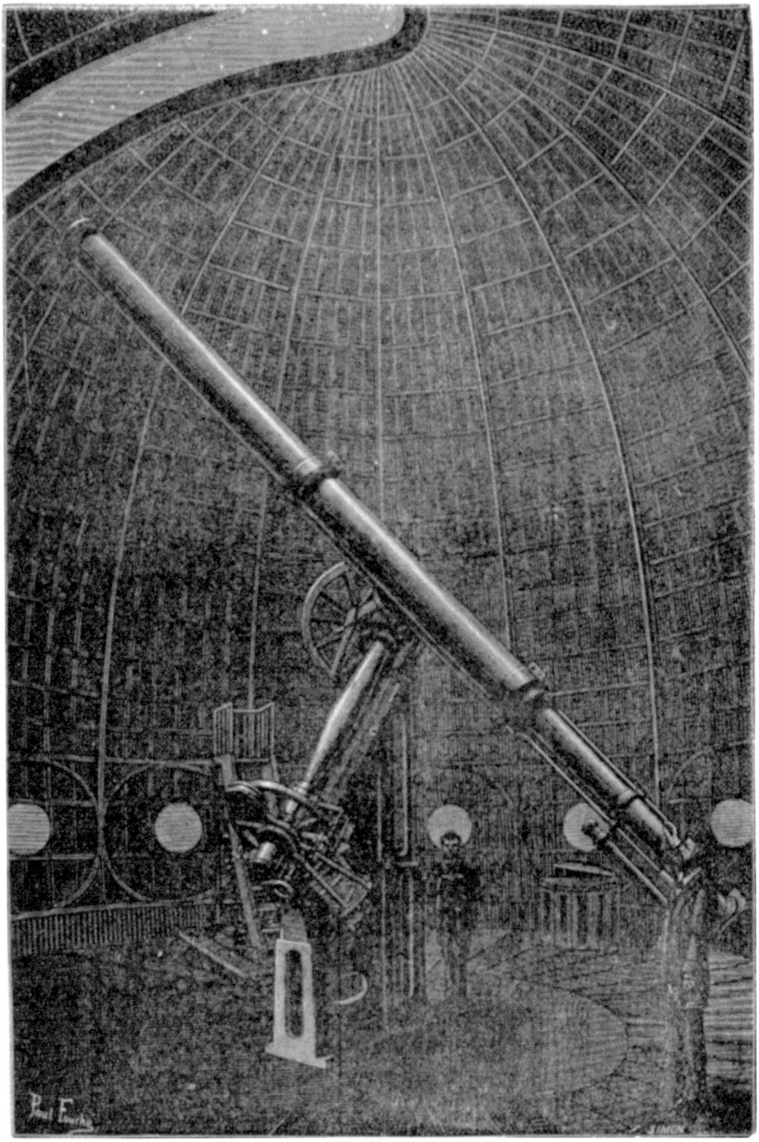

Fig. 7.9 The Brunner equatorial of the Paris Observatory at the end of the nineteenth century. From Flammarion's *Astronomie Populaire*, author's collection

the largest in the world, but no larger instrument was built in France until the 1.20- and 1.93-m telescopes of the Haute-Provence observatory, created in 1936. Indeed, the Paris 1.20-m telescope had been a failure, and as a kind of compensation one had installed in Meudon a very expensive double equatorial. During the gap in France, a 2.5-m giant was inaugurated in 1917 at Mount Wilson in California, allowing the impressive development of extragalactic astronomy in the US.

The Amphitheater

One of the duties of the Board of Longitudes was to give lectures in astronomy to "educate young men, either to fill up the positions of directors of the observatories of the different departments, or to supply astronomers for remote expeditions." Actually, during the Revolution the observatories in the provinces had suffered much more damage than the one in Paris, and it was a constant preoccupation of the Board to supply to them with astronomers and instruments, which were too often obsolete instruments from Paris. Arago was asked in 1812 to give these lectures, and he started in 1813. Despite its modest purposes, Arago's teaching was so good and given with such enthusiasm that it attracted many auditors. He first gave the lectures in the large meridian room of the Observatory, which has a very bad acoustics, then in Collège de France. But they deserved a special place worthy of their success.

In November 1836, Arago submitted to the Minister of Interior a project for an amphitheatre in the west wing of the Observatory. This project was accepted, with an exceptional subvention of 120,000 francs in 1839, and an equal one in 1840 (the total is roughly equivalent to $1 million). The amphitheater, designed by the Observatory architect de Gisors, was finished in February 1840. "Vast, elegant and convenient," according to a contemporary account, "of a scandalous luxury," as Arago said with humor, it could accommodate 800 auditors. Arago and the Board took special care of its decoration. At one angle, they put a statue of Cassini I finished by Jean-Guillaume Moitte (1746–1810) in 1810, and at the other a statue of Laplace by Gabriel Joseph Garraud (1807–1880). There were also busts of Newton, d'Alembert and Lagrange, and five medallions with the portraits of Clairaut, Picard, La Caille, Delambre and Bouguer. A map of the Moon decorated the ceiling, and six other medallions located on each side of the blackboard, itself well decorated, represented instruments of astronomy.

Arago gave his lectures in this amphitheatre until 1846. After his death, Le Verrier, who succeeded him as director of the Observatory, had it destroyed to put his own apartment in its place. It is a miracle that some of the elements of its decoration survive: the two statues, the blackboard and four of the medallions depicting instruments. The rest is lost.

Life at the Observatory

The Astronomers and the Arago Clan

The scientific personnel of the Observatory are well known thanks to the registers and publications of the Board of Longitudes. With the exception of retired sailors, geographers and to some extent artists, all the members of the Board were dealing with astronomy and geodesy, which was up to 1820 an integral part of astronomy. Figure 1.4 in Chap. 1 gives the list of the scientists of the Board, from its establishment to Arago's death in 1853. Aside from prestigious names, others on the list are

Box 7.2 Claude-Louis Mathieu

Born in 1783 to a modest family, Mathieu was helped financially in his studies by Delambre, who obtained for him a room at the Observatory in 1801. He entered the École Polytechnique in 1803, and started to work (like Fresnel later) in the bridges and roads administration. But he left this job in 1806 and in 1807 he was named assistant secretary (previously secretary-librarian) of the Board of Longitudes, replacing Arago, who had left for the Balearic Islands. He was named adjunct in 1817, was elected at the Academy of Sciences and replaced the aging Delambre at the Collège de France. He also became "extraordinary Inspector of cadastral survey." A discreet and erudite man, indefatigable worker, he took a large part to the editing of the *Connaissance des temps* (the French astronomical ephemeride) and of the yearbook of the Board of Longitudes. His scientific and political careers were more or less overshadowed by those of Arago, in spite of the fact that he was 3 years older. He survived him 22 years.

forgotten. However, one of these had an important role in the nineteenth-century astronomy: Alexis Bouvard (1767–1843).

In general, the astronomers were housed at the Observatory, either in separate buildings, which disappeared around 1815, or inside the Perrault building itself. One astronomer found it convenient to install apartments in the building, the observations taking place outside or in annexes. When he came to the Observatory in 1805 as secretary-librarian, Arago was accommodated in a small house on the Observatory grounds. In 1810, Humboldt joined him for some time, but after his marriage on 11 September 1811, Arago settled down in the main building (Fig. 2.6). He was to live there the rest of his life. It is there that his three children were born. His wife died in 1829 in this apartment, and himself in 1853.

In 1820, the Arago family was in a bad financial situation after the death of the François's father and the dismissal for political reasons of his brother Jean, who had succeeded his father as cashier of the Mint in Perpignan. For a time, Arago housed his brother Etienne and his sister Marguerite at the Observatory. Etienne did not stay for long, while Marguerite soon married an astronomer, Claude-Louis Mathieu (Box 7.2 and Fig. 7.10), 3 years older than Arago. This was one of the four unions that created what was sometimes called "the Arago clan," which included also Humboldt and the physicist Félix Savary (1797–1841), a close friend of Arago who went to Estagel to be nursed by Arago's mother and died there. Here are the three other unions:

- François's sister-in-law married the physicist Alexis Petit (1791–1820)
- In 1843, the daughter of Claude-Louis and Marguerite Mathieu, Lucie, married Ernest Laugier (Box 7.3), a pupil of Savary, *élève-astronome* at the Observatory since 1834.
- In 1840, Ernest Laugier's sister, Adèle, married Jacques Babinet (1794–1872), another physicist who was elected the same year to the Academy of Sciences and named the following year to the Board of Longitudes.

Fig. 7.10 Claude-Louis Mathieu (1783–1875) in 1867. Wikimedia Commons

Box 7.3 Ernest Laugier (1812–1872) and His Wife Lucie (1823–1900)

Unlike Mathieu, Laugier came from a prominent family of lawyers and scientists. After studies at the École Polytechnique, he entered the Observatory in 1834, and was elected to the Academy in 1843. He helped Arago in performing observations and experiments that the latter could not make because of his declining sight. His personal work was mainly about the sunspots and the rotation of the Sun.

His spouse Lucie, daughter of Mathieu and his wife Marguerite, had no scientific background, but she was intelligent and open-minded. She was the dominating personality of Arago's circle, the "Observatory angel" according to Humboldt, surrounded by a rather severe father and a stern husband, and by her uncle Arago, "who passed from the most violent storms to the most seducing serenities, and who did not always know how to spare his family from his angers".[24] She was his guardian angel at the end of his life, and, according to an unfounded legend, Arago dictated to her his *Astronomie populaire* (see Chap. 11). She kept his instruments, which she gave to the Observatory.

[24] From Jamin (1885).

The Visitors

The Observatory being one of the main centers of Parisian scientific life, it is not surprising that it was visited by important people. I will cite only two: the king of Prussia, Friedrich-Wilhelm III (1770–1840), and the writer Victor Hugo (1802–1885).

Soon after the fall of Napoléon's empire, many prominent foreign personalities wanted to visit the Observatory, but the astronomers, including Arago, were not too eager to receive them. However Humboldt, who wanted to introduce his sovereign the king of Prussia, asked him to disguise himself as a simple tourist. Arago describes this visit at the third person, not without some self-satisfaction[25]:

During the visit to the instruments, [...] politics mixes to the conversation, and Arago does not hide his opinion about the way the foreign sovereigns forced France to pay for the ambitious follies of Napoléon. Humboldt, who was turning anxious, took Arago aside while the other visitor was looking through a telescope: "Be careful, this is the king!—I was well aware, answered Arago, and this is why I expressed my opinion so neatly."[...] A letter from Humboldt told his friend a few days later that the king did not take the thing too seriously, and that he even approved the patriotic reflections of the French scientist.

The other visit dates from 1834. Victor Hugo came to the Observatory to observe the Moon under the direction of Arago, who was just promoted as Director of the observations. Very impressed by his visit, Hugo wrote in 1863, when he was in Guernsey, a rather exact although very romantic account [26] of which some extracts of "astronomical" nature are given in Box 7.4. The title of this little-known text alludes to the name of one of the lunar features that Hugo claims to have seen, the *Promontorium Somnii* (Promontory of Dreams). It describes first his observations, then undertakes a long and erudite philosophical reflection on dreams and myths. In spite of his exceptional memory, Hugo could not remember accurately the details of what he had seen 20 years after the fact. The names of the lunar features are given not in the order of their coming in the telescope, but in the order of their nomenclature in Cassini's map of the Moon that was glued on one of the sheets of the manuscript. Hugo describes as rapid, almost instantaneous, phenomena that were actually rather slow.

As shown by two astronomers of the Observatory, the large magnification of 400 of the telescope used by Hugo and its small field of view exclude that it was one of the instruments of the small observatory on the terrace.[27] It was rather one of the Cauchoix telescopes on parallatic mount purchased in 1827.

---

[25] *Hamy (1907), see p. 359–360.

[26] Hugo, V. (1863) *Promontorium Somnii*, in *Proses philosophiques de 1860–1865*, accessible via http://fr.wikisource.org/wiki/Proses_philosophiques/Promontorium_somnii.

[27] See http://www.obspm.fr/victor-hugo-et-le-promontoire-des-songes.html.

Box 7.4 Excerpts from Victor Hugo's *Promontorium somnii*

I remember a summer evening of 1834, a long time ago, when I went to the Observatory: that of Paris, where I was staying at that time. I went in. The night was clear, the air pure, the sky calm, the moon at its crescent; one could distinctly see with naked eye the dark side slightly illuminated: the earth-light. Arago was there, we climbed on the terrace; there was a telescope which enlarged 400 times [...]. Arago pointed the telescope and said: look.

I looked [...]

- I see nothing, I said.

Arago said: - You see the moon.

I insisted: - I see nothing

Arago went on: - Look.

After a moment, Arago said: - You have just taken a long trip.

- Which trip?

- Before, as all the inhabitants of the earth, you were at 90,000 leagues from the moon.

- So what?

- Now you are at 225 leagues.

[...] This was, in effect, the result of the enlargement by 400 times. [...] The field of view of the telescope was too narrow to cover the whole planet [sic], the sphere was not visible, and what I saw, if I could see anything, was only an obscure segment. Arago, as he explained to me later, had pointed the telescope to a part of the moon that was not yet illuminated. I resumed:

- I see nothing.

- Look, said Arago.

I did like Dante following Virgil, I obeyed.

Gradually my retina fulfilled its task. [...] My pupil dilated, my vision focused, as one says, and this darkness that I was staring at became clearer. I saw; what did I see? Difficult to say.[...] This impression cannot be described. Those who have not experienced it cannot realize what it is. [...]

Suddenly I caught my breath: a lightning flashed. This was wonderful and tremendous. I closed my dazzled eyes. I had seen the sunrise on the moon.

The lighting met something, a summit perhaps, and collided against it; a kind of snake of fire appeared in the darkness, coiled into a circle and stopped: this was an emerging crater. At some distance, another flash, another fiery snake, another circle: a second crater. The first one was the Messala volcano, Arago said; the second one, the *Promontorium Somnii*. Then, like flame crowns above a shadow, mount Proclus, mount Cléomèdes, mount Petsevius [sic, probably Petavius], those Vesuvius and Etnas of the moon, started to appear, resplendent; a tumultuous crimson ran on the darkness of this fantastic horizon, jagged teeth of charcoals jutted up and remained there, at a standstill, awesome. This is a chain of lunar Alps, Arago said to me. However, the circles

(continued)

Box 7.4 (continued)

grew, became larger and larger, their sides merging until they finally blended; valleys grew hollow, precipices opened, hiatuses gaped till a dark foam overflowed from their lips, deep spirals plunged, frightening the eye that followed them, huge cones of darkness were silhouetted, shadows moved, armies of rays of light lintel-like settled from one summit to another, crater nodes wrinkled around the peaks, profiles born out of the furnaces appeared in all sorts of confusion, some smoke-like, others radiant with light; capes, promontories, gorges, passes, plateaus, jagged surfaces, escarpments, cuttings tangled together, wrestled as rocky outcrops as so many beams; one could see the shape of the mountains. All this existed superbly. [...] It was if the dawn had suddenly set fire to this gloomy world.

As Arago explained to me as a rather obvious fact, while I was looking at the moon its motion had turned little by little to the sun the edge of the obscure part, and daylight invaded it progressively.

This vision is one of my deepest remembrances.

The Observations

In Arago's time, all observations were made locally. There was no question of going to Chile or to Hawaii to observe, as we do today. Only the passages of Venus required expeditions to remote places, but there was none during the lifetime of Arago. However, total eclipses of the Sun offered another opportunity for travel. Sometimes, the astronomers went to the places where a total eclipse could been viewed. This was the case for Arago, Laugier and Victor Mauvais (1809–1854) for the eclipse of 8 July 1842; the Sun was totally eclipsed in Perpignan, and offered a good opportunity for Arago to visit his family and his electoral district. Frederic Petit (1810–1865) went to Montpellier, and Bouvard to Digne. On the English side, Airy went to Pavia in Italy. For the first time the expeditions had a real scientific flavor, because they were well prepared by Arago, who gave instructions about what to observe.[28] Each astronomer took a telescope with him and made interesting observations of the solar protuberances and corona.[29] However, the first important expeditions involving a really extensive preparation were only devoted to the total eclipse of 18 July 1860 in Spain.[30]

Otherwise, astronomy was a long and tedious routine. Every night, using the meridian telescope, an astronomer had to measure the time of passage of bright stars

[28] *Annuaire du Bureau des Longitudes* for 1842, p. 609–642, accessible via http://gallica.bnf.fr.

[29] **Œuvres complètes*, t. 7, 136–290.

[30] See Lequeux, J. (2013) *Le Verrier, magnificent and detestable astronomer*, Springer, New York, p. 158–160.

and planets at the meridian to determine the time and the right ascension of planets (see Figs. 7.2, 7.3, and 7.4). With mural quadrants or full circles, one measured the angular distance to the zenith of stars and planets to obtain their declination, once the latitude of the observatory had been determined. Also, astronomers continued to observe the eclipses of Jupiter's satellites in the shadow of the planet, in order to improve their ephemeris using Laplace's theory; for this, several telescopes were used when they were available. Observing the time of occultation of stars by the Moon supplied a precise position for our satellite, which could be compared to the theory. For this, the precise position of these stars had to be known, something that was allowed after 1811 by the Reichenbach circle.

After August 1800, when the observations resumed after the revolutionary turmoil, they were mostly performed by Bouvard. After 22 March 1805, they were shared between Bouvard and the young Arago. Then Mathieu, "who had practiced a lot," replaced Arago during his trip to the Balearic islands. Arago observed again soon after his return, starting at the autumnal equinox in 1809. Nicollet joined the three observers in 1815, and Adolphe Gambard (1800–1836) took part in the observations in 1818 and 1819. When Arago became Director of observations in 1834, Bouvard did not observe anymore (he was 67 and died in 1843) and Nicollet had been forced to resign in 1830. Most of the observations were made by the *élèves-astronomes* (student astronomers) and Arago only took part in special observations.

The unreduced (raw) observations made before 1809 were published in the *Connaissance des temps*, which also inserted observations made elsewhere when they were judged of interest. They were also collected in a special volume. Later, the observations were only published occasionally in big volumes. No effort was made to reduce these observations, for example to obtain the coordinates of the planets; this had to be made by the interested astronomers, like Bouvard or later, Le Verrier, who had his board of calculators reduce all the observations from 1800 to 1854.[31]

When the Gambey equatorial was available, it was used to measure the position of comets with respect to nearby bright stars. With much delay and with respect to what Wilhelm Struve and John Herschel had been doing for a long time, Antoine Yvon Villarceau used it from 1849, together with other instruments, to separate double stars and to measure their motion around each other through observations distributed over many years. Astronomers also observed nebulae with the hope of detecting their possible proper motions (with respect to distant stars) and then to have an idea of their distances. This is what Laugier did without positive result. The successful observation by Bessel of the parallax (distance) of the star 61 Cygni in

[31] One finds, in the notice preceding the first volume containing the observations from 1837 to 1846 (*Observations astronomiques faites à l'Observatoire de Paris*, 1837–1846, 10 vol., Paris, Bachelier), the following sentence: "One would have to add two collaborators to the Observatory personnel, unless one had wanted to impose this new work to those people which, today, are in charge of the observations. In transforming these distinguished young men into computing machines, one would have completely changed their situation, forbidding them to do new researches.[…] All the same, the problem is reserved and will be solved when circumstances allow." They never did during the life of Arago.

Fig. 7.11 Urbain J.-J. Le Verrier, Wikimedia Commons, Magnus Manske

1838, that I will detail later, sparked a great interest. Faye attempted to measure the distance of other stars without success. Laugier was responsible for observing the Sun and its spots. Other observations, of the diameter and flattening of planets, were to a large extent reserved for Arago. None of these observations excited much interest, mainly because the available instruments were not up to the standard of those used in England, Germany, Russia and the United States. This is why Arago asked so forcefully for the construction of a great equatorial telescope.

The Le Verrier Affair[32]

Urbain-Jean-Joseph Le Verrier (1811–1877, Fig. 7.11) attended the École Polytechnique and received his diploma in 1833, becoming an engineer of tobacco. He worked with Gay-Lussac, and applied for a position of *répétiteur* (assistant) of chemistry, but he was unsuccessful. However, a similar position of geodesy, astronomy and "machines" was available and he got it. This gave him an opportunity to meet Arago. Le Verrier then specialized in celestial mechanics, and presented to the Academy of Sciences on 16 September 1839 a *Memoir on the secular variations of the elements of the orbits of the seven main planets, Mercury, Venus, the Earth, Mars, Jupiter, Saturne and Uranus*. He knew that there were anomalies in the motion of

[32] See for more details Lequeux (2013).

Fig. 7.12 Charles Delaunay. Wikimedia Commons, Tabularius

Uranus, observed by Delambre and Bouvard, that they attempted in vain to account for. This was a disaster for the Newtonian mechanics, which was the great article of faith of science at the time. When publishing in 1821 his *Tables of Uranus*, Bouvard suggested that an unknown "disturbing planet" could affect the motion of Uranus.

In 1842, Le Verrier occupied himself with the planet Uranus. He was not the only one to work on this problem. In Germany and Denmark, Peter Hansen (1795–1874), who had developed new methods of celestial mechanics, applied them to Uranus. In France, Joseph Liouville (1809–1882) assigned a young man, Charles Delaunay (1816–1872, Fig. 7.12), to examine Hansen's work and check the results, a task that he successfully accomplished in 1841. But Delaunay also found that certain terms in the series expansions, which he thought Hansen had not considered, were significant. Le Verrier did not agree with Delaunay, and a rather technical controversy ensued that can be followed in the *Comptes rendus de l'Académie des sciences*. Le Verrier was correct, but he rather nastily mocked his still inexperienced colleague. Such conflicts between them would occur often in the future, and in the end they would become mortal enemies.

Time was ripe to attempt to solve in a definitive way the problem of the motion of Uranus. The nephew of Alexis Bouvard, Eugène Bouvard (1812–1879), calculated new tables of the planet and noticed[33]:

The discrepancies between the observations and the theory lead me to believe that it is very likely that my uncle was right in postulating the existence of another planet which disturbs Uranus.

---

[33] *Bouvard, E. (1845) *Comptes rendus de l'Académie des sciences*, 21, p. 524–525.

This confirms that Alexis Bouvard was the first to suggest the idea of a disturbing planet. This idea spread in the scientific world and even in the general public. However, Arago was not very confident in Eugène Bouvard, who had made poor measurements during his expedition to Digne to observe the total eclipse of the Sun in 1842. He asked Le Verrier to deal with the problem, even though he was not a member of the Observatory. In the great article of 1846, where he summarized his research on Uranus and the subsequent discovery of Neptune,[34] Le Verrier wrote:

During the course of this last summer 1845, Mr. Arago made clear to me that the great importance of this question imposed a duty on every astronomer to contribute, to the utmost of his powers, the clarification of certain points. In response to his plea, I abandoned, therefore, my researches into comets, of which several fragments have already been communicated, in order to occupy myself fully with Uranus. Such is the origin of that work that I have the honor to present to the academy.

On 1 June 1846, Le Verrier, who had just been elected to the Academy of Sciences and named adjunct astronomer at the Board of Longitudes, announced at the Academy that he had proven the existence of the perturbing planet. On 31 August, he predicted the position of the planet in the sky. Le Verrier wrote on 18 September to his colleague Johann Gottfried Galle (1812–1910) in Berlin, and the very evening of the day he received the letter, the 23 September, Galle, helped by Heinrich Louis d'Arrest (1822–1875), discovered the planet at less than a degree of close the predicted position. This discovery was triumphantly announced by Arago at the Academy on 5 October[35]:

M. Arago reads a letter to M. Le Verrier from M. Galle from Berlin, dated 23 September. The first lines of this letter say: "The planet of which you sent the position really exists.[…]" M. Arago found in the letter from the Berlin observer an opportunity to detail the history of the memorable discovery of M. Le Verrier. "M. Le Verrier has seen the new celestial body without the need of looking at the sky: he saw it with the tip of his pen. […] It was necessary to build new maps of the sky until magnitude 10 before attempting with good chances of success to search for the new body. This should have been, for the small region in which the planet was located, according to M. Le Verrier, a work undertaken in Paris. There was no need for this in Berlin; detailed maps of the sky were already there. Through a happy coincidence, the map of the 21th hour [of right ascension] where the planet moves presently had just been engraved and published." M. Arago told the Academy that, having received from M. Le Verrier the right of giving a name to the new planet, he decided to use the name of his discovered, thus to call it Planet Le Verrier […].

---

[34] Le Verrier, U.-J. Recherches sur les mouvements de la planète Herschel (dite Uranus), *Connaissance des temps pour 1849*, p. 3–254, accessible via http://gallica.bnf.fr.

[35] *Arago, F. (1846) *Comptes rendus de l'Académie des sciences*, 23, p. 659–662.

Arago defended Le Verrier's discovery against the claims of Airy, of James Challis (1803–1882), the director of the Cambridge observatory, and of John Herschel, in favor of their compatriot John Couch Adams (1819–1892), who had made a similar study to that of Le Verrier, but did not disclose his results before October 1846.

Although Arago was in general full of enthusiasm, one might be surprised by the excessively eulogistic character of his interventions at the Academy, concerning the discovery of the planet. In a letter written in 1869 to the Minister of Public Education in order to protest against the exactions of Le Verrier, Delaunay wrote[36]:

In 1846, in the course of the discovery of the planet Neptune, M. Arago, driven by certain hideous circumstances which it is not appropriate to lift the veil on here, had raised M. Le Verrier on a pedestal, and made him out to be an extraordinary man, one of the greatest geniuses that France had ever produced. Some months later, M. Arago recognized the enormous mistake that he had made, but the harm was already done, and it was out of his control to repair the damage. His final years were poisoned by the vision of the dark consequences which were inevitably to follow therefore.

What were these "hideous circumstances?" It's a shame that the veil had not been raised! One persistent rumor speculates that Arago had had an affair with Madame Le Verrier. Le Verrier, discovering it, might have taken advantage of the situation by using it to blackmail Arago into supporting a proposition that was clearly indefensible. The hypothesis is plausible, but difficult to verify.

Of course, Le Verrier's discovery raised the enthusiasm of the scientific world and the general public, but there were jealousies. Laugier and Mauvais attacked Le Verrier on the question of the periodic orbits of comets, and the following controversy was acrid.

The relations between Arago and Le Verrier, although good in appearance, were soon to deteriorate. Indeed, Le Verrier, intoxicated by success, showed a pretention that irritated everyone. Delighted with the fact that he succeeded in forcing Arago to give his name to the new planet, he attempted to freeze the situation, hoping that Uranus would keep forever the name of Herschel, which actually was sometimes used:

In my subsequent publications, I will consider it a strict duty to make disappear completely the name Uranus, and to only refer to the planet using the name HERSCHEL. I sorely regret that my already published writings do not permit me to follow the determination that I shall religiously observe henceforth.

He did not succeed: Uranus kept its name, and Planet Le Verrier became Neptune, conforming with the usage of astronomers who give names of Roman gods to the planets.

---

[36] This letter is reproduced in *Annuaire du Bureau des longitudes for 1933*, p. A.30–A.33, accessible via http://gallica.bnf.fr.

Politics now entered in the affair. François Guizot (1787–1874), the prime minister under King Louis-Philippe (1773–1850), would have very much liked to control the hot-tempered Arago, who was a problematic opponent. He could not discredit Arago as a deputy, so he attempted a maneuver that consisted of promoting a rival for the position of Director of observations, with the hope that the Board of Longitudes would choose him to replace Arago. Le Verrier was the ideal rival, and one does not have to be a great scholar to be persuaded that he would not have been displeased by the idea of replacing Arago at the Observatory. But the 1848 Revolution, which drove Louis-Philippe and Guizot from power, put an end to the plan. Amazed by the treachery of a man he esteemed and had supported, Arago, as well as his whole clan, broke definitively with him. Now, it was war, with no holds barred. Unskillfully, Babinet attacked the discovery of Neptune, speaking of "enormous errors" and even suggesting that another planet, that he named Hyperion, added its perturbing action to that of Neptune.[37] Le Verrier himself had an easy time refuting Babinet. He wrote:

1. *Is it true that the direction in which I placed Neptune had an enormous error apart from the epoch of the discovery by M. Galle for a very few years before or after? NO. This is false.*
2. *Is it true that there were enormous errors relative to its distance from the Sun? NO. This too is false. I simply placed Neptune a bit too far away.*
3. *Is it true that the theoretical mass of Neptune differs from the mass derived from the observation of the satellite to such an extent that it would be an irresistible argument against the identity of the theoretical Neptune with the observed Neptune. NO. This is false.*[38]

Le Verrier had to wait for Arago's death to replace him at the Observatory. Delaunay wrote in his 1869 letter:

At the death of Arago (end of 1853) the Emperor thought naturally to M. Le Verrier to succeed him at the Observatory. This was natural: the reputation of M. Le Verrier did not allow another solution. But the character of this man, which was only known by those who, like myself, have been close to him for years, soon showed up fully and produced its effects.

Four months after Arago's death, Le Verrier took possession of the Observatory. He had the amphitheater destroyed, to be replaced by his apartment: a great loss that is still felt today. Mathieu, Laugier and others were expelled from the Observatory and lost their housing and instruments. Other astronomers left as well. Lucie Laugier collected Arago's personal instruments, such as his polarimeter, which

---

[37] *Babinet, J. (1848) *Comptes rendus de l'Académie des sciences* 27, p. 202–208. The answers of Le Verrier are p. 208–210 and 273–279.

[38] The mass obtained by Struve from the orbit of Neptune's satellite Triton, discovered by William Lassell (1799–1880) soon after the planet, is 65/100 of the mass predicted by Le Verrier, which appeared to him within the errors.

were preserved thanks to her action. By way of compensation, the ministry offered successively to Mathieu and Laugier the chair of astronomy at the Paris University, which they both refused. Mathieu would occupy himself with the publication of the *Connaissance des temps* and of the *Annuaire du Bureau des longitudes*,[39] then turned his attention to industry and was to direct the manufacture of tobacco in Dieppe. Laugier was to live from funds that he received in his capacity as examiner at the *École navale*, and was to receive the satisfaction of being in 1862 promoted to titular membership of the Bureau of Longitudes at the same time as Delaunay. All three were to become reformers of the Paris Observatory after Le Verrier's dismissal in 1870.[40] But neither Mathieu nor Laugier was to have the means to carry out astronomical research to the high level that one might have expected from them under more favorable circumstances.

The Observatory at Arago's Death

Arago's last years were darkened by diabetes and the progressive blindness it provokes, and probably also by the failure of the Second Republic, which lasted only from 1848 to 1852, and by his problems with Le Verrier. One has the feeling that he somewhat neglected the Observatory to concentrate on his duties at the Academy and on his personal work, in particular the editing and the updating of many memoirs and of the *Astronomie populaire* (We will discuss this in Chap. 11). When Le Verrier took over the direction of the Observatory in 1854, he wrote an interesting report on the situation of this institution.[41]

For Le Verrier, the Paris Observatory was to become "an observatory of the first order," comparable to those of Greenwich and Pulkovo in Russia. The great telescope of Brunner should have allowed it to achieve this rank despite its foreseen weaknesses, in particular its sensibility to vibrations. But this was not all; the report said:

- The meridian telescope had big defects that needed repairs; it should later be replaced by a larger telescope, to rival Greenwich;
- For the meridian observations, the time should be read on several clocks instead of a single one, whose running cannot be easily checked;
- The telescope of Gambey's meridian circle was not powerful enough and must be replaced

[39] Although Le Verrier succeeded in separating the Observatory from the Board of Longitudes, the Board survived but with much reduced prerogatives. It still exists today.

[40] Le Verrier behave so badly with the Observatory personnel and the authorities that he was dismissed on 8 February 1870 by the Emperor, on the recommendation of the Minister of Public Instruction. He took over anew the direction of the Observatory after the accidental death in 1872 of his successor, Delaunay.

[41] Le Verrier, U.-J. (1855) *Annales de l'Observatoire de Paris* 1, p. 1–68, Mallet-Bachelier, Paris, accessible via http://adsabs.harvard.edu/historical.html.

- One should construct a meridian circle with an aperture of 10 *pouces* (27 cm) which would be at the same time a passage instrument (Greenwich has an 8-in. one), to be placed to the west of the building to replace the amphitheatre "which is and will remain purposeless"[42];
- One should also contemplate the construction of large telescope objectives.

Le Verrier also described the conditions of the observers, who carried out a painful labor and were poorly remunerated (3000 francs per year for the adjunct astronomers, 5000 francs for the titular astronomers), whereas their lodgings at the Observatory were "inhabitable." But he judged that they should not disperse their efforts by participation in other activities, such as teaching.

Apart from his comments on teaching and efforts to popularize astronomy, which were debatable, Le Verrier's report appears to us moderate and objective: The Observatory was visibly not healthy. His proposals were in fact reasonable, and it was an innovation to forcefully recommend the practical application of the electric telegraph to measure longitudes, to provide a time-keeping service and, especially, to create a true weather service. In the end, the astronomers of the Observatory would acknowledge, despite the innumerable disputes they would have with him, that "in general his reasoning was basically sound; his fault was in the manners."[43]

---

[42] We have seen that the amphitheater was finally replaced by Le Verrier's apartment (400 m^2, or about 4320 ft^2!).

[43] Interview of Gaillot by Bigourdan in 1888, Bibliothèque de l'Observatoire de Paris, 3567(3), folder AP.

Chapter 8
Arago Astronomer

The great comet of 1843 over the Paris Observatory, seen from the north. From Flammarions's *Astronomie populaire*, author's collection

J. Lequeux, *François Arago*, Astrophysics and Space Science Library 421,
DOI 10.1007/978-3-319-20723-0_8

Astronomy in France in Arago's Time

The Triumph of Newtonian Mechanics

When creating the theory of gravitation, Newton applied it to the orbits of planets and also of comets. His method was used in 1705 by Halley to calculate the orbits of 24 comets; he discovered the periodicity of the returns of the comet to which his name was given later, and predicted that it would be visible again in 1758. When doing his calculations, Halley did not take into account the perturbations of the motion of the comets by the gravitational attraction of the planets, although he was aware that they existed. As a consequence, the exact time of the return of Halley's comet was uncertain. In 1757 Alexis-Claude Clairaut (1713–1765) got to work in order to perform the new and difficult calculations that were necessary to fix this date. He was helped by Nicole-Reine Lepaute (1723–1788), a female mathematician who was married to a famous clockmaker, and by the young astronomer Joseph-Jérôme Lefrançois de Lalande (1732–1807). They predicted that the return of the comet would be delayed by 518 days due to the attraction of Jupiter, and by a further 100 days due to Saturn, and that it would reach its perihelion (the point of the orbit closest to the Sun) on 13 April 1759. The comet was actually early by a month, but nevertheless Halley's prediction revised by Clairaut and his collaborators was immediately and rightly considered as the triumph of Newtonian mechanics. It was one of the reasons for the acceptance in France of Newton's theory, which had been considered with suspicion for a long time because it did not fit with the ideas of Descartes. Only a few scientists, such as Pierre Louis Maupertuis (1698–1759) and Nicolas Louis de Lacaille (1713–1762), were convinced by this theory, as was the famous writer Voltaire (1694–1778); the latter was influenced by his mistress Émilie du Châtelet (1706–1749), who produced a remarkable annotated translation into French of Newton's *Principia,* published by Clairaut 10 years after her death.

Fortunately, France did better at the end of the eighteenth century, thanks to Clairaut, d'Alembert (1717–1783), Lagrange (1736–1813), Condorcet (1743–1794) and above all to Laplace, whose monumental *Traité de Mécanique Céleste* appeared in five volumes from 1799 to 1825. For Laplace, the Solar System was an immense mechanism moved by universal gravitational forces, and it should be possible to predict its behavior forever. The following passage of his *Exposition du Système du Monde* gives a good testimony of his ideas:

Amidst the infinite variety of the phenomena that occur continuously in the Sky and on the Earth, one managed to unravel a few general laws that matter must follow in its motion. Everything obeys those laws in Nature; everything is derived from them with the same necessity as the return of the seasons; and the curve described by the lightest atom that the winds seem to carry at random is defined with the same certainty as the planetary orbits.[…] The geometers have reduced the whole mechanics to general formulae: it is only necessary to improve their analysis.

This mechanistic and deterministic conception of the Universe dominated not only astronomy, but all the science of the nineteenth century, as we have seen in Chap. 3 for Fresnel's optics. In a practical way, the progress of analysis required by Laplace was sufficient to make the determination of orbits of comets perturbed by planets a routine in Arago's time, even if the calculations were long and painful.

William Herschel's Contribution

The influence of celestial mechanics at the end of the eighteenth century was so great in France that the trend was to reduce astronomy to the study of the motion of planets, satellites and comets, without paying much attention to their physical nature. If Charles Messier had published the first extensive catalogue of nebulae, this was not because of the interest of these objects but rather to avoid confusing them with comets. The situation was different in Great Britain, where some original scientists like John Michell and above all William Herschel (Fig. 8.1) were interested in a broader variety of subjects. With the big telescopes he had constructed (Fig. 8.2), Herschel made many discoveries: that of the planet Uranus, that of double stars whose components orbit each other, and that of the motion of the

Fig. 8.1 William Herschel (1738–1822). Wikimedia commons, ArtMechanic

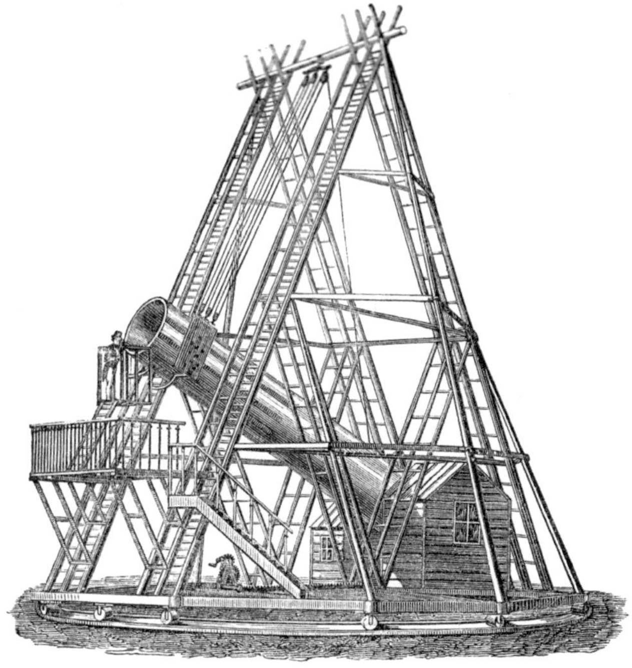

Fig. 8.2 The grand telescope of William Herschel (1.2 m in diameter), built in 1789. It was rather difficult to use, and most of Herschel's discoveries were made with smaller telescopes. From Arago's *Astronomie populaire*, author's collection

Sun with respect to nearby stars. He catalogued and studied nebulae; some of them seemed marginally resolved into many weak stars, and were perhaps the equivalent of our Milky Way, which Herschel himself was beginning to map. Other nebulae had a central star, and Herschel saw in them different steps in the formation of stars by contraction of nebulous matter. We know today that most of these nebulae were in fact made of material ejected by their central star, but no matter: a fruitful idea was launched. In 1811, Herschel wrote[1]:

We may conceive that, perhaps in progress of time these nebulæ which are already in such a state of compression, may be still farther condensed so as actually to become stars.

Curiously, the Germans and the French were more enthusiastic about Herschel's discoveries and ideas than the English. Herschel was not a clergyman, unlike most of the English scientists of his time, and his theories did not seem to conform to the standard ideas about Creation. In France, his works were very well known and religion had less weight. Twenty years after Herschel's death, Arago published an

[1] *Philosophical Transactions* 101 p. 269–336 (1811) accessible via http://www.jstor.org/stable/107348, see p. 318.

enthusiastic and well-documented eulogy with an extensive bibliography, to which he added many personal ideas.[2]

Herschel's observations and remarks were later considered by Laplace as supporting his own theory on the formation of the Solar System by gravitational collapse of a nebula. It is even possible that Herschel had been at the origin of Laplace's ideas. From Delambre[3]:

M. le Comte Laplace and M. Herschell [sic] came by different ways to consider that the Sun was surrounded in the past by a vast atmosphere: the first one, by deducing this state of the Sun from a consideration of the singular phenomena of the Solar System; the second one, by looking at the progress in the condensation of the nebular matter. This encounter, through the convergence of the proofs that they both gave and of their ideas, gives to the whole thing a probability that is almost a certainty.

Unfortunately, Laplace's influence was to divert French astronomers from observing nebulae for some time: The theory seemed to be enough!

The Program of the Board of Longitudes

Even after Laplace's death in 1827, admiration for Herschel did not influence French astronomy very much. Although the Board of Longitudes had promoted Arago as Director of observations in 1834, it did not intend at all to give up its prerogatives, and it decided on a program "which would serve as a rule for the astronomers of the Paris Observatory [...] for the year 1834." This program[4] was extremely classical; its emphasis was on observations of stars, the Moon, the Sun, planets and comets, of occultations of stars by the Moon and of eclipses of Jupiter's satellites. The purpose of these observations, that were essentially routine observations, was to keep the time, to favor the determination of longitudes and to verify the predictions of celestial mechanics, things that were normal for any observatory. It was also planned to build a new stellar catalogue to replace one that Lalande published in 1801, which included some 40,000 stars but whose accuracy was probably considered as insufficient. This new catalogue never came out. But, even if it had existed, it would have been of no help in the discovery of Neptune; its purpose was to accurately measure the position of the brightest stars, rather than to give approximate positions for many fainter ones, something that was needed for the discovery of the planet and was only done in Berlin.

At the end of its instructions, "the Board reserved itself to decide by separate programs the works to be done with the [Gambey] equatorial and the instruments

[2] **Annuaire du Bureau des longitudes pour 1842*, p. 249–608.

[3] Anonymous (almost certainly Delambre) **Connaissance des temps pour 1815*, p. 215–221, see p. 220.

[4] **Annuaire du Bureau des longitudes pour 1931*, p. A122–124.

Fig. 8.3 Lord Rosse's Leviathan in Parsonstown, Ireland, present state. Wikimedia commons, Niall McAulay

for magnetism and meteorology." These programs were soon forgotten, but Arago took care that the magnetic and meteorological observations be made regularly. The astronomers were thus free to do what they wanted with the Gambey equatorial and the other available telescopes, but none was really powerful enough to give important results. For instance, Félix Savary and Antoine Yvon Villarceau observed bright double stars like ξ Ursae Majoris or ζ Herculis, but there was no way to rival the "immense observational works of MM. Struve [father and son] and Sir John Herschel," who benefitted from larger instruments in Russia and England.

It had been an error to abandon the large reflecting telescopes after the failure of Dom Noël's (see the preceding chapter); in the meantime, Herschel and his successors, in particular his son John, made superb observations with their giant instruments. In 1845, Lord Rosse (1800–1867) discovered the spiral structure of galaxies with his "Leviathan" of 6 ft in diameter (Figs. 8.3 and 8.4). Arago did not ignore these works, but he did not foresee the construction of such telescopes. There was no large reflecting telescope in France like those in England, and no refractor like those of Russia, Great Britain and even the United States. Arago tried to fill the latter gap with the 38-cm equatorial, but this instrument would also be a failure. It was only after Léon Foucault had built his 40- and 80-cm reflectors at the end of the 1860s that French astronomers could undertake novel research. In Paris, Charles Wolf and

Fig. 8.4 The Whirlpool Galaxy (M 51) seen from *top* to *bottom* by John Herschel, Lord Rosse with his Leviathan and today with the Canada-France-Hawaii telescope. From Arago's *Astronomie populaire*, author's collection, and Jean-Charles Cuillandre

Georges Rayet (1839–1906) discovered in 1867 the emission-line Wolf-Rayet stars, while in Marseilles, Édouard Stéphan (1837–1923) made a catalogue of 800 new nebulae and galaxies, including the famous galaxy quintet that bears his name. Then, in 1914 Charles Fabry (1867–1945) and Henri Buisson (1873–1944) for the first time measured the temperature of a gaseous nebula, the Orion Nebula, with an interferometer invented by Fabry and Alfred Perot (1863–1925). It is significant that the latter observation was published in an American journal, the *Astrophysical Journal*; the French probably did not consider it very interesting. In any case, all this effort towards astrophysics remained marginal.

Astrometry at the Paris Observatory

Arago made his first observations with Bouvard on 22 March 1805, just after he was named secretary-librarian. These were routine observations of time and positions of stars, and also of occultations of stars by the Moon and of eclipses of Jupiter's satellites. They used not only the meridian instruments, but also the reflecting telescopes of Dollond and Short, and that of Caroché with its platinum mirror. Arago did not observe the following year because he was sent to Spain. He resumed observations when back, in September 1809. His participation to observing seems to have been regular, and he still observed himself from time to time when promoted in 1834 as Director of observations.

These observations have no particular interest, except for those on the velocity of light that I described in Chap. 3. In 1809, Arago worked on a new determination of the latitude of the Observatory with the Fortin repeating circle, while Mathieu and Humboldt did the same thing; but their results did not agree, and they spent a lot of time trying to understand why. According to Arago, the differences came from an "individual collimation," i.e. differences in focusing due to the defects of the eye of each observer. Apparently, this was a loss of time, but it is in this way, by trials and failures, that one becomes a true astronomer.

Proper Motion and Parallax

In 1812, Arago and Mathieu started a new series of observations to determine the distance of the star 61 Cygni. They noticed that this star was moving rapidly with respect to its neighbors (this is called the *proper motion* by astronomers), so that it could not be very far. They could only obtain a lower limit for its distance, and their new observations of 1825 did not do any better.

To understand what they were doing and why they did not succeed, let us see what are the proper motion and the parallax.

The proper motion (Fig. 8.5) is a slow, regular motion of a star with respect to its neighbors. It is expressed in arc seconds per year. It is determined by comparing positions of the star obtained at different epochs, after correcting the observations for precession, nutation and aberration. When the proper motion is large, it can be detected directly by meridian observations of the star[5]; it is in this way that Jacques Cassini (Cassini IV) determined the first proper motion, that of Arcturus (α Bootis), by comparing his 1738 observations with those made in 1672 by Richer in Cayenne, French Guiana. In 66 years, Arcturus moved by 150 arc sec, while a nearby star, η Bootis, had not changed place. Needless to say, this did not agree with the idea of "fixed stars" and made some noise at the time. William Herschel discussed 14 well-determined proper motions and showed that they were essentially due to the motion of the Sun (and of all the Solar System, of course) towards a point belonging to the Hercules constellation, a very remarkable result confirmed by all further observations. It then became possible to take the problem backwards, and to obtain some idea of the distance of celestial objects by determining their proper motion, assuming

[5]However, it is always better to observe its position relative to the largest possible number of distant stars.

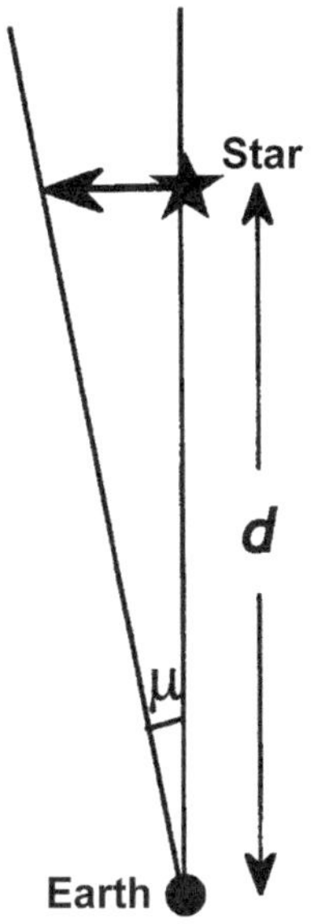

Fig. 8.5 The proper motion of a star. The star moves with respect to distant stars, assumed to be fixed. The annual angular displacement is μ, expressed in arc seconds per year. The lateral velocity v of the star with respect to the Sun is related to μ and to the distance d by the relation $v(\text{km/s}) = 1.45\mu(''/\text{year})\ d(\text{light years})$

that it was mainly due to the motion of the Sun; it would be smaller if the object was more distant. This method was used to try to get the distance of nebulae, which was completely unknown. Arago suggested in his eulogy of William Herschel that one could compare the proper motion of the central star of some nebulae to that of the nebula itself in order to determine if they were physically associated, a rather impractical method indeed. More seriously, Laugier started in 1849 to build a catalogue of the positions of nebulae obtained with the Gambey equatorial to compare them with those measured by Messier and John Herschel.[6] He wanted to see if some of these nebulae, which would not take part in the apparent systematic displacement due to the solar motion, could be very distant as suggested by John Herschel. This was a good idea in principle, but its practical implementation was impossible because the position of these diffuse objects could not be measured with sufficient accuracy. Laugier published his catalogue, but did not draw any conclusion from its comparison with the previous ones.

Let us come now to the parallax. Those stars with a large proper motion have, as I said, a good chance of being nearby; these are good candidates for distance measurements. The most direct way to obtain this distance—and the only possible one until the beginning of the twentieth century—was to measure the geometric parallax of the star (Fig. 8.6), which is the amplitude of its annual motion due to the orbital displacement of the Earth. One can locate the motion of the star with respect to much more distant stars that can be rightly considered as fixed. This method was

[6] *Laugier, E. (1849) *Comptes rendus de l'Académie des sciences* 28 p. 573–576; 37 (1853) p. 874–879.

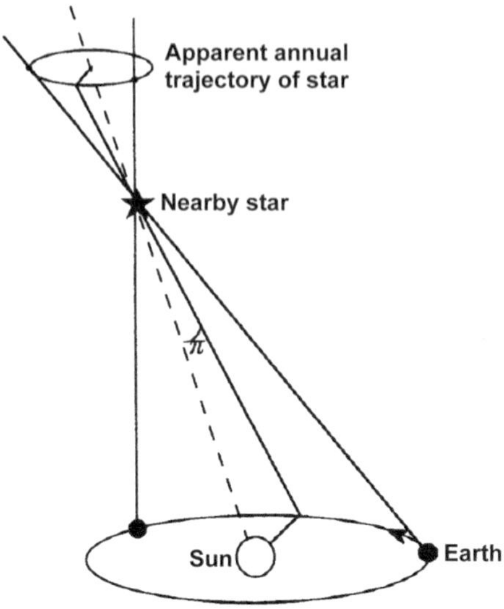

Fig. 8.6 The annual geometric parallax of a star. During the annual revolution of the Earth around the Sun, a nearby star seems to move with respect to the distant ones on an ellipse. The angle π of the semi-major axis of this ellipse is the parallax, usually expressed in arc seconds. It is inversely proportional to the distance of the star. A star with a parallax of 1″ would be at a distance of 1 pc, or 3.26 light years, or 3.08×10^{16} m. But no star is as close as this

suggested for the first time by Galileo. Many astronomers, including William Herschel, tried to detect the parallax of bright stars with a large proper motion, by measuring their position with respect to faint stars at different times of the year. None of them succeeded. However, in 1803, Herschel made a fundamental discovery: that of binary stars whose components orbit each other. This discovery cast doubt on the widespread idea that the stars were all similar, because the components of those double stars could have very different brightness, while being, of course, at the same distance. Since fainter stars were not necessarily more distant, it seemed dangerous to measure the position of a bright star with respect to its fainter neighbors in order to obtain its parallax.

This is why, in 1812, Arago and Mathieu inaugurated an absolute, direct method to obtain parallax that did not use the position of the studied star with respect to its neighbors. They measured with a repeating circle (probably that of Fortin) the variations of the zenith distance of several stars at different times of the year, when they crossed the meridian. Unfortunately it was difficult to measure parallaxes larger than about 0″.5 in this way because of measurement errors, variations in the atmospheric refraction and possible instabilities of the instrument. This method was fatally flawed, and indeed gave no positive result. Still, it was the method recommended in 1834 by the Board of Longitudes, perhaps under pressure from Arago.

The first successful parallax measurement was by Friedrich Wilhelm Bessel in 1838 and concerned 61 Cygni, the star studied before by Arago and Mathieu. Bessel found a parallax of 0″.31, revised later as 0″.375, using the heliometer of Fig. 8.7.

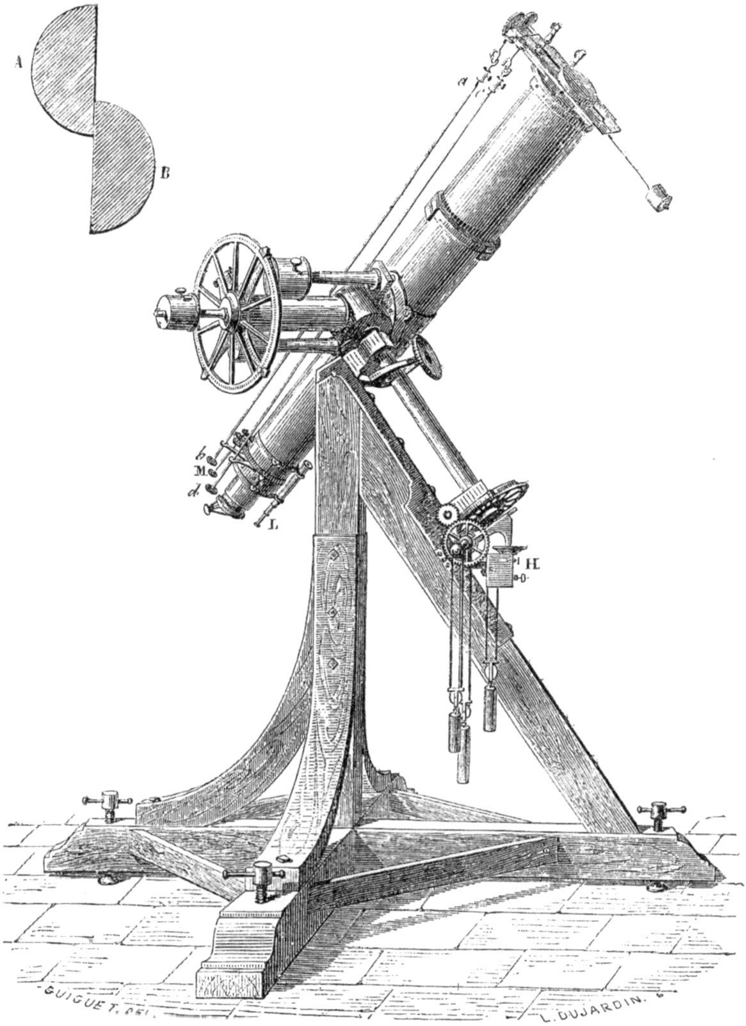

Fig. 8.7 The heliometer built by Fraunhofer for the Königsberg observatory, and used by Bessel for the first successful parallax measurement. As can be seen at the *top left*, the 157-mm diameter objective, which could turn around the axis of the telescope, was cut in two halves that could slide with respect to each other. Each star gave two images that could be oriented and separated at will. If one of these images was made to coincide with an image of a nearby star, one could accurately obtain the angular distance between the two stars. From Arago's *Astronomie populaire*, author's collection

Admiring, but at the same time vexed by this discovery, Arago recalled his own measurements[7]:

Astronomers will not be unpleased to see that the method of absolute zenith distances, observed with the repeating circle, gives approximately the same result as the heliometer measurements of M. Bessel. The latter, certainly very exact, have however the inconvenience to assume that the comparison stars have no appreciable parallax.

The "same result" is in fact an upper limit of 0″.5 for the variations of the zenith distance of 61 Cygni during the year. The measurement by Bessel made a big impression on the public, because no one had imagined that the stars could be so distant. The writings of the time are full of amazed considerations on the enormous distance of the star and on the long time its light takes to reach us. Still, Arago did not mention Bessel's measurement in his "Chronological table of the principal astronomical discoveries".[8] Was this a sign of rancor?

Geometrical parallaxes remained for a long time very difficult to measure. In 1854, only a handful were determined, and a comparison with recent determinations shows how large the errors could be. It is only with the Hipparcos European satellite, launched in 1989, that good parallaxes became available for many stars.

Arago's Micrometer

Noting large discrepancies between the published values for the diameters of planets, Arago put much effort in re-measuring them. He started as early as 1810 with the Rochon telescope, which gives a double image of an object, the separation of which can be adjusted by moving the bi-refringent prism along the telescope axis (Fig. 3.12). If one of the edges of the planet in one of the two images is made to coincide with the other edge of the other image, the angular distance between the images gives the angular size of the planet. But Arago soon realized that the results were not reliable. He decided to modify the telescope so that the Rochon prism (Fig. 3.11) was now placed between the eye and the eyepiece (Fig. 8.8). This device also gave a double image with an angular separation α/g, where α was the angle between the two rays coming out of the prism and g the enlargement of the telescope.

This device still had some problems, which disappeared with the final arrangement of 1840 depicted in Fig. 8.9. Now, the lenses of the eyepiece were fixed so that the enlargement was constant, but one of the prisms of a series was placed between the eyepiece and the eye. The separation between the two images was thus fixed and

[7] **Comptes rendus de l'Académie des sciences* 7 (1838) p. 793–795.

[8] *Arago, F. *Astronomie populaire*, t. 4, p. 787–790.

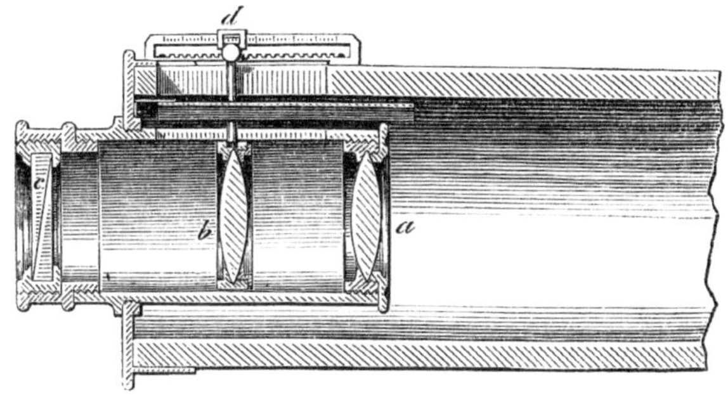

Fig. 8.8 The variable-enlargement micrometer of Arago. It is a telescope eyepiece, one of the two lenses of which can be displaced to vary the enlargement. A double bi-refringent Rochon prism *c* (*left*) is placed between this lens *b* and the eye. From Arago's *Astronomie populaire*, author's collection

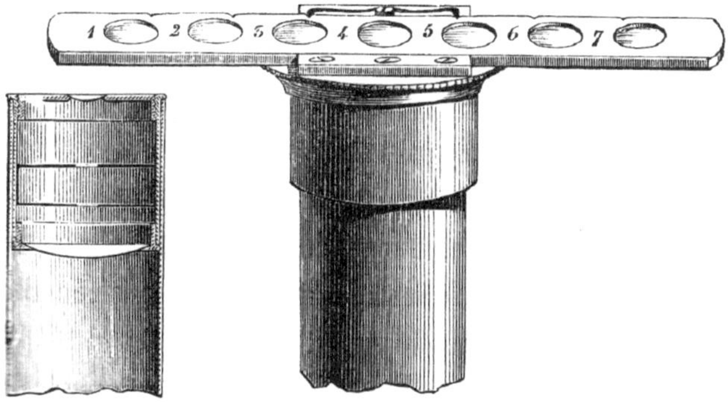

Fig. 8.9 The final ocular micrometer of Arago. A bi-refringent Rochon prism which can be chosen on the slide is inserted between the ocular and the eye. From Arago's *Astronomie populaire*, author's collection

reproducible. The collection of the Paris Observatory contains many such prisms and slides holding those prisms, all built by Soleil.

The observing logbooks of the Observatory are full of measurements made with these instruments. There were not many measurements after 1840, and they were most often made by Laugier, as Arago could not observe anymore because of excessive work and progressive loss of sight. They were concerned with Saturn's rings, the diameter and flattening of planets etc. These are good measurements, but not as good as those made in England, in Germany and in Russia with larger telescopes and heliometers. Furthermore, they were never reduced and published. But let's leave this topic for a more interesting one.

The Polarization of Light and the Physical Nature of Celestial Bodies

Let us recall that immediately after the discovery of polarization by Malus, Arago was very interested in and soon became a specialist in polarization. He discovered the rotational chromatic polarization produced by a quartz plate cut perpendicularly to the optical axis (see Chap. 3). Using this property, he built a polarimeter (Fig. 3.16) which allowed him to easily detect if a light was polarized. It produced in this case two images colored with complementary colors, but they were not colored if there was no polarization. The polarimeter was also able to measure the percentage of polarization of the light. We should not be surprised to see Arago pointing this instrument to various celestial bodies in order to detect the possible polarization of their light.

The Moon

The first of these bodies was the Moon. In October 1811, Arago discovered the first extra-terrestrial phenomenon of polarization by observing with his polarimeter the light of our satellite. His observations are reported in detail in his *Astronomie populaire.*[9] Curiously, the explanation that looks obvious to us at present—the polarization is due to the vitreous reflection and diffusion of solar light by the lunar soil—was not so clear for Arago, who wrote:

The light of the Moon is, it seems, more polarized at the time of the first quarter. This time would also be that in which the polarization of light reflected by some atmosphere similar to the terrestrial one would be at its maximum, if such an atmosphere exists around the Moon. Assuming that most of the polarization comes from a lunar atmosphere, one would naturally explain that it is larger in the direction of the lunar black spots.

Arago had indeed discovered the polarization of the blue sky due to the diffusion of solar light by the terrestrial atmosphere, and thought that an atmosphere around the Moon could also produce polarization. But to account for it, this atmosphere should be as thick as the Earth's. Arago was convinced that if the Moon had an atmosphere it was very tenuous. He wrote no fewer than eight pages on this in the *Astronomie populaire.*[10] There is a clear contradiction with his hypothesis on the origin of lunar polarization. I do not understand this contradiction. Arago wrote in several texts that the surface of the Moon is solid, and that the lunar seas … are not seas, and he knew that the light reflected obliquely on solids is polarized. He could

[9] *Arago, F. *Astronomie populaire*, t. 3, p. 463–467.

[10] *Arago, F. *Astronomie populaire*, t. 3, p. 434–442.

have understood that the polarization of the Moon comes from the diffusion and reflection of light by its soil, but he drew no clear conclusion from his observations.

The topic seems to have been forgotten until the French astronomer Bernard Lyot (1897–1952) started in 1920 an observing program on the polarization of the Moon and of planets (Araglo neglected the latter). This did not require big telescopes: Lyot's had only a diameter of 17.5 cm. These observations could have been done by Arago, but he missed the opportunity for lack of perseverance.

The Sun

In 1811 again, Arago observed the Sun with his polarimeter. He could not find any polarization even on the edges of the solar disk. He concluded rightly that the surface of the Sun was made of incandescent gas, and not of an incandescent solid or liquid. He had observed in the laboratory the light emitted by hot iron and liquid platinum and concluded that it was polarized if the surface was seen obliquely, like the edge of the solar disk. Thus, the surface of the Sun could not be solid or liquid. The summary of a communication of Arago given in front of the Academy on 14 June 1824 explains his approach[11]:

M. Arago gives an account of experiments he made a long time ago on the light emitted by incandescent bodies. He had seen that, if these bodies are solid or liquid, this light is partly polarized by reflection [sic] when the rays form an angle of a few degrees with the emitting surface. As to the light of burning gases, it is not polarized whatever the inclination. M. Arago deduces from these experiments the consequence that an important part of the light from the incandescent bodies is formed inside, until depths that he has not yet determined. He shows right now that the same means of observation can be applied to the physical constitution of the Sun: the results that he has obtained in this research confirm the conjectures of Bode, Schroëter [Schröter], Herschel, etc.

The importance of this discovery cannot be underestimated: it is, with that of the rotational chromatic polarization, the most important of Arago's[12] discoveries. This was indeed the first time that one could say anything on the nature of a celestial body from remote observations (the situation for the Moon was unclear as we have seen). According to the astronomer Félix Tisserand (1845–1896),[13] Arago "introduced physics in astronomy," creating a new discipline whose name would only appear much later: astrophysics. However, most historians of science consider that

[11] **Annales de Chimie et de Physique* 27 (1824), p. 89–90, accessible via http://gallica.bnf.fr.

[12] Arago was conscious of this importance as he proudly cites his discovery in his «Chronological table of the principal discoveries of Astronomy» at the end of **Astronomie populaire*, t. 4, p. 787–789.

[13] Tisserand (1893).

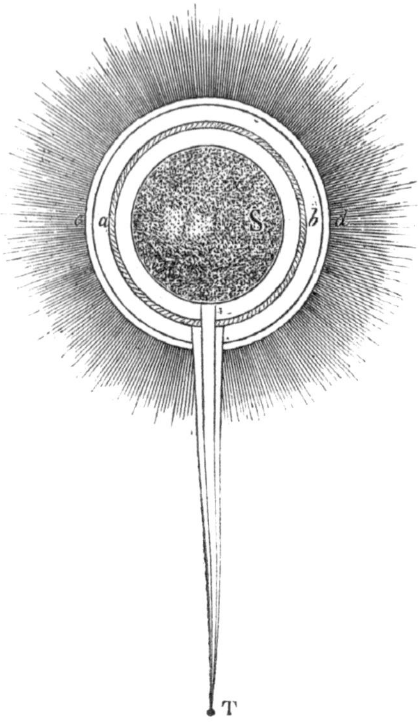

Fig. 8.10 The model of the Sun according to Herschel and Arago. A hole in the bright photosphere allows one to see a less bright deeper layer, which is the penumbra around the sunspots. A smaller, more or less concentric hole through the latter layer unveils the spot itself, which is a part of the relatively dark body of the Sun. From Arago's *Astronomie populaire*, author's collection

the beginning of astrophysics took place in 1860, the date when the German physicists Gustav Kirchhoff (1824–1887) and Robert Bunsen (1811–1899) had shown using spectroscopy that the Sun contains several of the chemical elements studied in terrestrial laboratories.

As recalled by Arago, several astronomers, such as Bode, Schröter and Michell, suspected that the Sun could have a gaseous atmosphere. Among the many models proposed in order to account for the properties of the Sun, William Herschel's (1795)[14] was particularly appreciated by Arago, who said that it was "almost generally adopted today." Figure 8.10 represents the model adopted by Arago for the Sun, which is an adaptation of Herschel's. For him, there were two layers in the atmosphere of the Sun: holes in the upper layer allowed seeing the lower one, which gave the penumbra around the sunspots, less bright than the surroundings. Holes in the lower layer allowed seeing the dark sunspots themselves. According to Herschel's[15] estimates, which were as good as it was possible at the time, the brightness of the penumbra was 469/1000 of that of the surrounding photosphere, and sunspots' brightness 7/1000 of the photosphere's. Still, as Arago said, "all the nuclei of the

[14]Herschel, W. (1795) *Philosophical transactions* 85, p. 46–72, accessible via http://www.jstor.org/stable/106944.

[15]Reported by *Arago, F. *Astronomie populaire*, t. 2, p. 168.

spots, as dark as they look on the Sun, would dazzle with their very strong light those who would look at them separately." Surprisingly, the extreme conditions on the solar surface did not prevent Herschel from thinking that the Sun could be inhabited.

Arago's conclusions about the gaseous nature of the solar photosphere did not convince everyone. He wrote in 1851[16]:

My hypothesis on the physical constitution of the Sun accounts for the facts in a very satisfactory way. However, it is not generally adopted: some authoritative books represented some time ago the sunspots as scoria floating on the liquid surface of the Sun and ejected by solar volcanoes much more powerful than the terrestrial ones.

Several contemporary textbooks correctly account for Arago's discovery, but others do not, including the treatise of John Herschel.[17] This discovery was also forgotten by Camille Flammarion, for which the Sun was "a globe of liquid fire".[18] This is surprising given the admiration of Flammarion for Arago, from whom he borrowed the title of his bestseller, the *Astronomie populaire*.

Solar Total Eclipses

In his 1851 notice on the physical constitution of the Sun and of the stars (see above note 16), Arago wrote:

We should ask to ourselves if the photosphere ends abruptly or if it is surrounded by a gaseous atmosphere, not luminous by itself or slightly reflecting. This third atmosphere would not be usually visible in the sea of light which seems to surround the Sun at all times.[...] There is a way to solve the problem: to chose the time when, in a total eclipse, the Moon covers the Sun completely. [...] We now know that this was the main purpose of the astronomers who went in 1842 to the south of France, to Italy, to Germany or to Russia to observe the total eclipse of 8 July. [...] The observers were quite surprised when they saw several pinkish protuberances with heights of 2 to 3 minutes, springing, so to say, from the edge of our satellite (Fig. 8.11).

Arago had prepared instructions for observing the 1842 eclipse and had himself taken part in the observations in Perpignan. This was certainly not the first time that protuberances had been seen, but not much attention was paid to them before. Arago estimated that the protuberances were "solar clouds swimming in a gaseous atmosphere," a conclusion confirmed by observations during the total eclipses of 8 August 1850 and 28 July 1851. Babinet had imagined that the protuberances were

[16] *Arago, F. *Œuvres complètes*, t. 7, p. 112–135, see p. 116.

[17] *Herschel (1853).

[18] *Flammarion (1892), see p. 67.

Fig. 8.11 Drawing of the Sun during the total eclipses of 8 July 1842. Note the protuberances which raise from the dark disk of the Moon, and the corona. From Arago's *Astronomie populaire*, author's collection

"igneous clouds" supported by centrifugal force while rotating around the Sun. But they would then move with a very great speed, which was not the case, and furthermore their height was variable. Arago concluded that an atmosphere was necessary to support them.

Was this atmosphere the corona seen during the eclipses? Before answering this question, one would have had to be sure that the corona was a real phenomenon, and not due to light scattered by the edge of the Moon. It took a very long time to decide that the corona was real. Arago had asked some observers to see if the corona was polarized, in which case it would have been diffused light, but nothing came out of the observations. Indeed the corona is polarized, because it contains dust mixed with a very hot gas. But this only became clear in the twentieth century.

Stars

At the end of his life—one does not know exactly when—Arago started a work on the polarization of stars to determine the nature of their surface as he had done for the Sun. He knew that their light was not polarized, but this was inconclusive because this was expected for symmetry reasons, the stars being uniform spheres. Cases where this symmetry is broken should be found. Noticing that there are stars which vary periodically, Arago indicated that there were only three possibilities to explain such variations: either a portion of the disk of the star is less bright than the rest and the star rotates, or the star has a darker companion that passes periodically in front, or, finally, the star is flattened and does not rotate around an axis of symmetry. The three cases indeed exist in nature and all three produce a break of symmetry. The best candidates were eclipsing binaries in which one of the two components passes periodically in front of the other, or behind. These were the ones that were observed by Laugier and others at Arago's request. Arago wrote[19]:

Thorough observations of several variable stars with the polarimeter at all phases of their variations have shown that their light is always of a perfect whiteness [thus is not polarized since polarization produces colors in this instrument]. One can thus state that this light comes from a material similar to our clouds [not polarized as the droplets are generally much larger than visible wavelengths] or to our burning gases.[...] One sees that it results from these researches, which have to be followed up, the complete confirmation of the hypothesis of the identity between our Sun and the innumerable stars which populate the sky.

The reasoning is clever, and Arago was clearly aware of the importance of the result, which unfortunately was only published in his complete works after his death. Anyway, the time was not yet ripe for astrophysics.

In 1949, the American astronomer William Albert Hiltner (1914–1991) discovered the polarization of the light of various stars, and understood that it was not intrinsic, but due to interstellar dust grains oriented by the galactic magnetic field. These observations were probably not possible in Arago's time because of the smallness of this polarization, and in any case no one would have been able to understand the cause of this polarization.

In his eulogy of Herschel, Arago drew other conclusions from his study of variable stars. He remarked quite judiciously that the absence of color variations implied that the different colors of their light propagate at the same velocity in interstellar space. He concluded that this could give an upper limit on the amount of interstellar matter, that he considered as a gas and, for him, was the famous ether. He knew from personal experience that gases disperse light, hence that the velocity of light in those gases depends on the wavelength (see Chap. 10). Wisely, he did not go as far as to specify an upper limit for the density of interstellar gas. This was another astute idea, which however only found an application in the 1950s to the propagation of radio waves in the ionized interstellar gas.

Finally, Arago made the statement that a thorough observation of the phases of Algol (β Persei, the prototype of eclipsing binaries) should allow determining the velocity of the light emitted by this star. Indeed, one expects that the apparent time that the smaller component of the binary star takes to come from the back to the front of the main star is shorter than the time taken in the other half of the orbit, due to the finite velocity of light. This is the same effect as for the eclipses of Jupiter's satellites (see Chap. 4). It should produce an asymmetry in the light curve of the binary star. If this asymmetry can be measured, then the time of propagation through the diameter of the orbit of the companion can be determined. If the distance to the star is known, so is this diameter, hence the velocity of light. Unfortunately the Algol system is more complex than Arago thought because there is a third component, making the determination impossible. Too bad, but we have seen here the qualities of imagination of Arago, and his interest in the velocity of light.

[19] *Arago, F. *Œuvres complètes*, t. 10, p. 261–264.

Comets

The study of comets was one of the main subjects of astronomy at the beginning of the nineteenth century. Arago devoted no less than 224 pages to comets in his *Astronomie populaire* (Fig. 8.12). He discussed the probability that a comet would fall on the Earth, and calculated that it was very small. He also tried to refute many beliefs about comets. His comments in general made good sense, but it is far from certain that they succeeded in convincing his contemporaries.

More interesting for scientists was the research on the physical nature of comets, which began to be active in Arago's time. On 3 July 1819, he discovered the polarization of the tail of Comet Tralles (C/1819 N1), but this was somewhat marginal and he had to wait for the return of Halley's Comet in 1835 to be sure. On 23 October of that year, he observed with certainty "the polarization of the light of the Halley's Comet, an observation confirmed by [Alexis] Bouvard, Mathieu and Eugène Bouvard, *élève-astronome* at the Observatory." Clearly this meant that the light was diffused solar light, but diffused by what? It was impossible at that time to decide whether it was dust or molecules that diffused solar light.

Arago and his contemporaries understood that matter was ejected from the nucleus of comets heated by the Sun, and that the comets disappear from our view when they are at a large distance from the Sun because no matter is ejected anymore. In 1836, Bessel reasonably assumed that the ejected matter occurred on the side of the nucleus heated by the Sun; then, this material was repelled by an unknown force of solar origin, forming the tail opposite to the direction of the Sun. This repelling force was only identified with certainty in 1900, by the Swedish physicist Svante Arrhenius (1859–1927), as the radiation pressure exerted on dust grains by the light of the Sun.[20] This confirmed that the tail was made of dust. Comets often show another tail made of ionized gas, as shown by spectroscopy. In 1951, the German astronomer Ludwig Biermann (1907–1986) understood that this gas is driven by the wind emitted more or less continuously by the Sun. The first solid comet nucleus, that of Halley's Comet, was seen by the European space probe Giotto in 1986.

At present, comets are of major interest to scientists because they are almost intact samples of the material that made the Solar System; they are preserved at a great distance from the Sun either in the Kuiper belt of frozen bodies outside the orbit of Neptune, or in the Oort cloud much farther away. From time to time, they are sent to the inner Solar System by gravitational perturbations. Then their trajectory can be deeply modified by the attraction of Jupiter and Saturn. Some comets become periodic, and will disappear after many revolutions when all their matter has been ejected or dispersed.

[20] This was indeed an old idea, as discussed by Arago himself in his **Œuvres complètes*, t. 7, p. 447–454. He considered with reason that an experiment to demonstrate the effect should be made in a vacuum with a torsion balance. This experiment was done with success in 1899 by the Russian physicist Pyotr Lebedev (1866–1912), confirming the theoretical study of James Clerk Maxwell (1831–1879).

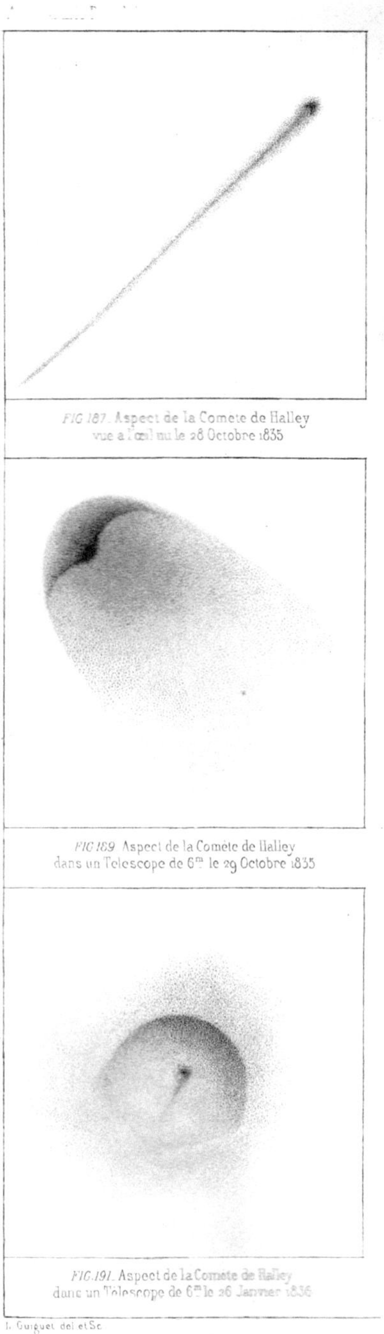

Fig. 8.12 Various aspects of the coma and of the tail of Halley's Comet, seen by John Herschel. Curiously, all the drawings of comets reproduced like these ones in Arago's *Astronomie populaire* come from foreign observers. From Arago's *Astronomie populaire*, author's collection

Photometry

Photometry is the art of measuring the intensity of light. It took up a major part of the late scientific activity of Arago. He invented new experimental methods using polarization, and had the corresponding instruments constructed, with which he or, more often, his collaborators made reliable measurements.

Three successive steps were necessary to reach this goal. The problem was that only visual observation was possible, and the eye can only compare equal luminous intensities. It was thus necessary to attenuate in a known way the comparison light, which had the higher intensity so that it was equal to that with the fainter intensity. This is what Arago's predecessors like Pierre Bouguer (1698–1758) or Johan Heinrich Lambert (1728–1777) had done with some success. The steps of Arago's work and his results are described in an introductory text of 1833 followed by seven Memoirs.[21] Six of these Memoirs were read at meetings of the Academy of Sciences from March to June 1850. By then, Arago was feeling his strength declining and wanted to present all of the unpublished parts of his work before it was too late. He had complementary observations made by Laugier and Frédéric Petit (1810–1865), his pupil who was now the director of the Toulouse observatory. Arago wrote in the first Memoir:

My first photometric experiments date from 1815. I did them at this time with a mobile apparatus that I held in my hand [the polarimeter or some adaptation]. However, the principle was so good that several of the results I obtained served to Fresnel to verify his theoretical formulae.

Unfortunately, there is no information on these early experiments. The first three of the later Memoirs explain the methods used by Arago, and the last four give the results, with some supplementary details. We should not be surprised that Arago, who was a specialist in polarization, used it for photometry. He used the attenuation of light by a set polarizer-analyzer, as shown earlier on Fig. 3.8. He went to great pains, with a special instrument preserved at the Paris Observatory, to verify the cosine square law that gives the attenuation of light as a function of the angle α between the planes of the polarizer and of the analyzer: $I = I_0 \cos^2\alpha$, where I_0 is the intensity after the polarizer and I that after the analyzer.

Once this law was verified, he used the apparatus whose principle is shown on Fig. 8.13. This was a Rochon telescope (see Fig. 3.12) in front of which a polarizer was placed; this polarizer was the prism invented in 1828 by the Scottish physicist William Nicol (1768–1851). If the incoming light was not polarized, which was generally the case for celestial sources, the intensity after the Nicol prism was one half of the initial one, the light being entirely polarized. This light then crossed an analyzer, which could be either another Nicol prism or a Rochon double prism

[21] *Arago, F. *Œuvres complètes,* t. 10, p. 150–167 (the introduction), 168–183, 184–215, 216–230, 231–250, 251–260, 261–281 and 282–297 (the seven Memoirs).

(Fig. 3.11). One could then compare the reference light attenuated in a known way thanks to the square cosine law with the light under study, turning the polarizer so to equalize the two fluxes.

The Limb Darkening of the Solar Disk

With his photometer, Arago tried to solve the problem of the darkening of the limb of the solar disk. He was not the first: Galileo and Huygens had already considered the problem, and Bouguer pretended that the edges of the disk were less luminous than the center. Arago observed the Sun with the instrument of Fig. 8.13, set such that the edge of one image was superimposed on the center of the other (Fig. 8.14). To test the sensitivity of the method, he turned the Nicol prism so that the fainter image was just visible over the bright one: this was when its intensity was 1/40 of that of the latter. Then he put a quartz plate between the Nicol prism and the objective, creating in this way rotational chromatic polarization. The two images were then equal and vividly colored in complementary colors, but white in their common part. If the edge of the disk were slightly brighter or fainter than the rest of the disk, one would have seen it slightly colored in the common part. But no color was

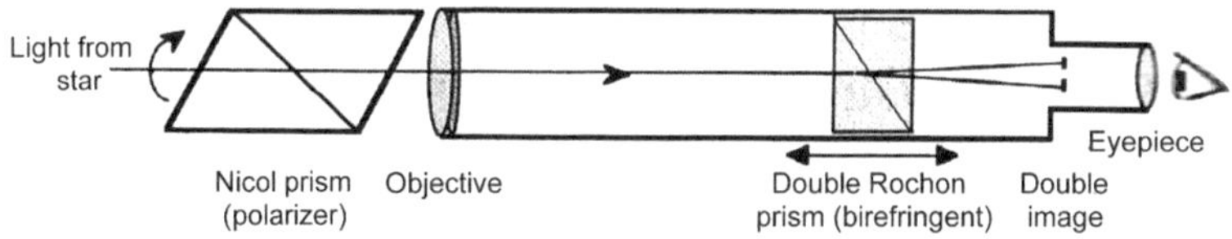

Fig. 8.13 Principle of Arago's photometric apparatus. This is a Rochon telescope with a polarizer in front. One of the polarized components of the light of the star, whose intensity is half of the initial one, enters the telescope and passes through a double Rochon prism, which produces two images polarized in perpendicular direction, hence are attenuated in different ways. One could change the separation of the images by displacing the Rochon prism along the axis of the telescope, and the relative intensity of the images by turning the polarizer by a known angle

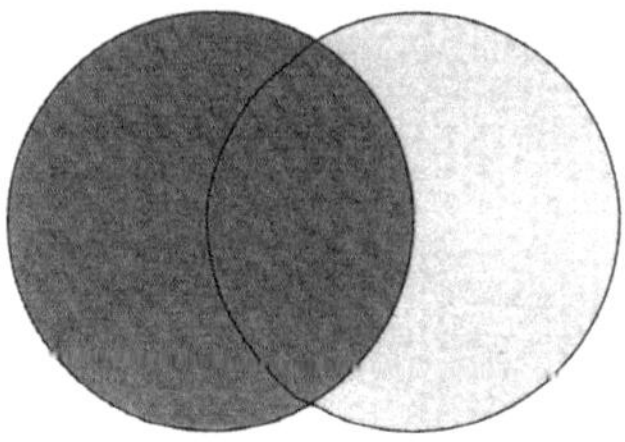

Fig. 8.14 The two images of the Sun as seen in the instrument of Fig. 8.13. The polarizer was turned to that the fainter image was just visible above the brighter one in the common part. Its intensity was then about 1/40 of that of the latter image

visible, and Arago concluded that the limb was neither brighter nor fainter than the center by more than 1/40. Curiously, this upper limit was transformed in *Astronomy populaire* into the positive detection of a limb darkening. Is it an error by Barral, who edited the book, or repentance?

The daguerreotypes of the Sun gave a clearer result. Arago was one of the first to photograph the Sun in collaboration with Louis-Mandé Daguerre (1787–1851), the inventor of the process. He wrote in his fourth Memoir on photometry:

Soon after the publication of the bright discovery of M. Daguerre, I noticed, when looking at the image of the Sun on a silver plate, that the rays from the central part of the disk had a larger photogenic action than those from the edge.[22] *This experiment has been repeated later by MM. Fizeau and Foucault who obtained in 1844 and 1845 a large number of photographic images of the Sun.*

The darkening of the limb of the solar disk is obvious on the latter daguerreotypes, in particular on the remarkable one which is preserved at the *Musée des arts et métiers* (Museum of Arts and Crafts) in Paris (Fig. 8.15). Arago wrote about these photographs that they confirmed his own conclusions about the edge darkening, a rather daring assertion to say the least!

When commenting on the photographs of the Sun, Arago regretted that no one had had the idea of placing the slit of a spectrograph radially through the edge of the Sun to see if the spectral lines would remain the same on the limb. This was his fault, because he had dissuaded the astronomers from doing this during the eclipse

Fig. 8.15 Engraving of a daguerreotype obtained on 2 April 1845 by Fizeau and Foucault. (The original daguerreotype and several smaller ones are reproduced by Tobin (2003) as Figs. 4.20 and 4.21.) The limb darkening is quite visible. The engraving is very faithful to the original, except that contours have been added to the sunspots to make them more conspicuous. From Arago's *Astronomie populaire*, author's collection

[22]This probably refers to the first successful daguerreotype of the Sun, taken by Daguerre and Arago during the partial solar eclipse of 15 March 1839; see Chap. 11.

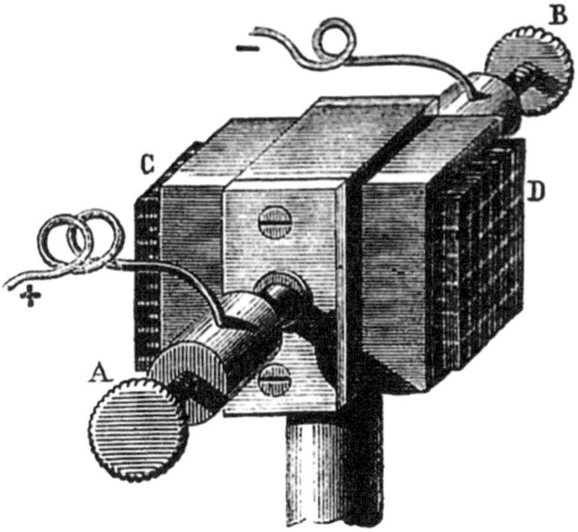

Fig. 8.16 The *thermo-multiplier* of Nobili and Melloni. 36 bismuth-antimony thermocouples were mounted in series. The instrument could detect the infrared radiation of a hand at a distance of 1 m. Private collection

of 1842, on the pretext that there were more interesting observations to perform! He would have found that the spectrum changes radically when crossing the limb, the chromosphere outside the limb producing emission lines rather than the absorption lines seen on the disk itself.

Arago also tried to detect a possible limb darkening in the "calorific rays," i.e. in the infrared. For this, he put a thermometer on the meridian in the large meridian room of the Observatory, and let the image of the Sun cross this thermometer. Nothing clear came out of this, and he proposed that someone make a better observation. This was done in 1852 by Father Angelo Secchi (1818–1878) at the Vatican observatory.[23] Secchi used "a special thermometer, based on the electricity produced by heat," actually a bolometer made of many thermocouples in series built by Leopoldo Nobili (1784–1835) and adapted to a galvanometer by Melloni, who named this device the *thermomultiplier* (Fig. 8.16). Secchi found a limb darkening in the infrared. He also found the limb darkening in visible light by a comparison of different regions of the disk projected on a screen.

Modern measurements have shown that the limb darkening of the Sun depends on wavelength, being stronger in the ultraviolet, where the daguerreotype has its maximum sensitivity, than in the visible and the infrared.

Arago did not try to explain the limb darkening, although its existence confirmed that the solar surface was an incandescent gas. Secchi assumed that the emitting layer was surrounded by an absorbing one, which produced the limb darkening because the thickness of the absorbing layer crossed by the light was larger near the edges of the Sun. This layer would also have produced the absorption lines seen in

[23] *Secchi (1875) *Le Soleil*, t. 1, p. 121–138. Secchi is one of the authors who did not cite the works of Arago on polarization: he only mentioned his photometry of the Sun.

the spectrum. Secchi even assumed that the layer would absorb a large fraction of the solar radiation, something that looks to us contrary to the conservation of energy but was probably acceptable at this epoch. On his side, Arago explained that the faculae—brighter, localized regions of the disk—were brighter on the edge than in the center due to the larger amount of incandescent gas crossed by the light. This seems to contradict Secchi's statement, and it is not surprising that the contemporaries got lost…

A long time was necessary to understand what really occurs in the atmospheres of the Sun and stars. This was the work of the German astronomer Karl Schwarzschild (1873–1916) and of his successors, from 1906. They showed that the temperature of the solar atmosphere decreases with altitude. It emits light over all its thickness and is only partly transparent, so that the radiation from the inner layers is partly absorbed by the outer ones.[24] The light we receive comes mainly from the layers whose opacity is important as seen by the observer. This region is more superficial, hence colder and less luminous, at the limb than at the inner parts of the disk, because it is seen more obliquely; this is the cause of the limb darkening.

The Brightness of the Sky, and Stellar Photometry

Arago's fifth Memoir on photometry describes the measurement of the brightness of the sky around the Sun. For this, Arago again used the photometer of Fig. 8.13, but he managed that the two images of the Sun were separated, and he turned the polarizer until one of the images disappeared in the sky light. He derived in this way a luminance of the sky near the Sun of about 1/500 of that of the solar disk, and noticed that this luminance was rather uniform up to 30 arc min from the edge of the Sun.

The fifth Memoir exposes more interesting results: It contains the discussion we have already seen about the absence of polarization of the light of stars and their nature, but also their photometry. At this time, the flux of the stars was estimated visually; they were classed into six magnitudes in an approximately logarithmic scale. Arago was one of the first, with William and John Herschel (Fig. 8.17), to do real measurements of their flux. For lack of photometric standards, only relative measurements were possible. Some scientists compared the fluxes of stars to that of a candle reflected on a polished sphere. John Herschel compared them to that of the Moon, from which he made a quasi-point image. Doing this comparison for several stars, he could obtain their relative fluxes.

One could also compare the flux of the star under study to that of another star taken as a standard, for example Sirius, the brightest star of the sky, by observing them successively with the same telescope. One put a diaphragm on the objective when looking at Sirius, so that it would look to be of the same brightness as the star

[24] This is indeed what Secchi said, but his model with two separate layers was far too simplistic. It remained, however, in use until the middle of the twentieth century.

Fig. 8.17 John Herschel in 1873. Wikimedia Commons, Materialscientist

under study seen without a diaphragm; the ratio between the area of the diaphragm and that of the full objective gave the ratio of the fluxes. This is what William Herschel did. Arago rejected this method because of the difficulty in comparing the fluxes observed one after the other, and also because the ratio of the areas was only a poor approximation of the flux ratio. He decided once again to use the apparatus of Fig. 8.13, and to turn the polarizer until the fainter image of the star disappeared into the sky background. Repeating the observation for another star, he could get the ratio of their fluxes provided the sky background had not changed between the two observations.

The results of these observations, made by Laugier, to which Arago committed the measurements, are somewhat erratic. This results from the fact that the sky background is variable in time and space, and also that atmospheric turbulence more or less spreads out the images; if the image is sharp, it is more easily detectable over the background. In spite of these weaknesses, the results are not bad when compared with modern measurements, as can be seen on Fig. 8.18. Arago, who was perhaps suspicious about Laugier's observations, concluded his Memoir in writing: "I have no other merit than to have given to the astronomers an exact and convenient instrument."

If this method is actually very convenient and fast and can reach relatively faint stars, down to the fifth magnitude, it cannot be qualified as exact. However, it had

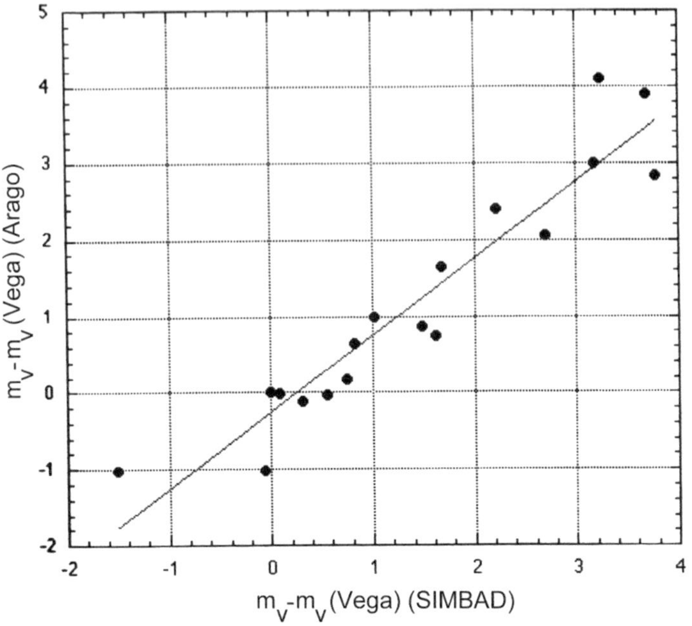

Fig. 8.18 Comparison of the photometric results of Arago and Laugier, with modern results collected in the SIMBAD catalogue of the *Centre de données de Strasbourg*. The scales are in magnitudes such that $m = -2.5 \log F + \text{constant}$, F being the flux from the star. The constants are adjusted such that the Vega (α Lyrae) has the same magnitude $m = 0$ in both systems. There is no systematic effect, but the dispersion of the measurements is rather large

the advantage of allowing the comparison of stars very distant from each other in the sky. Unfortunately, there were no other photometric measurements in France; everything was discontinued at Arago's death and the arrival of Le Verrier, who was not interested, and Arago's measurements are unjustly forgotten.

Most of the nineteenth century photometers after the death of Arago used his principle to attenuate the flux of the standard by a polarizer-analyzer set. The German astronomer Johann Karl Friedrich Zöllner (1834–1882) built in 1859–1861 a rather sophisticated photometer in which the standard was a kerosene lamp. In the United States, Edward C. Pickering (1846–1919) developed between 1875 and 1877 another stellar photometer in which he sent via a total reflection prism the image of the Pole star, attenuated by a polarizer-analyzer set, to the same field as the image of the star under study; this set-up allowed determining the stellar magnitudes in the great Harvard catalogue.

Like the sixth Memoir, Arago's seventh Memoir, on photometry, is a mixed bag. Arago was probably too ill to present it himself, and it was read by another academician. It contains the description of a rather clever method of estimating the height of clouds, which will be discussed in the next chapter. The Memoir also contains a discussion of the brightness of the different parts of the lunar disk, as measured with the photometer. The next topic is a measurement of the earth-light on the part of the Moon that is not lit by the Sun; Arago found a brightness of the order of 1/4000 of that of the luminous part, and noticed that it varies with time as a function of the cloud cover of the Earth. Finally, Arago reported about photometric measurements of Jupiter and satellites, which were not very successful.

Scintillation, Seeing and the Diameter of Stars

As any visual observer, Arago was troubled by *scintillation,* which is a fluctuation of the apparent flux and color of stars, and by *seeing* which blurs the images. The two phenomena are linked, and are due to the effect of turbulence of the terrestrial atmosphere, which affects the propagation of light. Arago had thought for a long time about this phenomenon, to which he devoted a long notice[25] in 1852.

Arago's Explanation

Of course, the degradation of astronomical images by the effect of atmosphere had been observed for a long time by astronomers, but without a clear explanation. Arago is the first to have produced a reasonable theory of the phenomenon; according to him, it was "given to M. de Humboldt in 1814,[26] inserted at the end of t. 4 of the *Voyage aux régions équinoxiales du nouveau continent.*[27]" The 1852 notice cited above shows by which tortuous, but logical, ways Arago came to his explanation.

He first recalled his interference experiments, in which he had inserted a glass plate in one of the optical paths, which produced a displacement of the fringes. He saw similar effects in his experiments on gases, which will be described in Chap. 10. Then he considered the objective of a telescope with which a star is observed, and noticed that by definition the lengths of the optical paths of the different rays that converge to the focus are the same (Fig. 8.19). He now assumed that a change in the refraction index of the air affects one of these rays, producing a phase shift with respect to the other rays; the interference between all the rays now would produce a blurring of the image and colored effects. Arago found by experience that an important phase shift on only 1/20 of the area of the lens produces a noticeable change of color. He also noticed that such changes are produced by atmospheric fluctuations at a speed that can reach 30 changes per second.

[25] **Annuaire du Bureau des longitudes pour 1852*, p. 363–504, accessible via http://gallica.bnf.fr; reproduced with a few changes and some additions in *Arago's *Œuvres complètes*, t. 7, p. 1–111.

[26] This was before the encounter of Arago with Fresnel, but Arago was already aware of the ideas of Thomas Young about interferences of light.

[27] *Humboldt (1816) t. 4, see Note B p. 285.

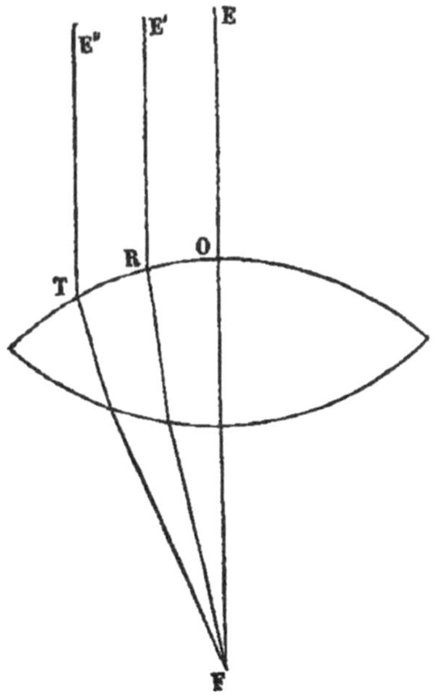

Fig. 8.19 Arago's explanation of scintillation. The three rays from a star come in phase at the focus F. But if one is affected by an irregularity of the atmosphere, its phase shift produces a variation of color of the image and a change on its intensity. From Arago's *Astronomie populaire,* author's collection

This theory is correct, but Arago's explanations are rather confusing and difficult to follow. One might also consider the deformation of the wave surface of the radiation coming from the star. If there were no atmosphere, as in observations from space, this surface would be a perfect plane. But the turbulence of the atmosphere produces deformations with a typical scale of 10 cm. If we observe the star with a telescope whose diameter is smaller than this, the deformation of the plane would increase or decrease the convergence of the lens so that the image moves continuously in front or in the back of the theoretical focus. For an observer who looks at this focus with an eyepiece, the image is deformed, displaced and spread out, with chromatic effects. The same is true for an observation with the naked eye. One can also reason like Arago in considering the interference between the different rays that cross the objective: the result is the same.

With a larger telescope, we have to consider the wave surface over all the diameter of the objective. It presents many zones with a typical size of 10 cm, inside which the rays are approximately in phase, and the image results from the interferences between all these zones. At a given time, it consists in small spots called speckles, which go on and off all the time and move so rapidly that the eye or the detector only sees a blurred image that looks approximately steady. At present, it is possible to correct for the deformations of the wave surface by placing on the light beam a deformable mirror, driven by actuators; a computer analyses the image and sends the relevant information to these actuators, usually 200 times per second. This is called the *interactive optics*. This benefits from the full resolving power of the telescope, at least in the infrared where the correction is easier; better images can now be obtained with giant telescopes from the ground than with the Hubble space telescope.

Arago's Scintillometer

To estimate qualitatively the importance of scintillation as a function of the elevation of the star over the horizon, a subject for which there was contradictory information, Arago imagined a "scintillometer." He had remarked that when one could de-focus a telescope by moving back the eyepiece, there is a position for which there is no more light in the central part of the image of a star. He interpreted correctly this observation by splitting the surface of the objective into two concentric zones. When one considers the image out of focus, there is a position for which the light of the central zone comes in phase opposition with the light of the outer zone, so that the central spot is dark due to interference between these two lights. If one puts the eyepiece in an intermediate place between this position and the focus, one increases the sensitivity to scintillation. There is still a dark center in the image, but from time to time a bright spot appears in the middle. This means that there is a part of the objective for which the phase shift due to the irregularities of the refraction index of the atmosphere compensates the geometrical phase shift. By counting the number of appearances of this bright spot per unit time, one can obtain qualitatively the importance of scintillation; the more frequent are these appearances, the larger is the scintillation.

In 1851, Jean-Jacques-Émile Goujon (1823–1856) and Mathieu made these observations with a small telescope. The result was that there were between five and 30 appearances of a bright spot every 5 min, and that scintillation was the same for all observed stars and depended only on their height over the horizon. Thus the question raised by Arago was answered. He suggested other means for measuring scintillation, but they were never implemented. His scintillometer was forgotten, another consequence of the reorganization of the Observatory by Le Verrier.

The Diameter of Stars

Arago devoted two small chapters of his *Astronomie populaire* (t. 1, p. 364–371) to this question. They contain a well-documented history in which Arago noticed strong contradictions. Before the invention of the astronomical telescope, Tycho Brahe (1546–1601) gave for Sirius an angular diameter of 4 arc min, while fainter stars seemed to be smaller, down to 1 arc min for the faintest ones visible with the naked eye. The stars seen through a telescope seemed even smaller: For Sirius, Pierre Gassendi (1592–1655) gave a diameter of only 10 arc sec, and Cassini even gave 5 arc sec. Arago concluded that these diameters were not real, but corresponded to the resolving power of the eye or of the telescope. He knew that this resolving power was limited by diffraction: he had observed the Airy diffraction rings around the image of stars given by a small telescope.[28] This resolving power is of the order

[28] *Arago, F. *Astronomie populaire*, t. 2, p. 137.

of 1 arc min for the human eye and of 1 arc sec for a 10 cm-diameter telescope. However, the atmospheric turbulence is such that the resolving power of telescopes is rarely less than 1 arc sec whatever their diameter; with a big telescope one gains in luminosity but not in resolution. Moreover, the saturation of the eye is such as to broaden the images of the brightest stars. This explains why Tycho Brahe found a larger diameter for Sirius than for fainter stars. Did Arago know all this? This is probable but not certain. Anyway, he wrote with good common sense:

In general, one should take for the diameter of a star the smallest quoted value, without being certain that it represents its real diameter.

He recalled that Herschel measured an apparent diameter of 0″.36 for Vega (α Lyrae) and even through a light smog, 0″.2 for Arcturus (α Bootis), with his reflecting telescopes (whose optical quality must have been outstanding). He calculated the corresponding linear diameters assuming that these stars were at the same distance as 61 Cygni, the only star for which it was known, and concluded that they would be incredibly large. Then he reasoned as a physicist, comparing the stars to the Sun to estimate their real diameters. He knew that Huygens, Michell, Lambert and finally Wollaston in 1829 had compared the flux received from Sirius with that of the Sun, a very difficult task indeed. Wollaston found that the flux from the Sun was 2×10^9 times that from Sirius (the real value is 1.53×10^9). If Sirius were similar to the Sun, its linear diameter would be the same and its distance would be $(2\times10^9)^{1/2}$ times that of the Sun, since the received flux is inversely proportional to the square of the distance. Then the angular diameter of Sirius would be equal to that of the Sun divided by the ratio of the distances: This gave 0″.04. But Wollaston thought that Sirius was at a distance ten times greater, hence had an intrinsic emission 100 times stronger than the Sun. Assuming that its surface brightness was similar to that of the Sun (wrong: it is larger), its apparent diameter would be only 0″.005. Being careful, Arago only stated that its diameter would be smaller than 0″.05. It is strange that neither he nor Barral, when editing the *Astronomie populaire*, took into account the parallax of Sirius measured by Thomas Henderson (1798–1844) in 1832 and by Thomas Maclear (1794–1879) in 1835, although it is cited in *Astronomie populaire*, t. 1 p. 435; they would have found a still smaller diameter.[29]

The preceding reasoning can be found almost literally in Wollaston's remarkable paper.[30] Arago did nothing original, but it is interesting that he had given attention to this article. This "astrophysical" behavior was exceptional in France at the time, while it was more common on the other side of the Channel. Unfortunately all this vanished after Arago's death: French astrophysics was stillborn, and reappeared only shyly with Jules Janssen 10 years later.

[29] The true angular diameter of Sirius is 0″.00585.

[30] *Philosophical transactions* (1829) 119, p. 19–27, accessible via http://www.jstor.org/stable/107764.

Chapter 9
Arago Geophysicist and Meteorologist

A polar aurora. Wikimedia Commons, US Air Force Zarex

J. Lequeux, *François Arago*, Astrophysics and Space Science Library 421,
DOI 10.1007/978-3-319-20723-0_9

Nothing escaped Arago about the physics of the Earth.[1]

While Arago is well-known as an astronomer and physicist, one often ignores that he was also an acknowledged specialist of meteorology, geophysics and oceanography. These topics were indeed among the attributes of astronomers; measurements of terrestrial magnetism and meteorology were done at the Observatory right from its beginnings. Arago continued this tradition, especially with magnetism, which was one of his favorite subjects. But he went much further, being interested in the temperature of the Earth, the currents of the seas, the winds etc. Perhaps his early travels through the Mediterranean Sea sparked his interest for the sea. But, overall, he was strongly influenced by his friend Humboldt, who was a great traveler and explorer. Arago collected old and contemporary observations, made a lot more himself and took advantage of his preeminent position and of his impressive knowledge to urge sailors and travelers to make even more observations. His purpose was to build a synoptic view of the properties of the Earth, its seas and its atmosphere. His *Instructions pour la physique du globe* ("Instructions for the Physics of the Earth")[2] of 1835, written in 2 weeks for the preparation for the mission around the world of the ship *La Bonite*,[3] are a summary of Arago's knowledge and ideas about geophysics and oceanography. They gave a serious framework for scientific observations at sea, and were generally followed by the sailors and scientists on this ship. Arago's instructions were however preceded by a much earlier English text on the same subject.[4]

Arago's ideas were strong and precise. Two years after he wrote the instructions, he declared in a speech on 5 June 1837 to the Chamber of Deputies[5] that the new expedition projected by Jules Dumont d'Urville was poorly conceived, with badly adapted ships, and he expressed his "reluctance for travels made only for curiosity." He wrote new instructions in 1838 and 1853 for the scientific exploration of Algeria.[6]

Arago was also interested in balloon flights. He listed the scientific equipment for a flight that Barral and Jacques-Alexandre Bixio (1808–1865) were planning in 1850; the equipment, prepared by the physicist Victor Regnault (1810–1876), was used effectively.

Let us now explore the various aspects of Arago's research in geophysics, beginning with meteorology.

---

[1] **Comptes rendus de l'Académie des sciences* (1854) 38, p. 1039.

[2] **Comptes rendus de l'Académie des sciences* (1835) 1, p. 380–410.

[3] *La Bonite* was not an oceanographic ship, but had "as a special mission to carry consular agents to the most remote points of the globe." She even went to Hawaii, an exceptional destination for a French ship.

[4] *Philosophical transactions* (1666) 2, p. 433–448, accessible via http://www.jstor.org/stable/101197.

[5] *Arago, F. *Œuvres complètes*, t. 9, p. 433–488. The expedition was to the Antarctic from 1837 to 1840; Dumont d'Urville discovered Terre Adélie during this trip, a non-negligible feat.

[6] **Comptes rendus de l'Académie des sciences* (1838) 7, p. 206–224.

Arago and Meteorology

The Birth of Scientific Meteorology

In Arago's time, meteorology was generally not considered a science. Until 1730–1740, thermometers were not even graduated and standardized. This standardization was achieved simultaneously by René Antoine de Réaumur (1683–1757) in France, Anders Celsius (1701–1744) in Sweden and Daniel Gabriel Fahrenheit (1686–1736) in Germany, each with his own system. It was possible in 1776 to compare the minimum temperatures in 24 French cities and 25 foreign ones, but there was not yet any regular publication, and the Paris Observatory made money by selling the results. At the beginning of the nineteenth century *Connaissance des temps* and other journals gave minimum and maximum temperatures and pressures for each day, as well as hygrometry (moisture) and wind direction at noon. New instruments were installed in 1810 at the Observatory, but Arago, who was very interested in meteorology (Fig. 9.1), considered what was published as insufficient. Starting in 1816, pressure, temperature and hygrometry were measured each day at 9 a.m., noon, 3 and 9 p.m., and cloud coverage and wind direction (not its strength) at noon. These data were published in *Annales de Chimie et de Physique*, the journal of Berthollet and Arago, and in the *Comptes rendus* after its creation in 1835. A general summary was published at the end of each month, giving also the total amount of rain. Joseph-Marie Bouvard, the brother of Alexis Bouvard, made most of the observations. None of the instruments could record automatically the data until 1850.

When an observation was missed, the Paris astronomers took the average between the next ones, as remarked aggressively by Le Verrier when he became director of the Observatory. Laugier replied that such cases were exceptional.

What could one do with all these data? Arago and Bouvard were particularly interested in the diurnal variations of atmospheric pressure. They were discovered during the geodesic expedition to Peru in 1735–1744, and well studied by Humboldt. But nothing really new was obtained.

Meteorological Forecast

Of course, we all want to know the weather in advance. Such was the case in Arago's time, but he refused to use the measurements for any forecast despite public pressure. Thus nothing happened in France until Le Verrier developed a meteorological network in which the information was sent to the Paris Observatory by electric telegraph from stations scattered first throughout France, then throughout Europe.[7] This was a great success, and some primitive forecasts could be made as early as 1857. Almost at the same time, the physicist Joseph Henry developed a similar

[7] See Lequeux (2013), chapter 9.

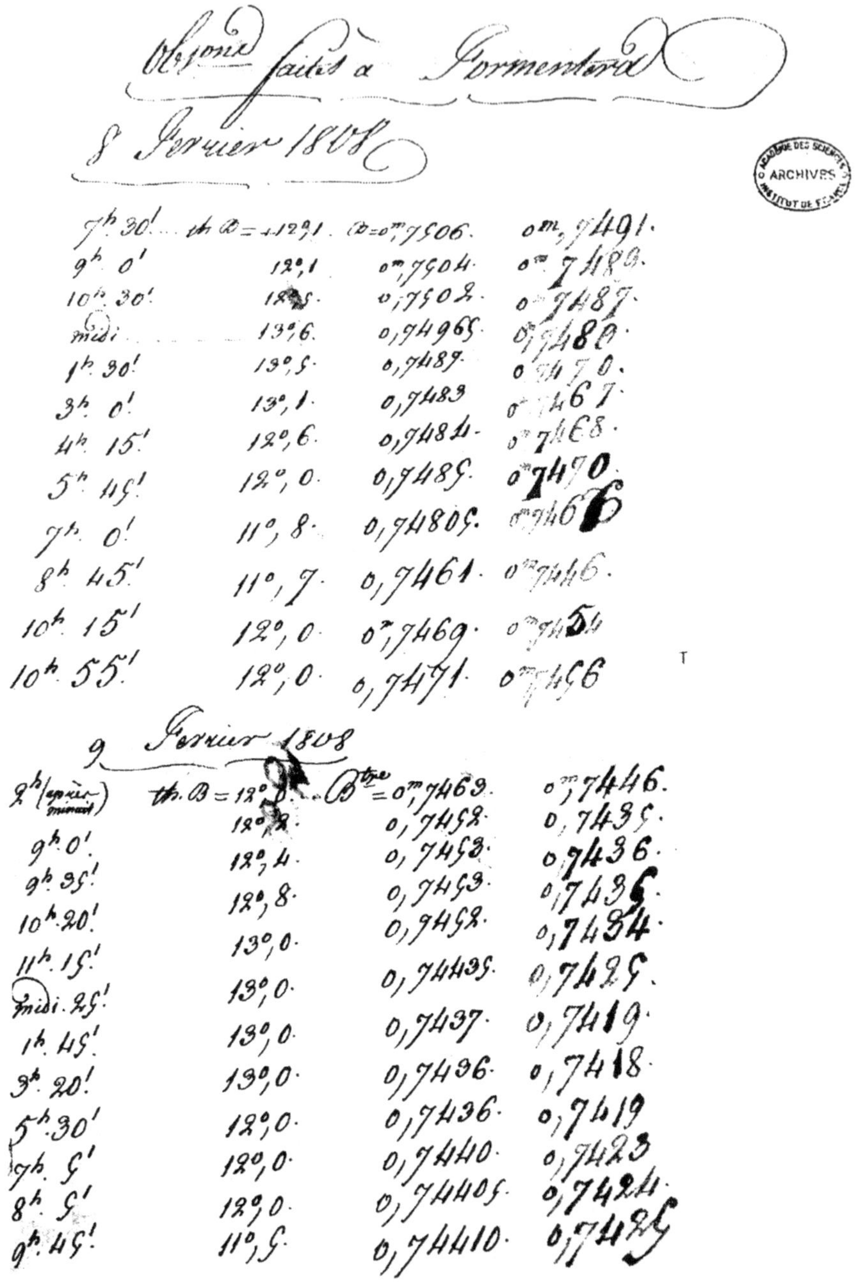

Obsons faites à Formentera

8 Février 1808

7h 30' ...	th. B = +12°,1 .	B = 0m,7506.	0m,7491.
9h 0'	12°,1	0m,7504.	0m,7489.
10h 30'	12°,5.	0,7502.	0m,7487.
midi	13°,6.	0,74965.	0,7480.
1h 30'	13°,5.	0,7489.	0,7470.
3h 0'	13°,1.	0,7483	0m,7467.
4h 15'	12°,6.	0,7484.	0m,7468.
5h 45'	12°,0.	0,7485.	0m,7470.
7h 0'	11°,8.	0,74805.	0m,7466
8h 45'	11°,7.	0,7461.	0m,7446.
10h 15'	12°,0.	0m,7469.	0m,7454
10h 55'	12°,0.	0,7471.	0m,7456

9 Février 1808

8h (après minuit)	th. B = 12°,[illegible]	Btre = 0m,7463.	0m,7446.
9h 0'.	12°,[illegible].	0,7458.	0,7435.
9h 35'.	12°,4.	0,7458.	0,7436.
10h 20'.	12°,8.	0,7453.	0,7436.
11h 15'.	13°,0.	0,9458.	0,7434.
midi 25'.	13°,0.	0,74435.	0,7425.
1h 45'.	13°,0.	0,7437.	0,7419.
3h 20'.	13°,0.	0,7436.	0,7418.
5h 30'	12°,0.	0,7436.	0,7419
7h 5'	12°,0.	0,7440.	0,7423
8h 5'	12°,0.	0,74405.	0,7424.
9h 45'.	11°,5.	0,74410.	0,7425

Fig. 9.1 Autograph register of meteorological observations by Arago in Formentera (Balearic islands). He measured temperature (in degrees Celsius) and atmospheric pressure (in meters of mercury) as read and then corrected for the effect of temperature, about 12 times a day. Archives de l'Académie des sciences

meteorological service in the United States; it was soon extended to the whole North American continent. The French and American services led to the founding of Météo-France and the National Weather Service, respectively.

A Synoptic Approach

While Arago had been a strong promoter of the electric telegraph in France, he did not imagine that it could be used for forecasting the weather. Together with Humboldt, he dreamed of a synoptic (global) approach to meteorology and climate covering the whole world, and thought that the electric telegraph would be useful for gathering the necessary data. In the meantime, the two friends asked sailors to make meteorological measurements at sea and in harbors as frequently as possible. As early as 1810, this made possible the construction of a map of the world where lines of equal mean annual temperatures were drawn (Fig. 9.2). Although published in *Astronomie populaire* "from the projection of Arago", this map came from Humboldt, as Arago indeed acknowledged it.

Arago was also interested in winds and their origin, especially the trade winds that blow from the northeast in the northern hemisphere, and from the southeast in the southern hemisphere. In his instructions for the *La Bonite* expedition, he asked that, taking advantage of a planned stopover in Hawaii, someone climb to the summit of the Mauna Loa volcano at 4200 m elevation, in the hope of detecting an "upper wind" that would compensate the lower trade wind "according to the explanation which is generally given of the trade winds." This explanation was given in 1735 by George Hadley (1685–1768), and is still considered as valid today (Fig. 9.3): hot, humid air from the equatorial latitudes rises to the tropopause at about 10 km and moves to higher latitudes on each side of the equator. Its motion is deviated to the right (hence towards the east) by the Coriolis acceleration due to the rotation of the Earth. When this air reaches the 30° latitude, it becomes cold enough to fall and to form a region with high pressures. It then moves back to the equator at low elevations, being also deviated to the right, this time towards the west, forming the trade winds. The existence of the counter-wind at high altitudes was, according to Arago, proven by the motion of very high clouds, by the fall on Barbados of dust ejected by the volcano on Saint-Vincent island located to the west, and finally by direct observations at the summit of Tenerife by Humboldt and Captain Basil Hall (1788–1844).

Storms, Lightning and Waterspouts

Arago gathered an enormous database on storms, lightning and waterspouts (Fig. 9.4). His notice on thunderstorms takes no less than 394 pages in *Annuaire du Bureau des longitudes,* with 11 extra pages in *Œuvres complètes*.[8] This required a

[8] **Annuaire du Bureau des longitudes pour 1838*, p. 221–614; *Arago, F. *Œuvres complètes,* t. 4 p. 1–404.

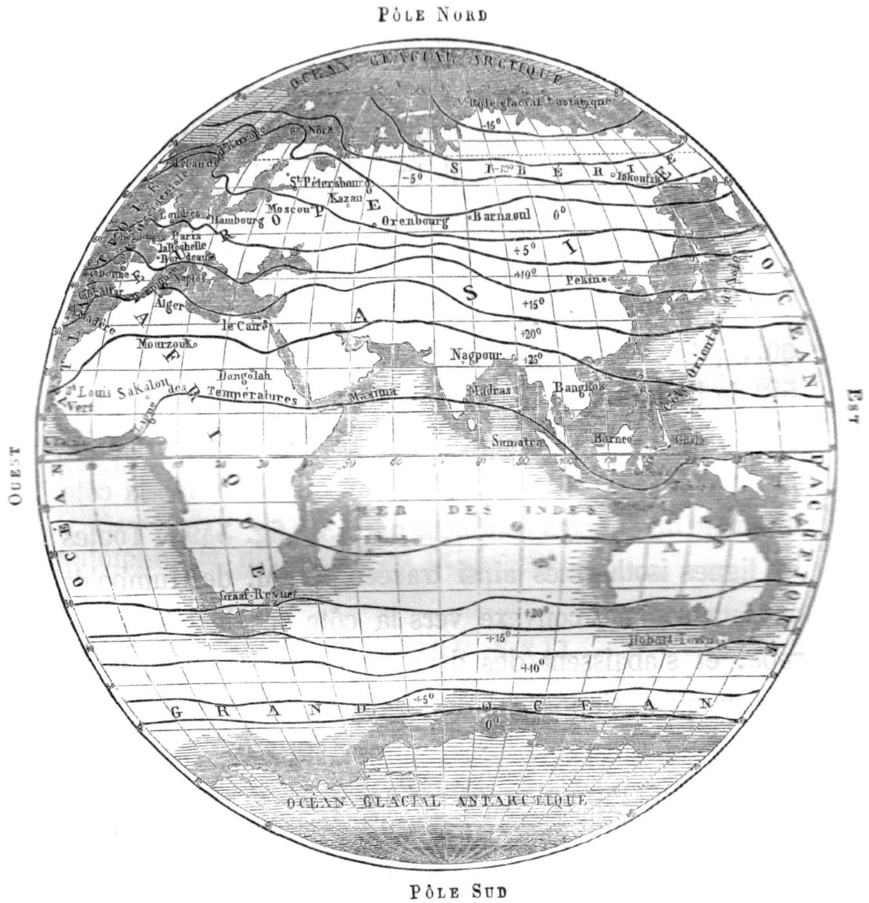

Fig. 9.2 Isothermal lines of one half of the terrestrial globe, according to Humboldt and Arago. The average annual temperatures are indicated on each line in degrees Celsius. From Arago's *Astronomie populaire,* author's collection

considerable amount of work. However, the result is disappointing. There is a complete description of the phenomena, but no attempt for explanations. It is true that the subject is a difficult one. Strangely, electricity is barely discussed for lightning or thunderbolts, although Arago said that he had demonstrated "the identity of ordinary electricity with the atmospheric one" by noticing that thunderbolts, like sparks in the laboratory, produced nitric acid from the nitrogen, oxygen and water vapor of the atmosphere. But he could not explain the origin of thunderbolts. This was not for lack of observations: he made no less than 2047 observations of atmospheric electricity scattered over 237 days in 1829–1830, using an electrometer connected to a lightning rod.[9] Presumably, he considered that no really interesting

[9] *Arago, F. *Œuvres complètes,* t. 11 p. 619–653.

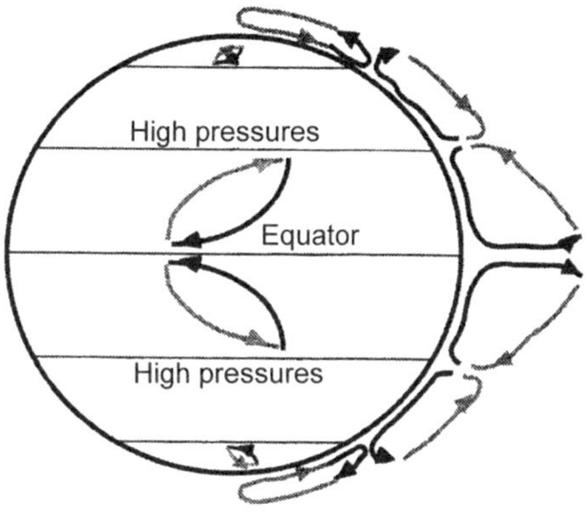

Fig. 9.3 Scheme of the general wind circulation in the terrestrial atmosphere. There are three closed cells in each hemisphere, seen in cut to the right outside the circle that represents the Earth: that of Hadley at low latitudes, that of William Ferrel (1817–1891) at intermediate latitudes, and the polar cell. The lower winds of the Hadley cell (*thick arrows*) are the trade winds. *Light arrows* schematize the counter-winds at high elevation. Note the deviation to the right of those winds due to the Coriolis acceleration

result was obtained, because nothing was published during his lifetime. As a matter of fact, it is only in the twentieth century that more sophisticated instruments were built and that these phenomena began to be understood.

The Red Moon

Arago was more successful in explaining the thermal equilibrium of the surface of the Earth. He was helped by several Memoirs of Joseph Fourier and by a book devoted to dew by the English physicist William Charles Wells (1757–1817).[10] He understood that the soil cools by emitting *calorique rayonnant* (radiating heat), actually infrared radiation with wavelengths around 10 μm; it can then be colder than the atmosphere by as much as 6–8 °C. Interposing a screen would avoid this cooling, because the outwards radiation is roughly compensated by the inwards radiation of the screen if it is at the same temperature as the atmosphere. If there are clouds, they also limit the cooling of the surface. This relatively simple physics was discussed by Arago in several notices gathered later in a single, long paper.[11] Arago showed in this paper his interest in agriculture, exploring various consequences of his observations. In particular, he fought against the prejudices of peasants about the supposed influence of the Moon on vegetation; but this did not convince most of his contemporaries, and these beliefs are still vivid today. However, some of the popular beliefs were not without basis, in particular the phenomenon of the red Moon.[12]

[10] Wells (1814), This book was analyzed by Arago in **Annales de Chimie et de Physique* 5 (1817) p. 183–216; he added interesting details.

[11] *Arago, F. *Œuvres complètes*, t. 8, p. 83–146.

[12] Ibid, p. 120–124.

Fig. 9.4 A waterspout over the sea. Wikimedia Commons, Ju TA

The story of the red Moon, told by Arago himself,[13] is amusing. After the fall of Napoléon in 1815, a delegation of the Board of Longitudes used to present to the king its publications, the *Connaissance des temps* and the *Annuaire*. During this ceremony, King Louis XVIII showed interest for unexpected subjects: "I am delighted that you are around me [...], because you will explain to me clearly what is the red Moon and its effects on the crops." Laplace, who was leading the delegation and to whom these words were addressed, was terrified. He could only answer that the red Moon had nothing to do with astronomical theories. Later, he asked Arago to inform him on the topic. Arago contacted the gardeners of the *Jardin des plantes* (the king's garden) about a phenomenon attributed to the full Moon at the end of April or May, which was supposed to produce some frost that would make the buds turn red. Arago had no doubt about the existence of this phenomenon; if one sees the Moon, this means that the sky is clear, so that the bodies on the surface of the Earth can freeze, damaging the buds that are particularly sensitive to freezing.

Arago would not go any further. But when reading the remarkable *Mémoire sur les températures du globe terrestre et des espaces interplanétaires* (Memoir on the Temperature of the Earth and of the Interplanetary Space)[14] published in 1824 by Joseph Fourier, one realizes that he had understood that the atmosphere itself limits the cooling of the terrestrial surface through the greenhouse effect. In 1861, the British physicist John Tyndall (1820–1893) showed that water vapor and carbon dioxide are responsible for this effect, which was analyzed quantitatively for the first time in 1896 by Svante Arrhenius (1859–1927).

---

[13] *Arago, F. *Astronomie populaire* t. 3, p. 497–503.

[14] **Mémoires de l'Académie royale des sciences* 7 (1827), p. 569–604.

Arago had some ecological preoccupations. He was among the very first who suspected that human activities could affect the climate. He wrote, for example[15]:

It is certain that the forests exert a noticeable influence on the temperature of the surrounding regions, because, for example, snow remains there much longer than in the surrounding countryside. The destruction of forests must therefore cause a change in the climates.[...]

The climate of Manchester suffered gradual and important changes following the development of industrial activities. Now that this city has become a great furnace, so to say, it rains more or less every day there.

Atmospheric Optics

We should not be surprised that Arago, a great observer, was interested in the phenomena of atmospheric optics. He devoted to them a long notice that is of little interest for us today. More interesting is his discovery of the polarization of the diurnal sky, which dates from 1811. He made many measurements in 1812, 1814 and 1815 and derived from them its empirical laws.[16] Polarization is obviously a property the solar light diffused by the atmosphere, and its maximum is approximately at 90° from the direction of the Sun. The light of the nocturnal sky lit by the Moon is also polarized. Arago could not understand the origin of this phenomenon, which was only elucidated by Lord Rayleigh (1842–1919) at the beginning of the twentieth century: It is the diffusion of light by the molecules of the air, the Rayleigh scattering (Box 9.1).

Box 9.1 The Rayleigh Scattering
When falling on an isolated atom or molecule, the oscillating electric field of light, which is an electromagnetic wave, produces a periodic deformation of its electronic cloud, so that the barycenter of the negative charges of the electrons oscillates with respect to the positive charges of the nuclei. The oscillating dipole created in this way radiates in all directions at the wave frequency, producing a diffusion of the incoming light. The diffused intensity depends much on the wavelength; as a consequence, the scattered light is bluer than the incident one. This is why the day sky is blue. The transmitted light is diminished and redder than the incident light; this produces the absorption and the red color of the rising or setting Sun, because the thickness of the atmosphere crossed by the light is then very large. The scattered light is partly polarized, as discovered by Arago. The Rayleigh scattering is only efficient if the sizes of the diffusing particles are small with respect to the wavelength, hence if these particles are atoms, molecules or very small grains of dust.

---

[15] **Annuaire du Bureau des longitudes pour 1846*, p. 574–608, see p. 595 and 606.

[16] *Arago, F. *Œuvres complètes*, t. 10, p. 548–560.

Arago also noticed that the light of a cloud illuminated by the Sun is not polarized, a property that he asked Barral and Bixio to check during their balloon flights. This is because the scattering water droplets are considerably larger than the wavelength, but Arago could not know this explanation. He then had the idea that this property could serve to measure the altitude of the clouds.[17] The diffused solar light originating in the part of the atmosphere located in front of the cloud is partly polarized, but of course less than in the free region outside the cloud, because the length crossed in the scattering atmosphere is smaller. A comparison between the two degrees of polarization allows a measurement of the ratio of the thickness of the atmosphere in front and outside the cloud, hence the elevation of the cloud. This was an excellent idea, but it seems that it has never been applied.

Finally, Arago has observed the polarization of the halo that sometimes surrounds the Sun.[18] He recognized that this polarization was not due to diffusion, but to the refraction by ice crystals in the upper atmosphere. This offered the means to determine if the temperature was low enough for the formation of these crystals, but once again there were no known observations after Arago, and his study was forgotten.

Terrestrial Magnetism

Since medieval times, sailors used a magnetic compass for orientation at sea. However, it was found that the magnetic needle does not really indicate the direction of the north, but differs by an angle called the magnetic declination, which varies slowly with time. It was thus important for navigation to measure the declination and to follow its variations. These measurements were done at the observatories of Greenwich and of Paris from their foundation and almost without interruption.[19] The magnetic inclination—the angle between the direction of the magnetic field and the horizon—was also measured with a needle moving around a horizontal axis. In Paris, these measurements lacked accuracy until Cassini IV renewed the observing methods in 1777, using principles developed by Le Monnier and Coulomb. This allowed good measurements of the diurnal variations of the declination, and also of the perturbations of the magnetic field linked to the polar aurorae, which will be discussed later.

---

[17] *Arago, F. *Œuvres complètes*, t. 10, p. 282–289.

[18] **Annales de Chimie et de Physique* 29 (1825), p. 77.

[19] For a history of magnetic observations at the Paris Observatory from 1667 to 1872, see Rayet, G. (1876) *Annales de l'Observatoire de Paris, Mémoires*, 13, p. A.*1–A.*40, accessible via http://cdsads.u-strasbg.fr/cgi-bin/iarticle_query?journal=AnPar&volume=0013&type=SCREEN_THMB.

Arago's Measurements

In 1810, soon after he came back from his long odyssey in the Balearic Islands and through the Mediterranean Sea, Arago took an active and long-lasting interest in the terrestrial magnetic field; he made a very large number of measurements himself and devoted many thoughts to this topic. He wrote as an introduction to the big Memoir that summarizes his researches[20]:

Nothing, in the vast domain of the physics of the Earth, is more mysterious than the causes of the variations of the three elements of the terrestrial magnetic field: declination, inclination and intensity. For the moment, we can only resign ourselves to collect the measurements that will serve as the basis for the researches of our successors.

Around 1820 in the large meridian room, Arago installed four new, improved instruments built by Gambey. They were:

- the declination compass (Fig. 9.5), a superb instrument derived from that of Cassini IV.
- the diurnal variations compass: It was similar to the declination compass, but its orientation was fixed and the variations were observed with a microscope looking at the tip of the long magnetic needle;
- the inclination compass (Fig. 9.6); and
- the intensity compass: like in the declination compass, the magnetic needle is suspended by a torsionless silk wire; it is made to oscillate around its equilibrium position; the retracting force being proportional to the magnetic field intensity, the observation of the period of oscillation allows measuring this intensity, at least in relative value.[21]

Contrary to a common belief, the Paris Observatory building contains iron, which spoiled the measurement of the magnetic declination and inclination. In 1823, Arago was obliged to build in the garden, 60 m south of the main building, a magnetic observatory in stone and wood, without steel or iron. It was replaced in 1830 by two other rooms, one of which is depicted on Fig. 9.7. Arago made himself a measurement of declination each year from 1818 to 1835; then Laugier and others took over. The other instruments were less or not sensitive to perturbations by iron, and remained in the more accessible meridian room. Before 1835, Arago made himself 52,599 measurements of declination with the diurnal variations compass. It is difficult to imagine how such a busy man found the time to do all these measurements in person. But there is no doubt that he did so: all his observations are reported in six in-folio autograph registers.

---

[20] *Arago, F. *Œuvres complètes*, t. 4, p. 459–541.

[21] It made little sense at this time to try to give an absolute value to the magnetic field; its unit was only defined at the end of the 1830s by the German physicists Carl Friedrich Gauss (1777–1855) and Wilhelm Weber (1804–1891).

Fig. 9.5 A Gambey's declination compass, ca. 1824. The box protects a long, thin magnet suspended by a torsionless wire. The direction of the magnet can be read on the horizontal graduated circle, and its angle with the north can be found thanks to the small telescope on the *top*, with which one can look at some fixed target or at the Pole star. Musée de l'École polytechnique

Diurnal Variations of the Earth's Magnetic Field

Arago worked on all the aspects of the terrestrial magnetic field. His systematic measurements of magnetic declination and inclination are precious in the study of the long-term variation of this field, the secular variation. They were, of course, very useful to navigation at their epoch. He also studied the small annual variation of declination discovered by Cassini IV; roughly, the magnetic needle is slightly to the east of its average position between the spring equinox and the summer solstice, and to the west during the nine other months. However, his measurements did not fit well with those of English and American scientists. The reason for these differences was not understood at the time, and even today we do not understand well what is this annual variation, whose behavior is somewhat erratic.

More interesting are Arago's observations of the diurnal variations, to which he devoted considerable effort; he made himself, as a mean, 11 observations each day, starting at 7 a.m. and finishing at 11 p.m. Sometimes, he even made extra observations

Fig. 9.6 A Gambey's inclination compass, ca. 1850. Musée de l'École polytechnique

during the night, depriving himself of several hours of sleep. The only systematic gap was on Mondays from 2 to 6 p.m., the time of the weekly meetings of the Academy of Sciences. Humboldt wrote in 1830 about these measurements[22]:

The most precise and important work ever undertaken on the phenomena of terrestrial magnetism, in particular the hourly motions of the needle and the perturbations of its motion due to boreal auroras and earthquakes in the most distant countries, is that of M. Arago at the royal Paris Observatory, followed during so many years. The physicists are waiting impatiently for the publications of his results.

Arago synthetized his observations as follows:

Every day, the magnetic needle [indicates] two maxima and two minima in declination, as follows:

1. *After 11 p.m., the north pole of the needle moves to the east, reaches its minimum elongation at 8¼ a.m., and retrogrades towards the west to reach its maximum declination at 1¼ p.m.*
2. *After 1¼ p.m., the needle moves again to the east, reaches a second minimum between 8 and 9 p.m. and then goes to the west, reaching its second maximum at 11 p.m.*

[22] **Annales de Chimie et de Physique* 44 (1830), p. 231–243, see p. 238.

Fig. 9.7 The magnetic pavilion of the Paris Observatory in 1862. Arago's instruments were replaced by more modern ones attributed to Gauss. Bibliothèque de l'Observatoire de Paris

Arago defined the amplitude of the diurnal variation as the difference between the maximum and minimum declinations: it varied from 3 to 17 arc min. He took the monthly mean and saw that the diurnal variation was maximum in April, and minimum in December, with large variations from year to year. Unfortunately, he did not try to put these annual variations on a graph (no one was doing that in this period); he would have seen that their amplitude is well correlated with solar activity, that one was beginning to quantify by counting the sunspots (Fig. 9.8). Several other scientists have seen this correlation, and Barral finally discovered it in Arago's data. It is clear now that the diurnal variation is due to variations in the electric currents of the Earth's magnetosphere, which are strongly dependent on the solar activity.

The other measures of Arago gave less interesting results, and will not be described here except the following ones.

Magnetic Field and Polar Aurorae

It was known for a long time that the compass needle moves in an erratic way when there are polar aurorae (a few of them can be seen from the latitude of Paris). Cassini IV described this effect in 1784. Curiously, Arago claimed that he was the

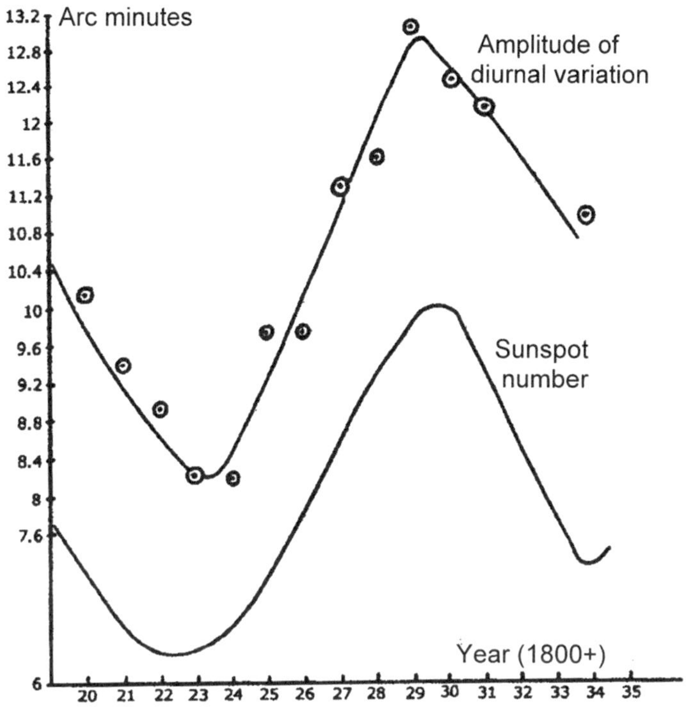

Fig. 9.8 Changes from 1820 to 1825 of the amplitude of the diurnal variation of the magnetic declination. They are shown in the *upper curve*, while the *lower curve* represents the variations of the number of sunspots (the Wolf number). The correlation is remarkable. From Le Mouël, 2006

discoverer[23]; it is true that his observations were better and that he made many more of them than Cassini. He searched for information on the aurorae that could not be seen from Paris, and noted that they also affected the magnetic needle. However, his publication was contested by several physicists. Arago recognized among them his Scottish enemy David Brewster, whose criticisms greatly displeased him. He replied[24]:

[My] work consists at present [in 1828] in more than 60,000 observations. Each time that I was obliged to leave the Observatory, friends were there to replace me.[…] If the critics want to compete with me, if they also want to produce forecasts, I will send them with pleasure the three talismans I have used, i.e. a thread of unbleached silk, a magnetized needle and a microscope. It would then remain to wish to them a good health, zeal and much patience.

---

[23] *Arago, F. *Œuvres complètes* t. 4, p. 571–594, see p. 571.

[24] **Annales de Chimie et de Physique* 39 (1828) p. 369–390, see p. 390.

The Origin of the Terrestrial Magnetic Field

Arago did not try to understand the origin of the magnetic field of the Earth. He had excuses, because science was unable to explain it at his epoch. At the end of the sixteenth century, William Gilbert (1544–1603) considered that the Earth was a giant magnet. Two centuries later, nothing had changed. However, as soon as electromagnetism was discovered, Ampère suggested that electric currents running from east to west at the surface of the globe could produce the magnetic field, an idea soon adopted by Barlow in England. Ampère also imagined that these currents could come from natural batteries formed by metallic veins. But the telluric currents, which indeed exist, are much too weak to generate the magnetic field.

We know now that most of the terrestrial magnetic field comes from electric currents running in the liquid part of the metallic core of the Earth, at radii between 1200 and 3400 km. They are generated by the differential rotation of this region, which works like a self-maintained dynamo. Poorly understood instabilities produce secular variations of the magnetic field, in particular of the magnetic declination. The magnetic field even changes polarity from time to time: the last inversion took place 800,000 years ago.

There are also electric currents in the environment of the Earth, in a vast region called the magnetosphere. This region (Fig. 9.9) is shaped by the solar wind, an ionized gas that travels and impinges on the Earth at 400 km/s. The rotation of the Earth at the center of the magnetosphere produces the diurnal variations so well studied by Arago. The time variations of the solar wind, which are directly linked to the solar activity, cause changes in the electric currents that run in the magnetosphere, hence in the magnetic field. This explains why the amplitude of the diurnal variations depends on solar activity. Finally, bursts of charged particles that come into the magnetosphere following solar eruptions produce deep changes of its structure. Electrons rush into the polar regions of the lower atmosphere, following the lines of force of the magnetic field. They ionize the gas, and the recombination of ions and electrons that follows produces light: these are the polar aurorae. We see why the aurorae are correlated with erratic changes in the magnetic field that disturb the needles of compasses. It took most of the twentieth century and many launches of artificial satellites and probes to reach a reasonable understanding of all these complex phenomena.

A Synoptic View of the Terrestrial Magnetic Field

Sailors know that the magnetic declination depends on longitude and latitude. At the end of the eighteenth century, they began to make systematic measurements of declination for the benefit of their successors. Humboldt and Arago recommended that sailors make such measurements as frequently as possible. Of course, these measurements cannot be as accurate as those obtained in a magnetic observatory but,

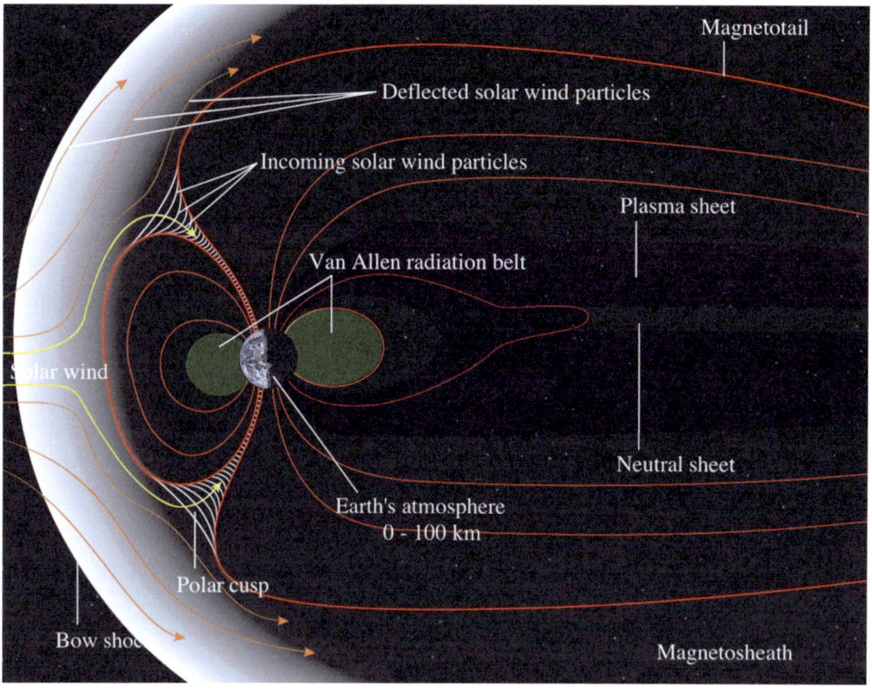

Fig. 9.9 The magnetosphere of the Earth. It is a complex ensemble of ionized media (plasmas) immersed in the terrestrial magnetic field. The solar wind, which itself carries a magnetic field, comes from the *left* of the figure with a velocity of 400 km/s, forming a shock with the material of the magnetosphere. The trajectories of the incoming ionized particles are schematized by *lines with arrows*. Some of these particles can reach the high atmosphere of the polar regions of the Earth as indicated. The other *lines* schematize the lines of force of the magnetic field. High-energy ionized particles are trapped by the magnetic field in the Van Allen "radiation" belt. Wikimedia commons, NASA-Medium69

if one is not too demanding, measurements made with a simple marine compass and related to the direction of the Pole star are sufficient, provided that the perturbations caused by the iron or steel of the ship are corrected for. In 1836, guided by Arago and Humboldt's observations, Captain Louis-Isidore Duperrey (1786–1865) built a world map of magnetic declination. Meanwhile, Gauss, Weber and their associates in Germany created a *magnetischer Verein* (Magnetic Society) with the purpose of making magnetic measurements throughout the terrestrial globe and making maps similar to that of Duperrey. Humboldt, who knew Gauss very well and had even introduced Weber to him, made many such measurements during his travels from the Ural mountains to China and his return via the Caspian Sea. He also built in his castle's garden at Tegel near Berlin "a small house completely ironless, similar to that of the Royal Paris Observatory." There, he started a regular series of observations of magnetic declination. The German geophysicist Adolf Kupfer (1799–1865), who worked in Russia, also built a magnetic observatory and obtained from various governments the construction of similar facilities. By the middle of the nineteenth century, magnetic observations were almost as common as meteorological ones.

The Temperature of the Earth

Arago certainly knew the works of Joseph Fourier (Fig. 9.10), his predecessor as Perpetual Secretary of the Academy of Sciences, even if he was probably unable to follow Fourier's very innovating mathematical developments, which were not really appreciated before the middle of the twentieth century. He found there an inspiration about a question that was a constant preoccupation of the scientists at that time: the temperature of the terrestrial globe. What interested Arago most was the long-term evolution of the climate. He wrote at the beginning of his very long "Memoir on the thermometric state of the globe"[25]:

Does the thermometric state of the globe vary with time? Do its variations affect the whole mass of the material of the Earth? Or, conversely, should we believe that they affect only its surface? In either case, are there any proofs that changes of temperature occurred in historic times?

The question of the age of the Earth is underlying. Of course, scientists did not believe any more in the Biblical chronology, but what should replace it? Buffon

Fig. 9.10 Joseph Fourier as an academician. Wikimedia commons

[25] *Arago, F. *Œuvres complètes*, t. 8, p. 184–652.

(1707–1788) and many others thought that the main source of heat of the Earth was fossil heat from the time of formation (the plutonian hypothesis). After experiments on the cooling of iron cannon balls, Buffon concluded that the Earth was 76,000 years old. This caused no problem at the time because no one had any idea about the relevant time scales. But Joseph Fourier (Fig. 9.10) demonstrated in his famous Memoir of 1824 (see note 14 of this chapter) that most of the heating of the Earth's surface came from irradiation by the Sun, although some residual heat was coming from the inside of the globe. To do this, he first noticed that the diurnal variations of the temperature of the soil were progressively damped with increasing depth; using his new theory of propagation of heat, he could derive from the scale length of this damping the thermal conductivity of the soil. Then, noting that the mean temperature rose with increasing depth, as could be seen in mines or drill-holes, he could evaluate the flux of heat coming from the inside of the Earth. He concluded that it was very small compared with the flux coming from the Sun. Thus, it was nonsense to calculate the age of the Earth as Buffon did, because the cooling of the Earth, although real, was considerably slower.

Arago was clearly convinced by this demonstration, and he was long active collecting measurements of the rate of increase of temperature with depth. This was not very useful, because he could not improve on Fourier's estimates, which were limited by the uncertainties in the thermal conductivity of the soil. Once again, Arago was driven by his love of history and of compilation.

Others had not read Fourier, or had not understood him. For example, in 1844, the geologist Léonce Élie de Beaumont (1798–1874) was still in the intellectual lineage of Buffon; forgetting the heating by the Sun, he revised downwards the age of the Earth to 38,000 years, which caused some problems because the time scale of geological phenomena now looked considerably longer.[26]

There was another way to estimate the cooling time of the Earth that called for astronomy, and here Arago was in his specialty ... although the reasoning was due to Laplace.[27] He noticed that the length of the day, when related to the motion of the Moon, which was expected to be stable, did not vary by more than 1/100 s since the time of Hipparchus (c. 190–c. 120 BC); Laplace even gave less than 1/380 s. Then, because of the conservation of angular momentum, the size of the Earth could not have changed much since that time.[28] Supposing that the thermal expansion coefficient of the Earth was the same as that of glass, Arago found that the temperature of its whole volume had not changed by more than 1/170 °C since Antiquity, or, more realistically because of the large uncertainties in this determination, by more than

---

[26] **Comptes rendus de l'Académie des sciences* 19 (1844) p. 1327–1331.

[27] **Connaissance des temps pour 1823*, p. 245–257 and 324–327.

[28] There was another motivation for Arago: As the unit of length was then defined from the size of the Earth, its contraction would cause a problem, so that another way of defining the meter would be welcome. As written by Tisserand (1893) p. 6: "Arago thought that from a thorough study of the light coming from the Sun or from stars one could find a more constant unit of length, no more related to the Earth but rather to celestial bodies, a *sidereal meter*." This meant that the unit of length had to be related to the wavelength of some atomic transition, or to the velocity of light: actually, this is what has been done in the successive new definitions of the meter of 1960 and 1975.

1/10 °C. This implied a time scale for the cooling of the Earth of many millions years, but Arago did not push the reasoning that far.

As a precursor of paleoclimatology, Arago asked himself about the long-term evolution of the parameters of the orbit of the Earth and its possible influence on the climate. He only considered changes in the eccentricity of this orbit, the existence of which was known, to conclude that their effect was weak. But he reasoned only on a time span of a few millennia. We now know that not only the variation of eccentricity, but also other important changes in the astronomical parameters (precession and variations of the obliquity of the ecliptic), have a great impact in longer periods of time.

In the short term, Arago tried to detect climatic changes through the determination of temperature. He was conscious of the difficulties of such studies due to the fact that the thermometers were only standardized during the first half of the eighteenth century. As a consequence, he tried to derive the temperature from the production of the harvests and from some natural phenomenon like the freezing of rivers. He compiled a large quantity of data that he examined often, but not always with a critical eye. They went from cold or mild winters in various points of the globe to the internal temperatures of different species of animals—a large accumulation of facts from which no serious conclusion could be derived.

Oceanography

Let us remember that there are the maritime successes of England and the existence of its Board of Longitudes that pushed Abbé Grégoire in 1795 to propose, with success, the creation of a French Board of Longitudes. Knowing the close ties of Arago with this Board and his constant interest for things of the sea, whose economical and military issues were so important, we should not be surprised to see the efforts he made to understand the physics of the sea. We shall see later what he did to promote nautical observatories, as well as lighthouses with their Fresnel's lenses. He understood the importance of maritime maps and strongly supported the body of hydrographic engineers, whose number was threatened by diminution after the maps of the French coasts would be completed. In a speech on 5 June 1837 at the Chamber of Deputies, he declared[29]:

M. Chairman of the Commission knows as well as me that the hydrographic works are like the Penelope needle works: they must constantly start over.[…] Amidst the bodies financed by the State, there is one which does much work but makes little noise: that of the hydrographic engineers.[…] There are in France twenty hydrographic engineers. Amongst them we find an engineer-in-chief, M. Beautemps-Baupré, and an adjunct engineer-in-chief, M. Daussy.[…] The engineer-in-chief receives 7,000 francs [per year, equivalent to about $30,000], and his adjunct 5,000 francs. As you can see, they do not cost much.

[29] *Arago, F. *Œuvres complètes*, t. 5, p. 635–644.

In the same address, he put the stamp of shame on the Navy, which showed no interest for scientific observations at sea, and he insisted that tides be measured regularly in many harbors with recording machines, as was done in England. He was successful in this.

The Detection of Reefs

As a practical mind, Arago discovered that polarization might improve the visual detection of reefs at sea. He knew that reflection on a non-metallic surface polarizes light, and that this polarization can be total for a certain direction called the Brewster incidence. For water, this occurs when the direction of sight makes an angle of 37° with the horizontal. As a consequence, when one looks at the surface of the sea through a polarizer, it becomes rather dark. An emerging reef would not be polarized and would then be seen with a good contrast with the sea. It is not known if this method has been much used in practice. However, polarized glasses are now used to attenuate the reflection of light on a road when the sun is in front of the car, so that obstacles like other vehicles are better seen.

Currents

Sailors are familiar with the Gulf Stream, the warm current that starts in the Sargasso Sea and disappears to the north of Norway (Fig. 9.11). Humboldt described another current, cold this time, which runs along the coasts of Chile and Peru, to which his name was given later. Other currents were identified by Portuguese, Spanish, Dutch and English sailors, but they were difficult to locate due to the uncertainties in the longitudes. Better descriptions became possible after the invention of marine chronometers, which allowed a good determination of longitudes. Worldwide maps of the currents were then published posthumously in 1832 by Major James Rennell (1742–1830) and independently in 1838 by Captain Duperrey.

However, Arago was still not satisfied with the situation. Overall, he regretted that the origin of the currents was so poorly understood. Some people thought they were pushed by the winds, and others explained them by a difference of level of the sea, which was supposed to be higher near the equator due to heating by the Sun (Jacques Babinet still used this hypothesis in 1847, in a rather naïve way). Arago, who had considerable knowledge of the different aspects of the problem, did not believe in these explanations, although we know that the winds actually play a role.[30] He proposed the following explanations, which in fact originate from Humboldt:

[30] *Œuvres complètes*, t. 9, p. 52–66.

The theory of currents has made little progress until now because one considered only what occurred at the surface of the seas. But currents created by differences in salinity or in temperature exist at all depths. There are for example currents flowing close to the bottom of the sea that carry to the equator the cold water of the polar zones.[...] [One can apply] to the sea the theory that accounts so satisfactorily of the trade winds.

The analogy with the trade winds shows that Arago understood that the warm currents, which flow towards high latitudes, are compensated by deep and cold currents running in the opposite direction. It is possible that Humboldt also understood this, but we lack proof. It was difficult at that time to directly detect those deep currents, but Arago recommended that sailors measure systematically temperature and salinity at different depths, foreseeing in this way modern oceanography.

Of course, we know much more today. It is clear that the surface currents, which are not deeper than 400 m, are principally driven by the wind (a wind blowing for ten consecutive hours drives the surface water at 2 % of its velocity). Arago did not believe in the effect of wind, as we have seen, but Humboldt had understood it.[31]

Fig. 9.11 The Gulf Stream, as drawn by Benjamin Franklin. Wikimedia Commons, ServiceAT

[31] Humboldt (1816), p. 73, wrote, "Weak winds acting continuously in a large zone like the trade winds cause a translation motion of the sea that we do not observe in the strongest storms, because they act only in a small area."

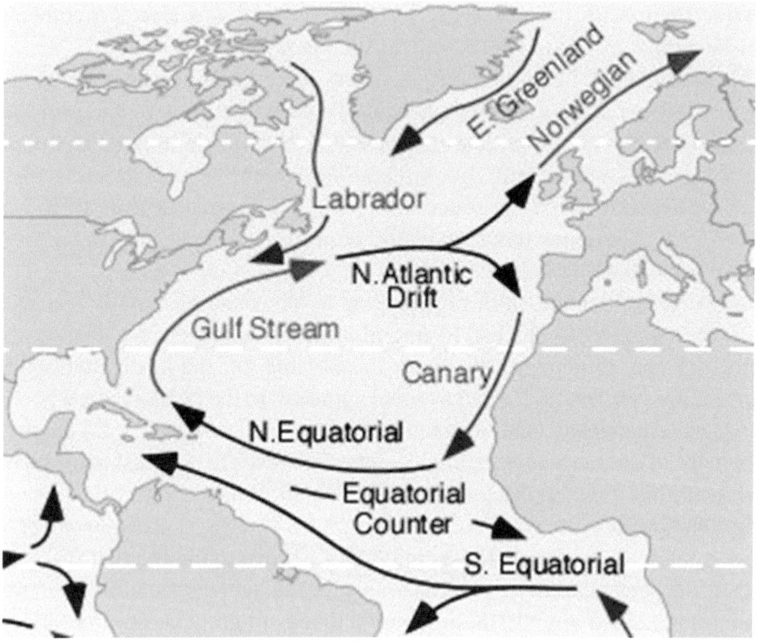

Fig. 9.12 The currents in the Atlantic Ocean. Wikimedia Commons

The Coriolis acceleration deviates the currents to the right in the northern hemisphere, and to the left in the southern one, so that a part of these currents form large circular loops, called *gyres*, in the main oceanic basins. But if a surface current like the Gulf Stream, warm and rich in salt comes to high latitudes, its water cools, with an increase in density, and falls to the oceanic depths. A deep current then takes place in a direction opposite to that of the superficial current, as described by Arago; his comparison with the trade winds is correct. However, the presence of winds along the coasts complicates considerably this simple description: there may be such that deep water rises to the surface, or inversely. The currents in the Atlantic Ocean give an illustration of these phenomena (Fig. 9.12).

A Method for Oceanography

There are many other examples of Arago's interest for the sea. But he was often limited by the lack of data. There were big gaps in the knowledge that could have been filled by systematic observations. However, the time was not ripe for oceanographic ships; the first specialized oceanographic ship, the fourth *Pourquoi-Pas* of Jean-Baptiste Charcot (1867–1936), was only launched in 1908, and one had to be content with observations from ships with other functions.[32] Arago was well placed

[32] The HMS Challenger, an adapted cruiser of the Royal Navy that sailed through the globe from 1870 to 1876, was however the first ship entirely devoted to oceanography.

to provide them with the necessary instructions, and was also a member of the commissions that examined the results of the observations.

In his instructions, he particularly recommended measuring as often as possible the surface temperature (every half hour in the case of currents), and also the temperatures at various depths with *thermométographes*; In addition, he recommended taking samples of water with Biot's *bouteille à renversement* to study dissolved gases. He also considered the reduction and scientific exploitation of these data. He knew from experience that they would require much time and effort, but he did not give a good example himself.

When we consider the work of Arago as a "physicist of the sea," as he called himself, we conclude that he had ideas and intuitions much clearer than most of his predecessors and contemporaries, with the notable exception of Humboldt, who influenced him very much. He had a global approach to the phenomena, a good sense of the measurements and analyses, and qualities of persuasion that did much for the development of oceanography in France. One also notes his interest in problems that are today of great concern, such as the evolution of climate and its relations with the oceans. This was far from the preoccupations of his contemporaries. However, Arago never was involved in theory. It was only after his death that the first mathematical models of the oceanic currents were developed. The general oceanic circulation was only well understood after 1948, principally thanks to the American oceanographer Henry Stommel (1920–1992).

Chapter 10
Applied Physics

The apparatus of Dulong and Arago to measure the vapor pressure of water. Water is contained in a boiler on the *left*, heated at a known temperature by a fire underneath. A safety valve with a weight is on the *top* of the boiler. The pressure is measured with a closed-tube mercury manometer on the *right*. Bibliothèque de l'Observatoire de Paris

J. Lequeux, *François Arago*, Astrophysics and Space Science Library 421,
DOI 10.1007/978-3-319-20723-0_10

The Optical Properties of Gases

The study of the refraction index of gases was Arago's first professional activity, started in 1805 under Biot's direction. Arago was 19. The experiments had an astronomical purpose. It was necessary to accurately know the refraction of light by the atmosphere in order to correct for the position measurements of stars. No one had studied the effect of water vapor contained in the atmosphere; Borda had made a few measurements on this before the Revolution, but could not reach any conclusion. The study needed to be resumed.

The First Measurements

Biot and Arago made their measurements with a hollow prism built by Borda, which is preserved in the collections of the Paris Observatory. They measured with a repeating circle the deviation of light through this prism filled with various gases, at known pressure and temperature. Figure 10.1 explains the principle of this measurement.

The first studied gases were dry air and humid air, whose humidity content was measured with a Saussure hair hygrometer. The refraction indices of several other gases were also measured: oxygen, nitrogen, hydrogen, ammonia, carbon dioxide, "hydrogen with carbon" (methane?), "hydrogen with more carbon" (ethane?). Biot and Arago had previously measured the density of those gases at atmospheric pressure by weighting a glass balloon with known volume, first evacuated and then filled with the gas.

The results were presented by Biot at the Board of Longitudes, then at the First Class of the Institute (the provisional name of the Academy of Sciences) on 26 March 1806. The paper should have been signed by the two men, but when Arago saw the proofs he discovered that his name was absent. He complained to Poisson and Thenard, who found his claim justified. Biot pretended "to have deleted [Arago's] name to comply to the academic uses, that according to him did not allow the name of an Academician to be coupled with that of a non-academician in the collection of Memoirs published by the scientific body." But he finally consented to add the name of Arago. This is a first example of the somewhat stormy relations between the two men. The Memoir was thus published under both names,[1] with a picturesque title: *Memoir on the affinities of bodies with light, and particularly on the refractive forces of various gases.*

One of the results, published solely by Biot in another Memoir, was that the water vapor produced only a negligible change in the refraction index of air. This pleased the astronomers enormously, in particular Delambre, who was enthusiastic about this work. But the conclusions that Biot and Arago derived from their measurements were a real surprise for their contemporaries, as we will see.

---

[1] Biot, J.-B. & Arago, F. (1806) *Mémoires de l'Institut national des sciences et arts—Mémoires de mathématiques et de physique* 7, p. 39–66.

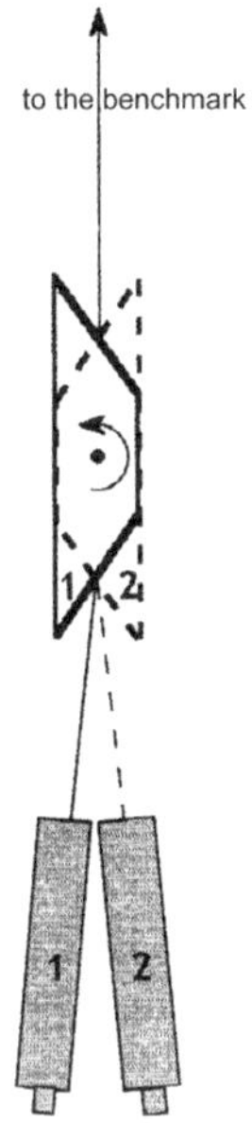

Fig. 10.1 The measurement of the refraction index of gases with Borda's hollow prism, filled with the gas under study at known pressure and temperature. A distant benchmark is sighted through the prism by the telescope of the repeating circle in *position 1*. Then the prism is turned by 180° and the telescope points to the benchmark in *position 2*. The difference of the readings gives twice the deviation by the gas in contained in the prism

The Refractive Power: A Stone in Newton's Backyard

Everyone at this time, with the exception of Thomas Young, was convinced of the validity of Newton's corpuscular theory of light. We saw in Chap. 3 that refraction was explained in this theory by an increase of velocity of the light corpuscles when they penetrated into a denser medium. One supposed (in modern terms) that a kinetic energy was added to that of the corpuscles, which depended only on the nature of the refracting body. This led the physicists to consider the increase of the square of the velocity of light corpuscles when they penetrated this material. It can be shown that in Newton's theory this quantity was proportional to n^2-1, n being the refraction index.

One also believed, logically but arbitrarily, that the increase of the square of the velocity of light was proportional to the specific mass ρ of the material, i.e. its mass per unit volume. Thus the quantity $(n^2-1)/\rho$, which the scientists called the *refractive power*, was supposed to be the same in all materials. Biot and Arago were satisfied to see that the refractive power was indeed constant when they increased the pressure of a given gas. But they were very surprised to discover that hydrogen had a refractive power 6.5 times greater than air. They noticed that the refractive power of a mixture of gases was the weighted sum of their individual powers, but this did not work for water vapor. Its refractive power was not the weighted power of its constituents, oxygen and hydrogen; they measured 1.72 instead of the expected value of 1.54. They were so perturbed by this result that they doubted their measurements, and even about the well-established proportions in which hydrogen and oxygen combine to make water.

Such was the motivation of new measurements made by Arago in 1815 with his brother-in-law Alexis Petit. They made a large number of measurements on many new gases.[2] These measurements remained unpublished until Barral reduced them in 1852 at Arago's request. Arago and Petit also made measurements of the refraction by liquid water, liquid carbon disulfide and other liquids, to conclude that their refractive powers were different from each other, and also different from the refractive powers of their gaseous forms. It was then necessary to conclude that, in the Newtonian hypothesis, "the attraction of a body for light is not proportional to its density."

Arago and Petit also made measurements of the dispersion of light by liquid and gaseous carbon disulfide, with similar unexpected results. They were forced to conclude that "these various suppositions perhaps diminished the simplicity and the probability of the Newtonian theory." It was a serious blow against this theory, which was soon to be replaced by the wave theory of Fresnel, as we have seen in Chap. 3.

Arago's experiments with the hollow prism were continued by Pierre-Louis Dulong (1785–1838) in 1825. He could only confirm the preceding results.

Interferential Measurements

In Chap. 3, we have seen that Arago and Fresnel inserted a glass plate in one of the beams of an interference experiment and saw a displacement of the fringes, due to the increase of the optical path in this beam. They understood from this experiment, although not immediately, that contrary to the expectations of Newton's theory light propagated more slowly in glass than in air. Arago realized that interferences could be used to compare much more precisely the velocity of light in dry and in wet air. For this, he had an apparatus built, probably by Jean-Baptiste François Soleil (1798–1878), consisting of two copper tubes, one meter long each, closed at their ends by optical-quality glass plates, into which different gases could be inserted. He produced two parallel light beams originating from the same source, each one running inside one of these tubes. Then he reunited them at the exit of the tubes and observed the interference fringes so produced. If the refraction index changed in the gas of one of the tubes, the fringes were displaced because of the change in the corresponding optical path. Fresnel and Arago made these measurements from 1816 to 1818, during Fresnel's stay at the Observatory. They introduced wet air in both tubes. Then they dried the air in one of them, and saw a small displacement of the fringes while the air became drier and drier; the full displacement was only by 1¼ fringe. They concluded that dry air has a refraction index slightly larger than wet air at the same pressure, but that the difference was so small that the astronomers could

[2] A summary of these measurements can be found in the first article published in **Annales de Chimie et de Physique* (1816) 1, p. 1–9.

well neglect the effect when they calculate atmospheric refraction. This confirmed with better precision Biot's results of 1807.

Arago and Fresnel's skill in these experiments was remarkable, and their contemporaries appreciated it. Arago noted with pride the opinion of Laplace, who spoke of the results to the journalists and wrote in 1824, in the fifth edition of his *Système du monde*: "M. Arago, using a means as accurate as ingenious, showed that the influence of the humidity of air on its refraction is negligible."

Much later, in 1850, Arago attempted to make interferential measurements of the variation with temperature and pressure of the refraction index of liquid water and described the necessary modifications of the equipment,[3] but these measurements were never performed. Also, he asked Soleil and Duboscq to construct a new interference apparatus to see if fog affects the refraction index of air. Laugier performed the experiments, and concluded that the index is slightly affected, but not enough to cause any problem to astronomical or geodesic measurements.

Finally, Arago, certainly impressed by the experimental skill of Fizeau, asked him to make new measurements on the refraction index of dry and wet air. Fizeau used a simple interferometric device, similar to Arago's in 1815, but compensated the displacement of fringes by inserting an inclined glass plate in one of the optical paths in order to improve sensitivity. The results of all the experiments of Arago and Fizeau are described in a long Memoir.[4]

Arago and the Lighthouses

A Short History of Lighthouses

Lighthouses have existed since Antiquity. In France, the first was erected by the Romans at Boulogne-sur-mer on the English Channel in 40 AD. There were six lighthouses on the coasts of France at the end of the seventeenth century, and 15 in 1775. A wood or charcoal fire was lit on the upper platform during the night, but not for the whole night because of the large quantity of fuel needed (typically 700 kg of wood per night, which had to be carried on a man's back). The fires were ignited at the approach of a ship when it was foreseen. The fishermen often had a fire made at the end of the pier in order to find the entrance of the harbor. Things were better after 1770, the date when a private company, Tourville-Sangrain, obtained a contract from the government; it installed the first oil lamp with a concave reflector in Sète, then at Pointe Saint-Matthieu near Brest and on the Planier lighthouse near Marseilles. The latter could be seen from 28 km when the weather was clear. The lighthouses now worked all night long and were maintained by two keepers.

---

[3] *Arago, F. *Œuvres complètes* t. 10, p. 298–311.

[4] *Arago, F. *Œuvres complètes* t. 10, p. 312–333: this is the last Memoir dictated by Arago before his death.

To increase the power of the lighthouses, several lamps with reflectors were often installed in parallel. For example, the Cordouan lighthouse at the estuary of the Gironde River had no less than 84 oil lamps in parallel, fed with whale oil. But the result was disappointing until the lamps were replaced by more powerful ones with hollow circular wicks, invented in 1784 by the Swiss engineer Aimé Argand (1750–1803). Some lamp systems could rotate around a vertical axis, thanks to a clockwork, to produce a blinking light; the first one was inaugurated in Dieppe in 1787. The first serious apparatus of this kind, built by Étienne Lenoir (1744–1832) on a design by Borda, was installed in 1791 in the new Cordouan lighthouse (Fig. 10.2). It consisted of 12 large parabolic reflectors, 81 cm in diameter, and was certainly the most powerful in the world.

During the Revolution, the lighthouses were nationalized, while the responsibility of the previous private company remained. Things deteriorated, especially because of bad relations with the Ministry of Navy, so that the lighthouses were placed under the responsibility of the Ministry of Interior, with the exception of those of military harbors. A *Commission permanente des phares* (Permanent Commission of Lighthouses) was created in 1811, with representatives of the Institute, the Navy and the Administration of Bridges and Roads.

Arago and Fresnel Intervene

The Commission of Lighthouses included Malus at its creation, but he died in 1812 and was replaced by Arago the following year. There was little money to build and maintain the lighthouses, and no new ones were erected during the Empire, i.e. before 1815. The Commission however made some tests, for example that of a large, 1-m reflector between Montmartre and Montlhéry, "in the presence of several members of the Institute and of many distinguished engineers." But complaints of sailors about the poor efficiency and bad condition of lighthouses piled up on the desk of the Director of Public Works, who was responsible for them. At his arrival in 1817, the new Director, a M. Becquey, worried about the situation and decided to revive the work of the Commission. Arago wrote[5]:

I was charged by my colleagues to preside to the experiments made by the Commission; but my many other occupations did not allow me to do this work with all the desirable strength, so that I asked for a collaborator at the beginning of 1819. I consider as one of the most happy events in my life that I remembered of an almost unknown engineer, who was to illustrate France with the most remarkable discoveries: he was Fresnel, and I obtained that he was named as the secretary of the Commission.[6]

[5] *Arago, F. *Œuvres complètes* t. 6, p. 1–56, see p. 7.

[6] In this late text, Arago remembered only Fresnel. However, he had also asked that Mathieu be named as his adjunct for a new series of experiments.

Fig. 10.2 The Cordouan lighthouse, present state. A first lighthouse was installed here as early as 1360, and replaced in 1611 by a new one. The decorated lower part dates from this time. The upper part is from 1790. This beautiful building, which has always been a test bench for innovations, is classed as a historical monument. Wikimedia Commons, Thibault Grouas

This famous scientist first imagined to use large lenses for the lighting of our coasts and to build them from small pieces [...]. This was done as soon as it was conceived. Numerous and delicate experiments also led to the construction of a lamp with several concentric wicks, which was twenty times more powerful than the ordinary ones with a double air flow. By combining those lamps with the Fresnel lenses, we reached astonishing results.

Let me say a few words about these oil lamps, whose power Arago and Fresnel together succeeded in increasing so well. Following an idea of Benjamin Rumford that was gathering dust, they made lamps with two, three, then four concentric wicks.[7] They were fed with oil under pressure according to a system invented by Bertrand-Guillaume Carcel (1750–1812) for his famous lamps. Captain Henry Kater, who had already collaborated with Arago and would again collaborate with him for the geodesic connection between Paris and Greenwich (see Chap. 6), attended the first test of the system in October 1819. The lamps, the largest of which used 750 g of oil per hour, worked very well and were progressively placed on all the lighthouses.

The Fresnel Lenses

Fresnel considered that converging lenses were more adapted for the lighthouses than mirrors; silvered mirrors were difficult to make, tarnished too easily and were cumbersome. However, the lenses must have a short focal length so that the optics keeps a reasonable size even with a large lens diameter. A single lens with these characteristics would be very thick, hence very heavy, would absorb light and would present a large chromatic aberration; hence, the idea of the multi-part lens made of thin concentric zones that all have the same focus, where the lamp is located (Fig. 10.3). This idea was in fact attributed to Buffon, but he imagined making the lens from a single piece of glass, which was then completely impossible. Condorcet, in his eulogy of Buffon of 1788, proposed making these lenses from separate pieces of glass. But nothing happened until 1820. One does not know if Fresnel knew about these precursors, but in any case he affirmed[8]:

This application of the lenses to the lighthouses could not be a new idea, as it came easily to one's mind.[...] What was my major preoccupation was their realization, for which I have been helped to perfection by the zeal and intelligence of M. Soleil, an optician who courageously undertook the construction of these big lenses.

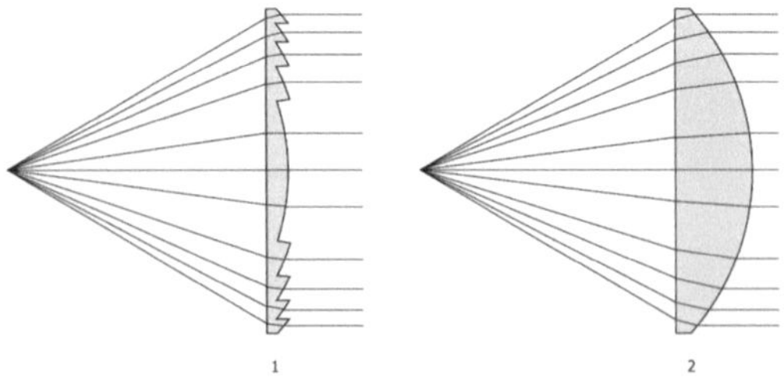

Fig. 10.3 The Fresnel lens (**1**) compared to a thick lens with the same focal length (**2**)

[7] See in image on http://www.cordouan.culture.fr/accessible/fr/uc/04_02_04-La%20première%20 optique%20de%20Cordouan.

[8] Fresnel (1866, 1868, 1870). Tome 3 is entirely devoted to the lighthouses, but is not easily accessible. It contains the fundamental paper *Fresnel (1822), see p. 2.

The first lens built in this way was tested on 12 May 1820 by the *Commission des phares*, whose members found it much better than the best reflectors of the time. It was made of many curved prisms glued together (one such lens survives at the Paris Observatory). Soon after, Soleil succeeded in producing annular prisms covering a part of the circumference of a circle. A system consisting of eight lenses constructed in this way was tested in August 1822 on the top of the Arc de Triomphe at the Place de l'Étoile in Paris, then under construction. Four members of the Commission went to a distance of 32 km from Paris and observed that the flashes of light were "very bright." In July 1823, this apparatus was installed on the top of the Cordouan lighthouse, replacing that of Borda and Lenoir. The comments of sailors, and of Britons who were attending the sea baths in Royan, were enthusiastic. The cost of the device was estimated as 28,000 francs (equivalent to some $125,000), including 13,800 francs for the optics. This magnificent instrument is preserved in the Museum of Lighthouses on the island of Ouessant, off the coast of Brittany.[9]

After this success, the Commission, whose the *de facto* chief was now Fresnel, decided to build as many lighthouses as possible with this new optics, which would take various forms according to the needs. Fresnel and his colleague Paul Édouard de Rossel (1765–1829) produced in 1825 an ambitious and well conceived program with 28 first-order lighthouses with a range of 60 km, five second-order ones with a 40-km range and 18 third-order lighthouses with a range of 28 km. This project, still improved with ten more lighthouses, was almost entirely realized in 1850; there were then 58 lighthouses on the coasts of France, requiring an annual budget of about 500,000 francs ($2.2 million). This number more than doubled during the following years.

Of course, Soleil could not supply the optics for all these lighthouses. Other makers had to intervene: Henri Lepaute (1800–1885), François jeune (?–?), and Louis Sautter (1825–1912, Fig. 10.4).

Fresnel's invention prompted great interest in foreign countries. In 1850, there were 126 lighthouses on the coasts of England, Scotland and Ireland, and no less than 138 on the coasts of the United States. The vast majority were equipped with Fresnel lenses. The improvement in the safety of navigation was outstanding. Arago noted that there were on average 163 wrecks per year on the French coasts from 1816 to 1823, a number reduced to 39 per year from 1824 to 1831, although only about ten lighthouses had Fresnel lenses by that time. The invention saved a large number of men and ships. As Arago wrote in an article of 1828, in which he protested again against Brewster, who claimed to be the first to have had the idea of making lenses with several pieces of glass [10]:

In writing this long article,[…] I wanted above all to prove that the friend that a cruel death separated from me so fast did not only attach his name to the most subtle and ingenious discoveries of optics, but also deserved through his invention of lenticular lighthouses the gratitude of sailors of all nations.

---

[9] See in image on http://www.cordouan.culture.fr/accessible/fr/uc/04_02_04-La%20première%20optique%20de%20Cordouan.

[10] **Annales de Chimie et de Physique* 37 (1828) p. 392–409. Brewster was terribly afraid of the possibility of being drowned in a wreck, hence his interest in lighthouses.

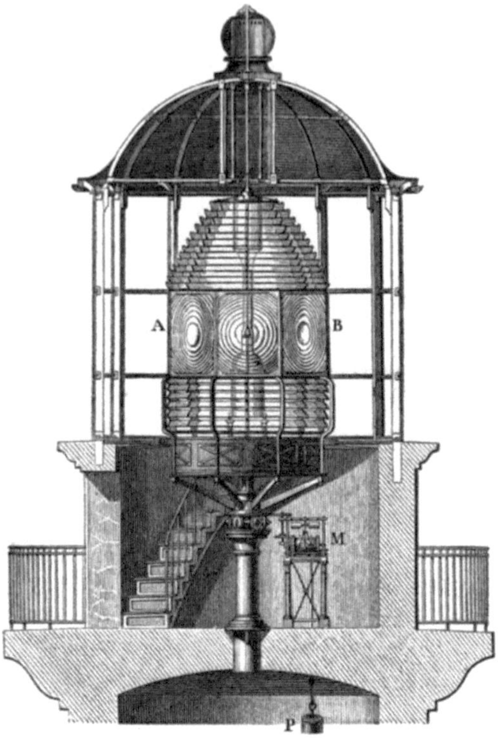

Fig. 10.4 A lighthouse presented by Sautter at the Universal Exhibition of 1855. The mirrors above and under the lenses used in the Cordouan lighthouse optics have been replaced by circular total-reflection prisms. Note the clockwork M for the rotation of the lighthouse. Wikimedia Commons, Chetvorno

The Velocity of Sound

Marin Mersenne (1588–1648) and Pierre Gassendi (1592–1655) appear to have been the first to measure the velocity of sound in air, during the first half of the seventeenth century. Other measurements were made later in France, England and Italy, with contradictory results. As a consequence, in 1738, the Academy of Sciences decided to make new determinations between various places around Paris, whose location was well known by triangulation.[11] A gun was fired at one of the sites, and one noted at the other site the time between the light and the noise. Some of these measurements were reciprocal, i.e. there was another gun at the second site and the record was made at the first site. This allowed freeing the determination from the effect of the wind. The result was quite good: 333 m/s at 0 °C versus the present value of 331.5 m/s.

What were the reasons that drew the Board of Longitudes to decide to make new measurements in 1822? Officially, it was the alleged poor quality of the 1738 measurements, and also the similarly alleged lack of "atmospheric circumstances."

[11] **Mémoires de l'Académie royale des sciences* for 1738, p. 128–146.

While reporting new measurements made in Chile, Arago gave the true reason[12]: it was to check the new formula given by "an illustrious geometer" (Laplace) in 1816, which allowed for calculating the velocity of sound.[13] Whatever the reason, under Laplace's proposal, the Board ordered Humboldt, Gay-Lussac and Bouvard to go to Montlhéry, and Arago, Mathieu and Prony to Villejuif to make the new measurements. Therefore, some of the best physicists of the time were in charge of the operation. Laplace was 73, perhaps too old to attend the experiment. There was a gun at each place; Laplace's son, who was an artillery officer, was in charge of the guns. The measurements took place during the nights of 21 and 22 June 1822. To measure the delay, one had chronometers from Breguet and a recording chronograph, a novelty built by another clock maker, Nicolas-Mathieu Rieussec (1781–1866). The distance between the two locations was already known from triangulation, but to be sure this triangulation was completed by measurements at the Pantheon, at the Observatory, in Montmartre and at the Invalides, with the best available instruments.

The results can be found in several publications.[14] They correspond to 331.2 m/s at 0 °C, closer to the present value of 331.5 m/s than the preceding result. One cannot say that the new measurement was a real check of Laplace's formula, because the parameters that take place in this formula were poorly known at that time. Furthermore, Laplace attempted to correct the result by taking humidity into account, and came to a figure that was too large[15]!

The "Elastic Force" of Water Vapor

Steam engines caused many serious accidents from the start, and we should not be surprised that this was a cause of concern for the Academy of Sciences. At the beginning of 1823, the Academy created a commission in charge of improving the safety of the machines. This commission made several reports; one of them contains a "Table of the elastic force of water vapor at different temperatures, up to 8 atm," i.e. the saturation vapor pressure as encountered in the boiler of a steam engine.[16] It asked for further measurements at higher temperatures and pressures. The Academy charged a new commission composed of Arago, Ampère, Dulong, Pierre-Simon

[12] **Annales de Chimie et de Physique* 7 (1817) p. 93–98, see p. 98.

[13] The Laplace formula that gives the velocity of sound v is $v=(C_p/C_v)^{1/2}\,(p/\rho)^{1/2}$, where C_p and C_v are respectively the specific heats of air at constant pressure and at constant volume, ($C_p/C_v = 1.402$), p the pressure and ρ the specific mass of air, which varies as the inverse of the absolute temperature T. The velocity of sound is thus proportional to the square root of T.

[14] For example in **Connaissance des temps* for 1825, p. 361–371. Note that the distances were still expressed in *toises*, while the official unit was the meter.

[15] **Annales de Chimie et de Physique* 20 (1822) p. 266–268. Laplace uses the meter as the unit of length.

[16] **Annales de Chimie et de Physique* 27 (1824) p. 95–101.

Fig. 10.5 Pierre-Louis Dulong. Wikimedia commons, Materialscientist

Girard (1765–1836) and Prony to make these measurements. Only Arago and Dulong would effectively work at the project, from 1825 to 1829. Dulong (Fig. 10.5) was familiar with dangerous experiments, and had lost an eye in one of them.

The measurements were indeed difficult and dangerous. Two steps were necessary: first, to build and calibrate a manometer able to measure high pressures, and then to measure the temperature and the pressure in a boiler. The manometer was a closed vertical tube containing air at the upper end and mercury below. It communicated with another vertical tube, long and open, which contained also mercury. When pressure was applied to the vessel containing the mercury, it raised on each side. The height reached in the long open tube allowed to estimate the pressure and to calibrate the height of the mercury in the manometer as a function of this pressure. The open tube was 26 m high and the mercury ascended to 20.5 m at the maximum pressure of 27 bars. The open tube was so tall that it had to be attached along a high tower, that of the Royal College Henri IV near the Pantheon, at the top of Montagne Sainte-Geneviève in the Latin quarter of Paris.

Once the manometer was calibrated, it was connected to the boiler as shown on the figure at the beginning of this chapter, and the pressure was read as a function of the boiler temperature. There was a considerable risk of explosion, given the very high pressures reached in the experiment, so that the administration of the College Henri IV feared an accident and required that the apparatus be transferred to some other place; it was moved to the Paris Observatory, along the northern part of the building. During the experiment, the boiler leaked so much that the pressure had to be limited to 24 bars. The apparatus was so terrifying that the readings on the manometer were done from a distance with a telescope. But the purpose was reached: the measurements were excellent, and were published in many places.[17] Arago and Dulong made a somewhat dubious extrapolation up to pressures of 50 bars, fortunately without interest in practice because no boiler could reach such high pressures. Now, it became possible for the first time to know the pressure in a boiler from its temperature, which was easy to measure: this was reassuring.

We have now completed the description of Arago's contributions to direct applications of physics, with a few unimportant exceptions.

[17] For example in **Annuaire du Bureau des longitudes* for 1830, p. 241–242, in the same issue as a Notice by Arago on the danger of steam engines and the ways of avoiding them. A full description of the experiment, with drawings, is in *Arago, F. *Œuvres complètes* 11, p. 13–54.

Chapter 11
The Promotion of Science and Technique

The Popularization of Science

Arago was one of the best popularizers of all time. In an epoch when there was none of the present audio-visual facilities, only conferences, speeches and articles could make science and technique known. Arago used all of them profusely, through his lectures on astronomy, discourses at the Academy of Sciences and at the Chamber of Deputies, obituaries of great scientists that often served as pretexts to explain some scientific topic and to promote science, and finally scientific notices in the *Annuaire du Bureau des Longitudes* and elsewhere. Let us give some details.

The Public Lectures in Astronomy and the Astronomie Populaire

We have seen in Chap. 2 that, on a proposition by Laplace, the Board of Longitudes created a public course of astronomy in 1812. It was committed to an enthusiastic Arago; he acknowledged that he had "a mania for giving lessons." His pedagogical qualities and his eloquence attracted a large public, much larger than expected. The course was first given in the enormous meridian room of the Observatory, whose acoustics are very bad, then in *Collège de France* and, finally, after February 1841 in the specially constructed amphitheater of the Observatory. Arago gave his lectures for 34 years until 1846, when he was forced to stop because of his many other activities and his poor health. A large fresco in the Sorbonne is a testimony of his fame as a teacher (Fig. 11.1).

After 1835, some auditors, anonymous or not, published lecture notes without Arago's permission. Some were translated and published in England and in the United States. There were many errors in these books, so that Arago decided that he

J. Lequeux, *François Arago*, Astrophysics and Space Science Library 421,
DOI 10.1007/978-3-319-20723-0_11

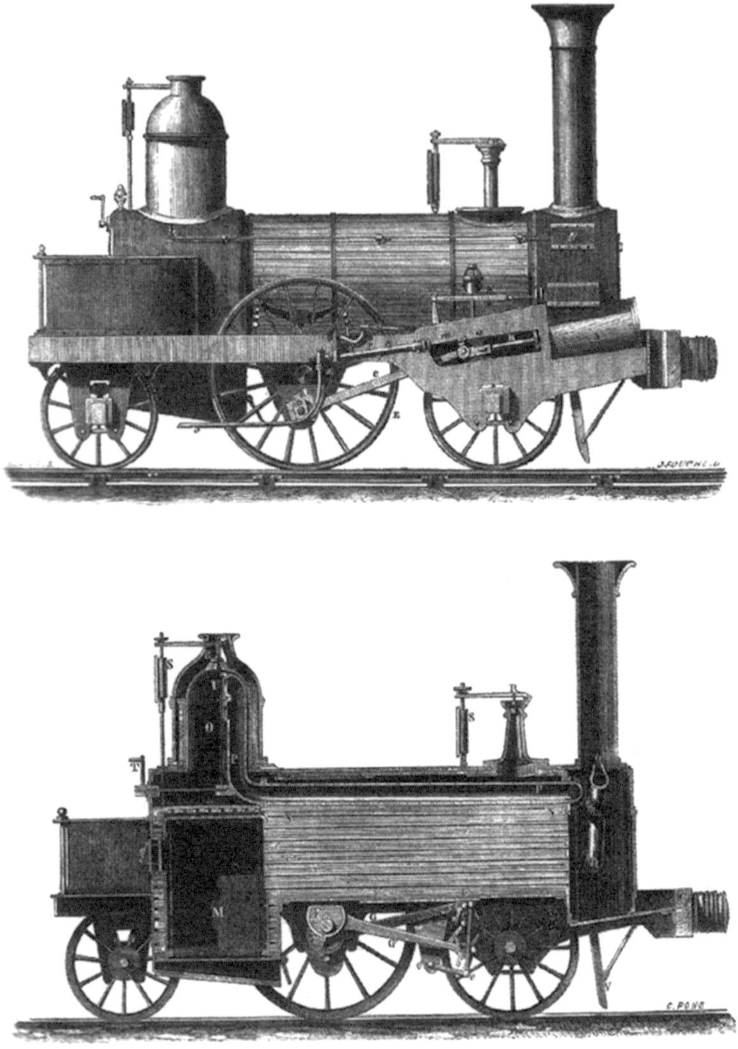

The Buddicom steam locomotive used on the Paris-Rouen line from 1843, external view and cut. From Delaunay (1850), author's collection

would himself write a book from his lecture notes and other documents; it was to be his *Astronomie Populaire*. This decision occurred rather late in the 1840s, and because Arago was almost blind he had to dictate the contents of the book. To whom? A legend suggests that it was his niece Lucie Laugier, but this was not confirmed by Barral, the editor of Arago's *Œuvres complètes*, who claimed that most of the dictation

Fig. 11.1 Arago giving his course of popular astronomy in the amphitheater of the Observatory. Note in the background the statue of Cassini I. Fresco by Theobald Chartran (1849–1907), ca. 1885, in the peristyle of the academic palace of the Sorbonne, in Paris. Rectorat de l'Académie de Paris, photo Gilles Vignier

was to Jean-Jacques-Émile Goujon (1823–1856), one of the *élèves-astronomes* of the Observatory. Barral wrote in his introduction to the *Œuvres complètes*[1]:

Prepared since a long time, corrected many times, all the chapters of the Astronomie Populaire *were revised and completed by M. Arago during the three last years of his life. None of the many figures of the pages dictated to M. Goujon was drawn. A singular and remarkable thing is that M. Arago, almost blind, was drawing in his imagination the most complicated figures. He called the lines by letters, as we do usually, and dictated as if the figures really existed, without any error. It is thus from a text mixed with letters representing straight or curved lines that I had to reconstitute the figures of the* Astronomie Populaire.

[1] *Arago, F. *Œuvres complètes*, t. 13 (Tables), p. VII–CCLXXIII, see p. XVI.

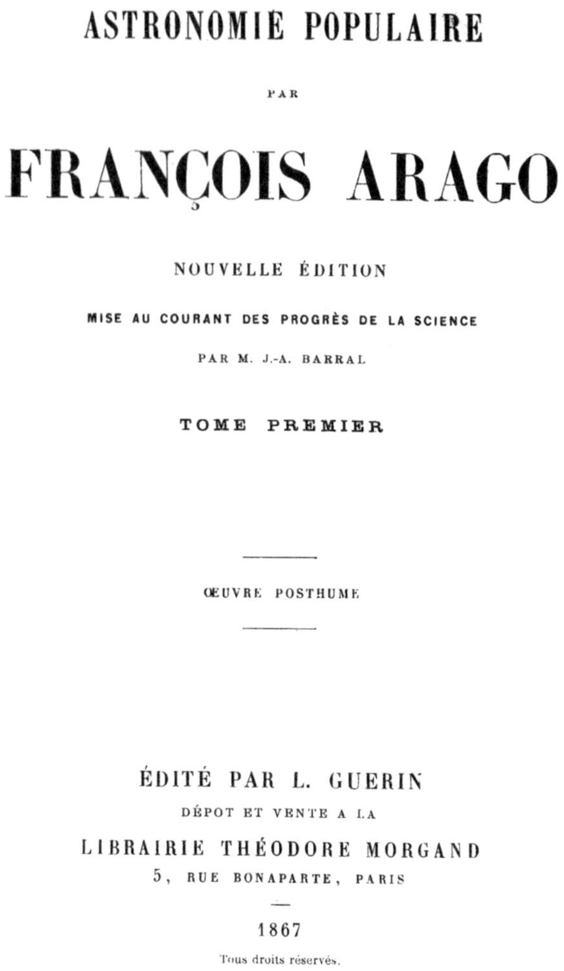
ASTRONOMIE POPULAIRE

PAR

FRANÇOIS ARAGO

NOUVELLE ÉDITION

MISE AU COURANT DES PROGRÈS DE LA SCIENCE

PAR M. J.-A. BARRAL

TOME PREMIER

ŒUVRE POSTHUME

ÉDITÉ PAR L. GUERIN

DÉPOT ET VENTE A LA

LIBRAIRIE THÉODORE MORGAND

5, RUE BONAPARTE, PARIS

1867

Tous droits réservés.

Fig. 11.2 Title page of the second edition of Arago's Astronomie populaire. Author's collection

Barral was probably right. We have a hard time imagining that Lucie, who had no scientific training, would be able to collect such technical texts and figures. In any case, Arago died before bringing his book to a close; it was completed and published by Barral between 1854 and 1857, and there were several further editions (Fig. 11.2).

The physicist Jules-Célestin Jamin, who knew Arago very well, wrote that "*Astronomie populaire* […] [was] only a pale attempt at a reconstruction, without the heat and life that Arago lavished around him." This judgment may look harsh, but it is true that the term "popular," although cherished by Arago, does not do justice to the content of the book, which is a real treatise of astronomy (without the celestial mechanics), although without any equations. Although slightly outdated on some points at its publication, *Astronomie populaire* remained an excellent introduction to astronomy for years. It was better than the 1880 book by Flammarion with the same title, which contains some rather whimsical topics.

Discourses and Interventions at the Chamber of Deputies

During the time he was a deputy, from 1831 to 1848, Arago delivered no less than 66 speeches at the Chamber, some quite long, as well as a large number of reports and interventions on a wide variety of subjects. He was a tall and impressive orator, with a very strong voice probably affected by a Catalan accent; the contemporaries said that he was able to impress the deputies even if they were not on his side. Of course, he seized these opportunities to be a promoter of science and technique, and also to make known or to defend the scientists and inventors he esteemed. Sometimes, he seemed to forget that he was not at the Academy and bored the deputies with technical matters that they could not understand. This was only a minor defect; most often, his speeches were well written and well organized, even if their inflated style, which was common in that time, can make us smile today.

Most of these discourses have been preserved in Arago's *Œuvres complètes*. A large number of speeches were concerned with steam engines and railways, which is not surprising because they were in full development at the time. In each of these speeches, Arago gave a history of the subject in question, always well documented, and a clear and precise account of the latest developments. A good example is given by his report on the electric telegraph.[2] On 29 April 1845, a vote concerned a subvention of 240,000 francs (about $1 million) for tests of electric telegraphs. Arago, who was chairman of the relevant committee, gave a history of the telegraph, then described a first test in which a magnetic needle was affected by an electric signal sent from Paris to Mantes-la-Jolie at a distance of 57 km (about 35 miles), first through two wires then by a single wire with return through the ground. He said that he would arrange another test between Paris and Rouen the next Sunday; finally, he described the telegraph systems of Breguet and of Morse (see Chap. 7). How could one resist a man who knew so well his documents and who was so convincing? The budget was approved.

Arago strongly defended the builders of scientific instruments. He tried to increase their protection by improving the legislation about patents,[3] of which he was a good specialist, and never missed an opportunity to give work to these builders. For example, he asked the Chamber to add 30,000 francs to the budget of the Navy for purchasing marine chronometers and reflection circles (sextants or others) for navigation.[4] In order to convince the representatives, he recalled that the ships that had only compasses were shipwrecked frequently because their position was uncertain, in particular due to the poor knowledge of magnetic declination.[5] He also presented a history of the prizes awarded for the determination of longitudes, to conclude that England had been more generous than France, and recalled that many more chronometers and sextants were built in England than in France, and that a

[2] *Arago, F. *Œuvres complètes* t. 5 p. 467–489.

[3] *Arago, F. *Œuvres complètes* t. 6 p. 677–698.

[4] *Arago, F. *Œuvres complètes* t. 5 p. 658–671.

[5] One may see here a justification of the very numerous measurements of magnetic declination that Arago made and collected during his entire career (see Chap. 9).

number of French ships did not even have a chronometer. Again, the Chamber could not resist Arago's eloquence and arguments.

Arago attempted, as expected, to convince the Chamber of his economic and social ideas. He had competence in economics because he taught this matter as "social arithmetic" at the École Polytechnique; because he prepared his speeches well, he had barely any opponents in this matter. He was clearly an adept of a liberal economy, but with social preoccupations. As a supporter of free trade, he was in principle opposed to the custom rights, except for the "big machines," because the iron and coal necessary for their construction were more expensive in France than in England.[6] However, his opinions changed somewhat, as we will see later when discussing the steam engines and the railways.

At the Academy of Sciences

Arago was at home at the Academy of Sciences, and his position as Perpetual Secretary after 1835 made him particularly powerful there. This was the best agora to convey his ideas. It would be fastidious to enumerate his many speeches and reports at the Academy, but it is of interest to discuss his obituaries of deceased colleagues.[7] Arago paid a great attention to these obituaries, which are still of interest.

In his first biography, the 1830 eulogy of Fresnel, who died 3 years earlier, Arago warned his colleagues as follows:

This eulogy [...] is not in the usual form. I ask you to consider it as an ordinary scientific Memoir in which, on the occasion of the works of our fellow, I examine the progress that several of the most important branches of optics made recently. At a time in which the lectures given at the Collège de France, *at the University or at the* Jardin du Roi *[the Museum of Natural History] attract so many people, it seems to me that the Academy of Sciences could also inform directly the friends of science who want to attend our meetings about some of the various questions that we would examine especially.*

One can see here the desire of Arago to explain scientific knowledge to the general public.

The next obituaries read by Arago were those of Volta, Joseph Fourier and Watt, 1 or 2 years after their deaths. Then, the situation became anarchical. The obituaries written by Arago were successively those of Lazare Carnot (1753–1823), Ampère, Condorcet, Bailly, Monge, Poisson, Gay-Lussac and Malus, not at all in the order of their respective deaths. Some of these eulogies could not be read before, often for obvious political reasons. That of Malus, deceased in 1812, was only written for the 1854 public meeting of the Academy, "by a special decision of this scientific body", and read on 8 January 1855. Arago had died more than a year earlier.

---

[6] *Arago, F. *Œuvres complètes* t. 6 p. 666–676.

[7] All the obituaries written by Arago can be found in *Arago, F. *Œuvres complètes* t. 1, 2 and 3.

Arago had the power to choose the men to be celebrated. His choices were not neutral; they reflected his political opinions, first democratic and later republican. This was for him the opportunity to make remarks about the past or present political regimes, which are not always straightforward for us but were transparent at the time. We have seen that this was the case in the biographies of Fresnel and Condorcet. In Lazare Carnot's eulogy, Arago's hatred for Napoléon, "who had seized the Republic for his own profit," is expressed in strong words.

Arago realized that journalists were the best possible mediators of science, because they reached a much larger audience than the orators of the public lectures. When he became Perpetual Secretary, he allowed them to attend the meetings of the Academy, and even to consult the documents discussed during these meetings. This was not to the taste of all his colleagues, so the privilege for journalists was canceled after his death.

The Scientific Notices

When Delambre became Editor-in-Chief of the *Connaissance des temps* in 1802, he made important changes in this annual journal. Before, it contained only tables and a "history of astronomy," i.e. a summary of the astronomical research of the preceding year. Delambre introduced critical reflections on the present state of science, detailed analyses of books and a few scientific articles. Arago wrote one on the repeating circles.[8] Some papers were translations from English. There were few changes after Delambre's death in 1822, but a new heading appeared in 1827:

The Board of Longitudes decided that every year the editors will publish in the Connaissance des temps, *under the title of* Notices scientifiques, *an abridged analysis of the astronomical, geographical and hydrographic work that it is aware of.*

Arago was put in charge of these notices. The first two were published in 1827 on problems of navigation. The next year, one finds a notice "On multiple stars," relating the old discovery by Herschel of binary stars. This had little to do with the preoccupations of the Board, and this is probably the reason why it decided that all the scientific notices would henceforth be published in its *Annuary*. There would be very few exceptions to this rule.

Arago was certainly very pleased with this decision, because the *Annuaire du Bureau des Longitudes* was a small-format, inexpensive book with a much larger readership than the *Connaissance des temps*. He had already written many articles there, so that it is difficult to understand why the Board had decided to publish scientific notices in the *Connaissance des temps*. This does not matter, because in any case these notices symbolized the best of Arago's popularization of science. They were the main reason for the considerable success of the *Annuaire*.[9]

[8] **Connaissance des temps* for 1816, p. 345–358.

[9] All Arago's notices, classed by subject, can be found in *Arago, F. *Œuvres complètes* t. 4–8.

The scientific notices reflect Arago's interests in all their variety; they went from astronomy, navigation, geophysics, oceanography and meteorology to different points of physics (optics, electric battery) and of technique (steam engines, drilling of wells, chalk and cements). One can also find extracts from biographies or obituaries presented at the Academy as well as a remarkable "Historical and critical analysis of the life and works of Sir William Herschel" (1842), where Arago exposed his own new ideas on stars. There was no notice in the *Annuary* for 1847. Note the coincidence with the end of Arago's public lectures, due to excess of work and illness. There were again some notices after the 1848 Revolution, but they were no longer concerned with timely subjects, but were parts of the material gathered by Arago for his *Astronomy populaire*. One also finds in the *Annuaire* the biographies of Monge and Bailly, written long before but unpublished. That of Bailly, published in 1853, the year of Arago's death,[10] contains a moving footnote on the first page:

Read in the public meeting of the Academy of Sciences on 26 February 1844. This biography has remained unpublished up to now, for good or bad reasons that it would not be of interest to give here. I have it printed today without changing a single word. I apologize for the misprints and perhaps even for grammatical faults that could have slipped into the following 282 pages. The public, which has always honored me with its benevolence, will have the kindness to remark that, being now almost blind, I could not revise the proofs myself.

I will now concentrate on a few topics exposed in Arago's scientific notices. For lack of space I will consider only the steam engine, railways and photography, setting aside topics like navigation (of which he was an expert), hydraulic lime, cements and mortars and various items concerning the city of Paris, with the exception of the water supply and the artesian wells. The French edition (Lequeux 2003) covers these and other topics that are mostly of interest for the French reader.

The Steam Engine and the Industrial Progress

Improvements and Dangers of the Steam Engine

The steam engine was the principal element of the industrial revolution of the end of the eighteenth century. Arago had a strong interest in it; one of the plans that he asked the students of his "course of machines" at École Polytechnique to draw was Watt's relatively advanced steam engine (Fig. 11.3). He wrote in the *Annuaire du Bureau des longitudes* for 1829 a long notice on the history of the steam engine, reproduced and enlarged in the *Annuaires* for 1830 and 1837. In the latter, we find an "Examination of the critical observations of which the preceding notice was the object," which is a discussion of priorities about the steam engine. We know how

[10] *Annuaire du Bureau des Longitudes for 1853, p. 343–623.

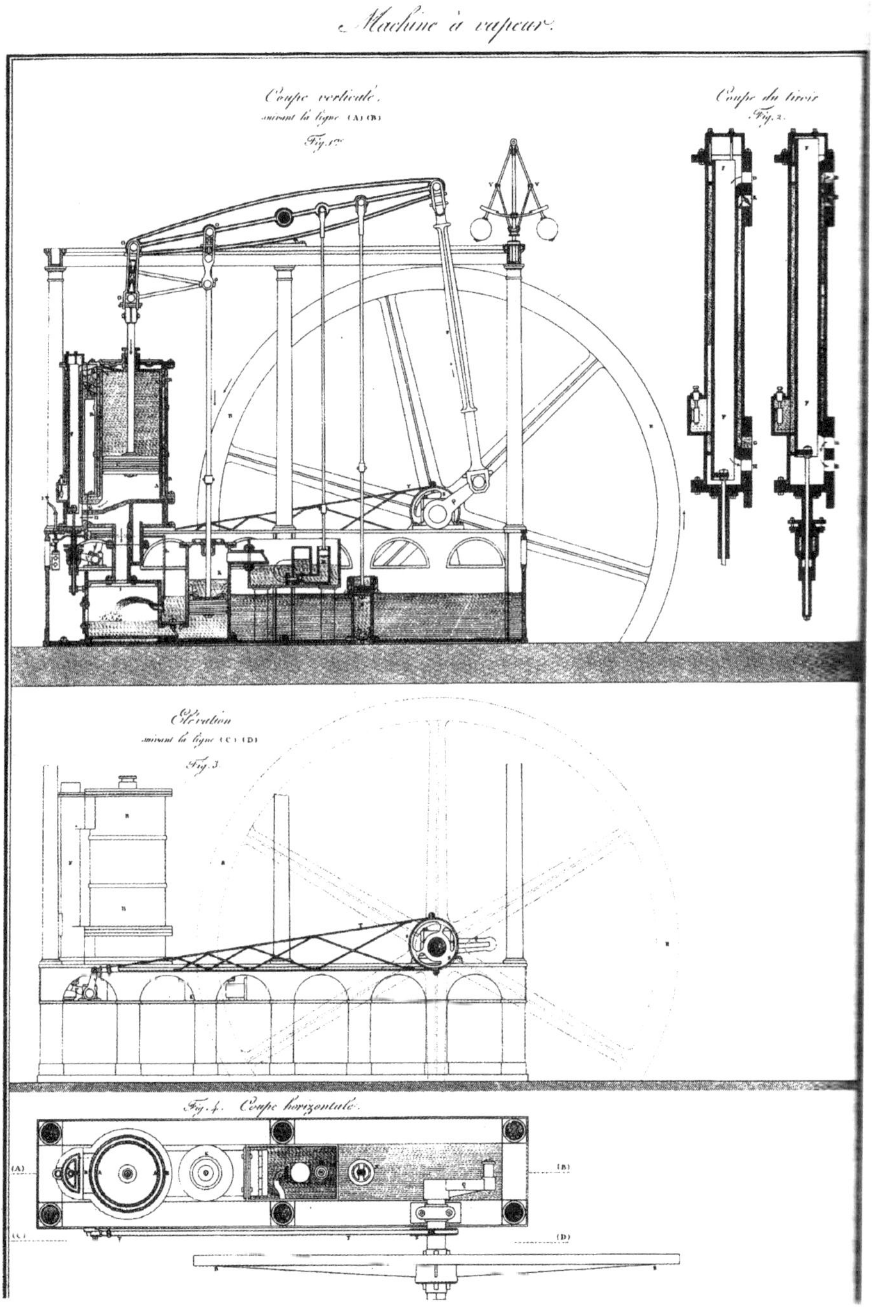

Fig. 11.3 Drawing of a Watt's steam engine by a student, for Arago's *Cours de machines* at *École Polytechnique*. Bibliothèque de l'École polytechnique

touchy Arago was on questions of scientific or technical priority, sometimes to the limit of bad faith when the contested inventions were French.[11] In his "Historical eulogy of James Watt,"[12] he reproduced a part of these texts and added a number of considerations about economy. Arago's documentation is impressive, and in spite of its nationalistic biases, his history is so well done that it has often been reproduced, for example by Delaunay in his *Cours de mécanique* of 1850.[13] Several of the illustrations that follow are taken from this book.

One soon had the intuition, verified by experience and by the theory developed by Sadi Carnot, that the high-pressure steam engines were more efficient than the first low-pressure ones, for example that of Thomas Newcomen (1664–1729), which used atmospheric pressure (Fig. 11.4), and the first machines of Watt. They were also less cumbersome and more adapted to locomotives. But of course they were more dangerous because of an increased risk of explosion. Arago was perfectly conscious of this risk after his terrifying experiments with Dulong (see Chap. 10), but he was such a strong advocate of high-pressure machines that he sometimes concealed their danger.

A first report on the safety of steam engines, elaborated by a commission presided by Baron Charles Dupin (1784–1873), was read at the Academy in 1822. It recommended two safety valves per machine, that the boilers be tested at a pressure 4–5 times greater than normal, and that walls 1 m thick be constructed around the machines close to buildings. As Arago was supposed to be particularly competent on the subject, the Academy asked him later to write a report on it.[14] One finds in this text a description of several explosions, in particular the one that occurred on 4 March 1827 in Lyons during the tests of the ship *Le Rhône* built by Aitkin and Steel, whose boiler had been "perfectly executed" in the *Ateliers de la Gare*. There were many casualties, including Steel himself, who had blocked the safety valves! This was not exceptional: Arago cited other explosions in which the safety valves had been overloaded, voluntarily or not. One often put disks of fusible metal on the walls of the boilers, but Arago said that "when the firemen want to push the fire more than usually, they know very well that, in order to prevent the fusion of the plate, it suffices to send on its surface a continuous flow of cold water." On the other hand, he also wrote some "remarks about the pretended dangers of high-pressure machines," in which he said that the boilers were tested with a hydraulic press up to three times the nominal pressure, and recalled his own tests with Dulong. He concluded with satisfaction: "We are pleased to say that most of the dispositions we recommended were adopted by the government."

---

[11] Arago was certainly displeased when reading the commentary of Sadi Carnot (1824, p. 6): "If the honor of a discovery belongs to a country where it was improved, this honor cannot be refused to England: Savery, Newcomen, Smeathon, the famous Watt, Woolf, Trevetick and several other English engineers are the true inventors of the fire machine; it got from them all the successive degrees of improvements. It is indeed normal that an invention is born and developed at the place where it is the most needed."

[12] *Arago, F. *Œuvres complètes* t. 1 p. 371–510.

[13] *Delaunay (1850).

[14] *Arago, F., *Œuvres complètes* t. 5 p. 117–180.

Fig. 11.4 The "atmospheric" steam engine of Newcomen (1712). The cylinder B containing the piston C is open at the top. Steam from the boiler A is sent to the bottom of the cylinder, then a small quantity of cold water from the tank G is introduced. The steam condenses, and the atmospheric pressure on the piston drives the machine. When the piston is at its low position, the water is evacuated and the cycle resumes. Here, the machine is used to pump water from a mine. The weight of the rod E and of the additional charge F is such that the piston goes up when steam is injected. The taps for admission and evacuation are driven by wheels run by the machine. From Delaunay (1850), author's collection

The Birth of Thermodynamics

The theory of the steam engine was much less advanced than its practice. It could only be established after the fundamental work of Sadi Carnot was understood and assimilated. I find it interesting to write a few lines about his *Réflexions sur la puissance motrice du feu*[15] (Reflections on the motive power of fire) of 1824, which is the founding text of thermodynamics (Fig. 11.5).

As most of his contemporaries, Carnot did not seem to be concerned with the nature of heat, on which there were two opposite ideas. For Lavoisier, heat was a subtle fluid that he called *calorique*, which flew from the hot bodies to the cold ones.[16] For a minority of scientists, Babinet for example, heat was a vibration of the "molecules" of the bodies; this is the present idea. This concept was introduced by

[15] *Carnot (1824).

[16] *Lavoisier (1862–1893) t. 2, p. 623–655.

Fig. 11.5 Sadi Carnot (1796–1832) in the uniform of the École polytechnique. Wikimedia commons, Materialscientist

the Swiss physicist Daniel Bernoulli (1700–1782), the creator of the kinetic theory of gases.[17] However, its impact was less than that of the alternative theory, which was defended by prestigious scientists like Laplace or Poisson.

Carnot conceived the functioning of a "fire machine" as the transport of *calorique* from the hot source (the boiler) to a cold source (the condenser or the open space), while there was simultaneously production of "motive power" in the cylinder. He showed, and this was his major discovery, that this power could not be larger than a certain amount that depended on the temperature difference between the two sources. He developed the analogy with a waterfall, whose power cannot be larger than what the height of the fall can supply. He also showed that "the diminution of the *calorique* produces more power in the lower degrees than in the upper ones," for example this power is larger when one goes from 100 to 0 °C than from 200 to 100 °C. We are not far from the law that says that the efficiency of a thermal machine cannot be larger than $(T_{hot} - T_{cold})/T_{cold}$, where T is the absolute temperature expressed in kelvins (°C + 273). However, for lack of data Carnot could not give this quantitative law, which had to wait for Rudolph Clausius (1822–1888) and Lord Kelvin (1824–1909). He concluded with the following principles:

1. *The temperature of the fluid [steam for example] must first be raised to the highest possible temperature, in order to produce a large diminution of the* calorique *and as a consequence a large production of motive power*;
2. *For the same reason, the cooling must be as strong as possible*;
3. *The passage of the elastic fluid from the higher temperature to the lower one should be due to an increase of its volume [this is the expansion in the cylinders*

[17] Bernoulli (1738).

of the steam engine], i.e., the cooling of the gas occurs through the effect of its decrease of density.

The novelty of Carnot's work is absolute, and it was probably difficult to appreciate it at the time. Carnot himself was probably aware of this, as he wrote: "The matter treated here being completely novel, we are forced to use new expressions which are perhaps not as clear as desirable." Probably, many contemporaries did not understand what he recommended; Carnot cites many realizations that do not obey his principles, for example, those that do not use the expansion of steam in the cylinders (the Watt's engine represented in Fig. 11.3 is an example).

A Few Curiosities

In his *Réflexions sur la puissance motrice du feu*, Carnot remarked judiciously that "the vapors of all the bodies able to pass into the gaseous state, of alcohol, mercury, sulfur etc. could play the same role as water vapor."

As a matter of fact, motors "with combined vapors," also called "binary machines," were built, using ether, carbon sulfide, carbon perchloride or chloroform. In these machines, the steam of an ordinary steam engine was cooled in a container where it caused the ebullition of a liquid more volatile than water. It was thus possible to recuperate a part of the latent heat of water. The vapor of the volatile liquid was sent in a cylinder working in parallel with the first one, increasing the total power. One finds for example in the *Comptes rendus* of 1848 a letter by Arago, then Minister of Navy and Colonies, with the title "Questions relative to the use of chloroform vapor in one of the cylinders of binary machines".[18] He proposed to replace ether, which was the most usual material, by chloroform judged less dangerous. Needless to say that these machines were soon abandoned because of the unavoidable losses of the volatile liquid, whose cost was very high.

The hot air machines were more interesting. Carnot indicated that any gas, air in particular, could be used to build a heat engine without using vaporization and liquefaction. The main interest is that one does not have to supply the energy needed to vaporize the liquid, an energy that is difficult to recuperate, as we have just seen. Figure 11.6 represents an example of hot air machine due to the American John Ericsson (1803–1889). It is an adaptation of the motor built in 1816 by the Scottish clergyman Robert Stirling (1790–1878). Stirling is generally considered as the inventor of the hot air machine, but it was already known for some time and he only perfected them. This type of motor, which can reach an efficiency of 30 %, was revived in the 1930s and is still used for some applications, while the classical steam engines have almost disappeared except in some locomotives in India and in a few other countries.

---

[18] **Comptes rendus de l'Académie des sciences* 26 (1848) p. 408–409.

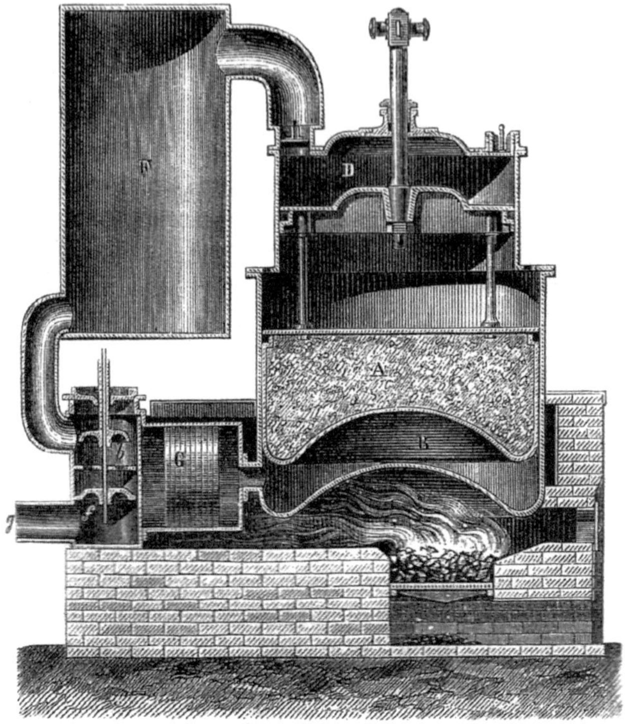

Fig. 11.6 The Ericsson hot air motor (ca. 1826). The cylinder contains two superimposed attached pistons, with different diameters. The lower one A is thick and filled with a thermal insulating material. The two valves *f* and *b* on the left being closed, the air heated by the fire below raises the two pistons, supplying the power to the rod above and then to a mechanism not represented. The upper piston pushes the air of the space D to the tank F, where the pressure is approximately constant. When the pistons are at their upper position, they descend thanks to an inertia wheel (not represented). The valve *f* is now open and the valve *b* closed, so that the air below the piston is evacuated in *g*. The valve to the top right of volume D opens, and fresh air is introduced in this volume. Then *f* is closed and *b* is opened; air from the tank F is introduced in volume B under the piston A, and the cycle resumes. In G are wire gauzes used as heat exchangers: they are heated by the air evacuated from volume B and pre-heat the fresh air coming from the tank F. From Delaunay (1850), author's collection

Carnot regretted that the heat source of steam engines was not hot enough because the pressure was limited (often to 6 bars, corresponding to 160 °C). The hot air machines could work at a much higher temperature and had a better efficiency. Now, if one could inject directly into the cylinder a flammable substance, realizing in this way an internal combustion engine, the temperature in the cylinder could be still higher, and so the efficiency. One of the first attempts in this direction was by Nicéphore Niepce (1765–1833), better known as one of the inventors of

Fig. 11.7 Nicéphore Niepce. Wikimedia Commons, Daderot

photography (Fig. 11.7), and his brother Claude Niepce (ca. 1763–1828). Carnot described this invention as follows:

This machine, very ingenious and interesting mostly by the novelty of its principle, had a major defect. The material used as a fuel, the lycopode powder,[19] *which is used in our theaters to produce flames, was so expensive that the advantage was canceled by this cause.*

And he added:

Instead of doing like MM. Niepce, I would consider it better to compress some air by pneumatic pumps, and to have it cross a closed box in which one would introduce the fuel in small quantities by a mechanism easy to imagine, and to develop its action in a cylinder with piston.

We see here the principles of several elements of the combustion engine: the carburetor, the compressor and, of course, the cylinder and piston. However, Carnot seems to have ignored that the first combustion engine was built in 1805 by the Swiss engineer Isaac de Rivaz (1752–1828). This motor, which used hydrogen, was never commercialized, nor the motor of the Niepce brothers. The first commercial combustion engines were only built in 1860 by Jean-Joseph-Étienne Lenoir (1822–1900), a Belgian-French engineer. He produced 400 two-stroke engines with a power of

[19] These are the spores of Lycopodium clavatum, a flowerless plant that grows in humid places of mountains. The lycopode powder was still used a few years ago in high schools to visualize stationary sound waves.

1.5 kW, using coal gas. He also built in 1863 the first four-stroke engines, whose principle had been proposed the preceding year by Alphonse Eugène Beau de Rochas (1815–1893). The same year, he also constructed the first working automobile, which took 3 h to go 18 km.

The Economic and Social Consequences of the Steam Engine

In general, the steam engine was considered as a factor of economical progress, and also of social progress. Among the many testimonies, one could cite that of Baron Dupin, who on 24 April 1821, in front of the four Academies of the Royal Institute of France, presented a "Discourse on some advantages of industry and machines in France and in England." According to the minutes of this meeting[20]:

[M. Dupin] shows how unfounded are the reproaches made to machines: to take away from the poor his means of work and existence. Conversely, we should realize that the masses are happier and are more affluent due to the larger abundance of products made by machines to satisfy the needs, the pleasures and the luxury of men.

Arago had a similar point of view in his eulogy of James Watt[21] at the public meeting of the Academy of Sciences on 8 December 1834. The style is characteristic of the time:

The products of industry, while perfecting themselves, will be less expensive.

The population, well fed, well clothed, well heated, will increase rapidly; it will cover with elegant houses all the parts of the territory.[...]

Installed on ships, the steam engine will replace a hundredfold the efforts of the triple and quadruple rows of oarsmen, from which our fathers required efforts considered as the punishment of the worst criminals.[...] The travels will be much more rapid; the time of arrival of the ships will be foreseen as accurately as that of the public stage-coaches.

Finally, steam locomotives drawing thousands of travelers will run on rails much faster than the best race horse only loaded with its light jockey.[...]

In a few years,[...] the sybarites, driven by vapor on the rails, will be able to visit rapidly the various regions of France. They will be able in a single day to see our Navy get under steam in Toulon; to lunch in Marseilles with succulent Mediterranean red mullets; to plunge at noon their tired limbs in the mineral water of Bagnères; and they will return from Bordeaux to Paris in the evening to dance at the Opera.

This is going a bit too far: despite our fast trains, we are not yet there!

[20] *Mémoires de l'Académie royale des sciences* 5 (1821–1822), *histoire,* p. 104–105.

[21] *Arago, F. *Œuvres complètes* t. 1 p. 371–510.

This enthusiasm was tardy: France was late with respect to England. According to Arago himself,[22] there were already in 1819 10,000 steam engines in England, "with a total power of 600,000 horses or three or four million workmen [sic]." But in 1835 there were in France only 1000 machines, equal to 14,000 horses. Therefore, Arago recommended that powerful steam engines be constructed in France, mentioning three possible builders. Unfortunately, the English machines were 30 % cheaper than the French ones, because of the lower cost of coal and iron. Custom tariffs of 33 % compensated for the difference.

In 1837, tariffs were cut to 15 %, but the English machines were still cheaper so that the protection became totally ineffective. Arago, who in principle opposed protectionism, did not ask for an increase of the tariffs, but presented at the Chamber of Deputies on 16 June 1840 the following amendment to a proposed law on the construction of railways[23]:

The nine-tenths at least of the locomotives used by the Company must be built in France. This would cease to be compulsory if the prices of the French machines become higher than those of the English ones by more than 15 percent.

At this date, the French railways owned 59 French locomotives against 97 English ones. The number of French builders had increased since 1835. Many arguments were used to stimulate the construction of locomotives in France; to illustrate the quality of the French products, Arago cited the success in England of the astronomical telescopes of Cauchoix, which had little to do with steam engines!

Arago's attempts did not have the expected results immediately; his amendment was transferred to the Commission of Finances of the Chamber, and was dead and buried. The protectionist measures were only voted in 1844, and English locomotives continued to be imported. In spite of interesting inventions of Marc Seguin (1786–1875), the French machines were generally mere copies of English ones. But the national industry was born, and two companies were going to dominate it for several decades: that of Nicolas Koechlin (1781–1852) in Mulhouse and that of Creusot, founded by Adolphe Schneider (1802–1845) and his brother Eugène (1805–1875).

The Railways

The Beginnings

As for the steam engine, France was late compared to England and even to the United States.[24] To be sure, there was a 20-km line between Andrézieux and Saint-Étienne (Loire) since 1828. It was used to carry coal and the carriages were drawn

[22] *Arago, F., *Œuvres complètes* t. 5 p. 206–214.

[23] *Arago, F., *Œuvres complètes* t. 5 p. 215–226.

[24] The first English railway, with horse traction, dates from 1801, and the first line with steam locomotives from 1825. In the United States, the first line was opened in 1827 by the Baltimore and Ohio Railroad.

by horses. In 1832, it was opened to travelers, but the horses were only replaced by steam engines in 1844. Another line was built from 1826 to 1832 between Saint-Étienne and Lyons; it necessitated building 14 tunnels. The first French line was extended to Roanne in 1834, with some slopes so steep that the trains had to be drawn by stationary steam engines and cables. These railways were essentially used for carrying heavy loads: transportation of passengers was not a priority. In any case, they were not faster than the stage coaches. The banks were not interested in financing these enterprises, probably judged too risky, and preferred to invest in canals and roads.

Things were going to change after 1832, a date when the railways found strong advocates, because they were considered as essential for the emerging industrial society. Paulin Talabot (1799–1885), an engineer who was to become one of the most powerful businessmen of the nineteenth century, associated with the wealthy Maréchal Nicolas Soult (1769–1851) to create a new line between Alès and the mines of La Grand-Combe (Gard). In 1833, the Chamber of Deputies voted a loan of 500,000 francs (equivalent to $2 million) for studies of railroads, and fixed a rather restrained juridical framework for their construction and exploitation.

The Parisian Railways and the Intervention of Arago

The first really important railway project in France started in 1835: the line between Paris and Saint-Germain, 19 km long. This line actually went only to Le Pecq, because the final slope to Saint-Germain was too steep. It was financed by several bankers, including the brothers Émile and Isaac Pereire (1800–1875 and 1806–1880) and James de Rothschild (1792–1863), under the form of 10,000 shares of 500 francs. The spacing between the rails was fixed to 1.435 m, the same as in England, and the minimum radius of the curves was 800 m in order to allow high speeds. Then followed two different lines from Paris to Versailles as the result of a conflict between two rival groups of interest. This gave Arago the opportunity for his first intervention about railways at the Chamber of Deputies on 13 June 1836. He was opposed to the line to Versailles running on the right bank of the Seine river, essentially because a long tunnel was required. He thought that the passengers might catch a cold due to the change of temperature and that, although the risk of explosion of the boiler was not large, it would have devastating consequences if occurring inside the tunnel. Arago was often reproached for his conservative attitude, which was actually shared by many people. In fairness to him, one should remember that the carriages had no roof except the first-class ones. Perhaps he found the existence of two lines useless, or he feared the creation of a monopoly because the right-bank line was to be given to the same group as the Saint-Germain line. In any case, his arguments were not sufficient to stop an irreversible impulse: the two lines were actually open, in 1839 for the right-bank one and in 1840 for the other.

On 8 May 1842, a very serious accident occurred in Meudon on the left-bank line. A broken axle of one of the two locomotives resulted in 55 persons killed, including the famous explorer Jules Dumont d'Urville (1790–1842) and his family, and more than 100 seriously injured people. This was not enough to stop the progress of the railways. However the Academy began to search for "the most appropriate measures to limit the badness of railway accidents." Many more or less realist solutions were proposed, in particular for improving the brakes which were then very primitive. Arago suggested an emergency brake that would consist in implementing in a special carriage two parachutes with a total effective surface of 16.5 m^2, designed to brake the train when deployed. Furthermore, the locomotive would be separated from the carriages. Arago calculated that with two such special carriages, braking would be more efficient at 50 km an hour (the maximum speed of the trains) than the usual one with a hand brake on each carriage. It would have been more efficient to inverse the steam in the locomotive cylinders, but Arago obviously did not think about this. Later, Arago refrained to propose such ridiculous ideas about the brakes; however, they remained a weak point of the railways until the adoption around 1880 of the Westinghouse brake.

Banks or State?

As early as 1838, the French government proposed a railway network consisting of seven lines linking Paris to Belgium, Spain, Le Havre, Nantes, Marseilles, Toulouse and Strasbourg, plus two transverse lines. This network would have had a total length of 4400 km, for a cost of one billion francs (some $4 billion). Other lines would be private. However, this magnificent project was not realized as foreseen. The major problem was, of course, financing. Would the State be able to build and exploit the main lines? A commission of the Chamber of Deputies presided by Arago was charged with examining the question. Its report,[25] presented on 24 April 1838 and in which one can easily recognize Arago's style, concluded negatively. It recommended that the lines be constructed in succession rather than simultaneously, a reasonable proposition at a time when the techniques were not stabilized. Then the commission recommended that the intervention of the government should be limited to the parts that the private companies would not be able to realize. It is true that the State did not want to create a sovereign bond for this, and that a lot of money was already being spent on roads and channels. Moreover, the profitability of the railways was far from insured, and it seemed better to leave the risk to private investors. Finally, the deputies decided against the project. This decision kept the State from building railways for years.

This is the heart of a controversy that is still present today in France and in other European countries: should the railways be private or owned by the government? It already started in 1837, when Lamartine violently attacked Arago's position in

[25] *Arago, F. *Œuvres complètes* t. 5 p. 254–301 and 301–356.

favor of private railways. But Arago was to change his mind. As written by Pierre-Joseph Proudhon (1809–1865)[26]:

In 1842 M. Arago was in favor of the construction of railways by private companies, and the majority of France was thinking like him. In 1846, he said that he had changed his opinion; and apart for the speculators of the railways one can say that the majority of the citizens have changed like M. Arago.[…] So that, today as well as previously, neither the public nor M. Arago in spite of their flip-flopping know what they want.

The controversy was revived in 1848, where Louis-Antoine Garnier-Pagès, the minister of finances of the provisional government, proposed the complete nationalization of railways, an impossible thing due to the low level of the finances of the State. Then it became clear that the big financial groups would win, in spite of the impediments that a meddlesome administration put in their way and of the State's desire to keep at least a partial control of the railways. It is not the place here to trace the chaotic history of the French railways. The debate would continue until the railways were nationalized in 1937, and could revive in the future.

The first important line of the French network was that from Strasbourg to Basel, finished in 1838 by Koechlin. Then came those from Paris to Orléans, and from Paris to Rouen, both financed by two different groups of bankers and completed in 1843. The locomotives of the latter line, built in Rouen by the English company Allcard, Buddicom & Co. were already quite sophisticated (see the figure at the beginning of this chapter). Both lines were a financial success, so that the investors were encouraged to build more. Thus, by 1850, and not long before Arago's death, the network was already well developed (Fig. 11.8).

Two False Trails: The Atmospheric Railway and the Arnoux System

One should not be surprised that, in the period of uncertainty about the technology of railways, alternative projects blossomed. One of these, which went rather far with Arago's blessing, was the atmospheric railway. The idea, which originated in England, was simple but difficult to implement: one of the carriages of a train was attached to a piston that circulated horizontally in a tube under the tracks. Vacuum was produced on one side of this tube by a stationary pneumatic machine moved by steam engines, and the atmospheric pressure on the other side pushed the train (Figs. 11.9 and 11.10). The major difficulty was, of course, to make the tube air tight, while it had a slit over all its length. This was achieved, albeit not exactly successfully, by a rubberized canvas or leather lip running all along the slit; a series of disks lifted this lip on the passage of the blade connecting the carriage with the piston.

[26] *Proudhon (1846), see p. 327.

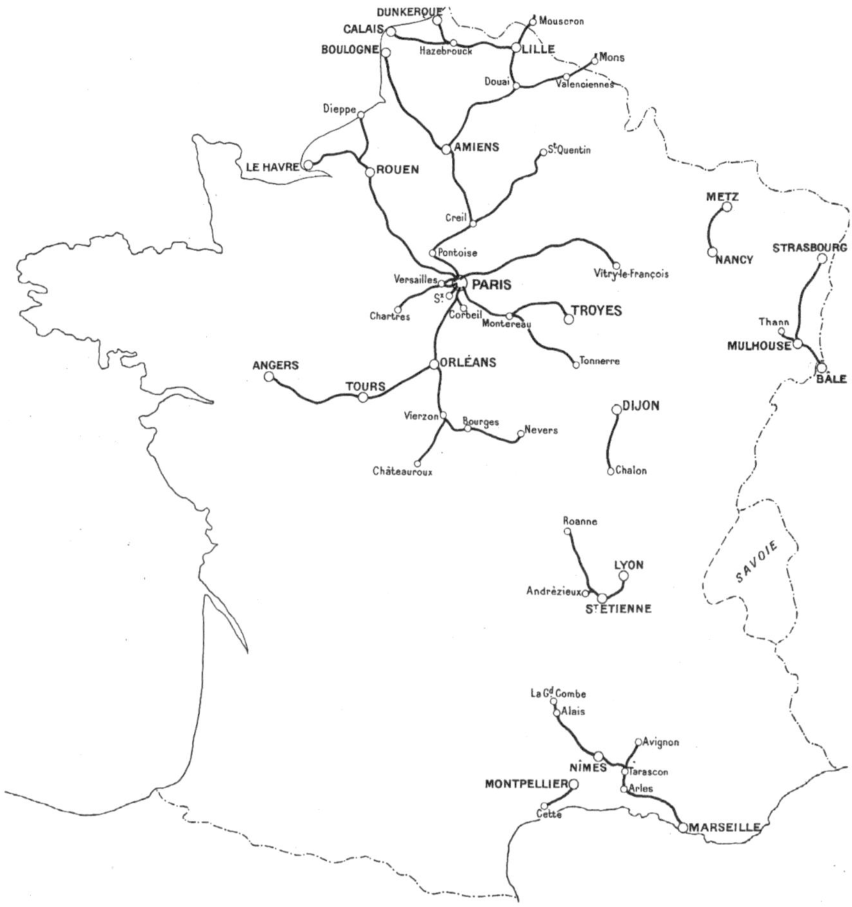

Fig. 11.8 The French railway network in 1850. Author's collection

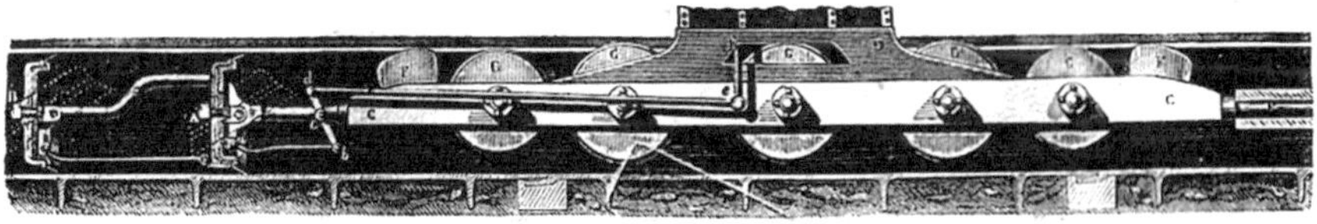

Fig. 11.9 The driver of the atmospheric train. The part of the tube to the left of the pistons is evacuated by a vacuum pump, and the atmospheric pressure on their right pushes them. The discs serve to lift progressively the lip so that the blade connecting the pistons to the carriage above can move. To stop the system the pistons could be rotated around a horizontal axis. From Delaunay (1850), author's collection

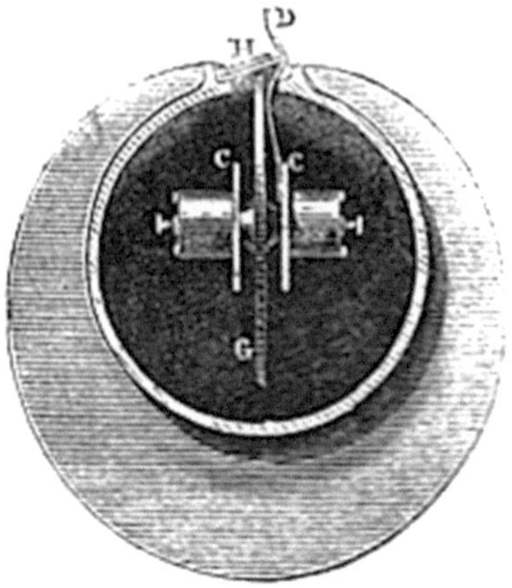

Fig. 11.10 Cut of the atmospheric tube. The blade D attached to the piston and to the carriage above was free to move thanks to the fact that the beveled disk G was temporarily raising the lip H. From Delaunay (1850), author's collection

The first railway built on this principle was installed in Ireland in 1843 by Samuel Clegg (1814–1856) and Joseph Samuda (1813–1885). It had a length of 2413 m and could draw a 30-t train at 64 km/h. Several others followed. A Frenchman named Mallet, honorary Inspector of Bridges and Roads, was sent by the French government in November 1843 to see the Irish railway, and his report was very favorable.[27] Others opposed the principle, remarking rightly that the atmospheric railway was only practical for climbing steep slopes, and that the running expenses would be very high, but they were a minority. The French builders set to work: the constructor Alexis Hallette (1788–1846) built a prototype 100 m long; Arago was enthusiastic, and asked from the Chamber on 16 July 1844 a credit of 1,800,000 francs to try the various systems proposed by the builders.[28] There were three candidates, but finally the Council of Bridges and Road chose another one: the Saint-Germain railway company. This company saw the project as a means to complete economically its railway, which had to stop before Saint-Germain as we have seen because of the excessive slope. It chose the English design of Clegg and Samuda, instead of that of the Hallette company, which went bankrupt in 1847, in spite of Arago's intervention. Eugène Flachat (1802–1873) was in charge of the construction. In order to carry the necessary material, he built a locomotive that could climb the 3.5 % slope between Le Pecq and Saint-Germain, which demonstrated the uselessness of the atmospheric railway! However, the credits were voted and the construction proceeded. It was finished in 1847, for a cost of six million

[27] An English version of this report is available via http://en.wikipedia.org/wiki/Dalkey_Atmospheric_Railway#cite_note-5

[28] *Arago, F. *Œuvres complètes* t. 5 p. 393–418. See also **Comptes rendus de l'Académie des sciences* 18 (1844), p. 226–227, where one can read: "Developing the idea of M. Hallette, M. Arago remarks that a system of pneumatic tubes fixed along the walls of the embankments of the Seine river would be much less expensive than a towing path, and that the use of steam to tow the barges would have several clear advantages compared to horses."

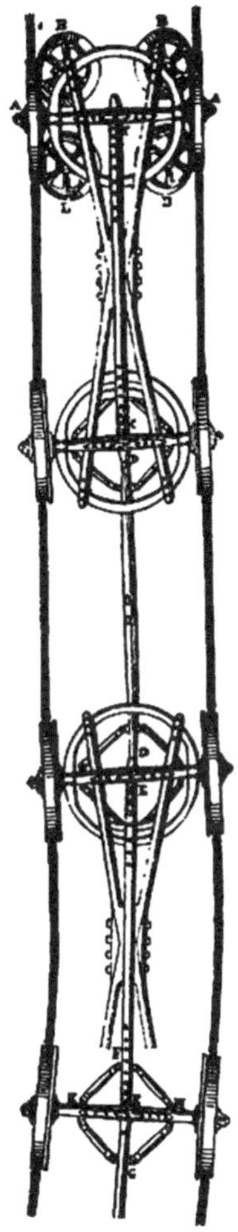

Fig. 11.11 Principle of the Arnoux system. In front of the train, four guiding wheels rolling along the inside of the rails forced the shaft to be perpendicular to the rails. A system of rods communicated this orientation to the following shafts. A similar system was placed at the other end of the train. Author's collection

francs (about $25 million), far higher than the initial estimate. This atmospheric railway operated until 1858, but it was not without problems, and was stopped because of an accident that had three casualties. In this case, the brakes failed during the descent, which was by gravity.

There were problems with carriages as well as locomotives. The most serious was adapting the carriages to small-radius curves of the tracks. Initially the engineers built carriages with parallel and rigid axles; they had to be very short and were quite uncomfortable for the passengers as a result. In 1838, Jean-Claude Républicain Arnoux (1798–1866) invented what was apparently the solution to this problem, a "system of carriages for railways of any curvature." This system of rods (Fig. 11.11) forced the axles, mounted on pivots, to stay perpendicular to the rails, which indeed allowed tackling short-radius curves.

A commission of the Academy, of which Arago was a member, examined small-scale tests: its report was favorable and recommended full-scale ones. Arnoux made full-scale tests in Saint-Mandé on a 1142-m-long track with many curves, including one with an 18-m radius.[29] It cost Arnoux 150,000 francs, but the Academy judged the tests so successful that he won one of its prizes. Thanks to a large extent to Arago, Arnoux obtained in 1844 the concession of the railway line from Paris to Sceaux, called the *ligne de Sceaux*, which still exists after some changes as a part of the RER B commuter rail line. His system allowed the train to make a half turn at both ends of the line on a 25-m radius track. This is the reason why the Parisian RER B station at Place Denfert-Rochereau, the oldest railway station in existence in the city (1846), has a curved shape.

However, the carriages of the Arnoux system could not circulate on a normal track of 1.435-m gauge, because its own gauge was 1.750 m. No rail switching was possible in this system: it was a deadlock. The *ligne de Sceaux* was the only one in which Arnoux's system was ever used, and it was replaced in 1857. However, the problem of the curves was still there, although it had already been solved in the United States, thanks to the "American carriage," whose body rested on two bogies with two shafts each. This system is still in use today. In France, it was tested on the line between Lyons and Saint-Étienne in 1842, then abandoned because the French industry felt unable to adapt to it. However, it was revived in France in the 1860s. It was only then that the passengers could find some comfort in their travels by train.

Arago was certainly well-intentioned in supporting these projects. He had taken a wait-and-see attitude in his 1838 report, letting foreign countries to develop new rail technologies before they were adopted in France. Historians believe they were bad recommendations, because the only way for France to catch up was to jump into the venture without losing time. Fortunately, the French industrials and bankers did it. If Arago's strategy had been followed, France would have depended for a long time on English techniques.

A Pioneer of Photography

Photography was born during Arago's period of activity, and its development owes a lot to him. We should not be surprised of his motivations: as a specialist of optics, it is natural that he was enthusiastic about this discovery. At the end of his life, in 1853, Arago wrote an important text on photography[30] that incorporates his preceding writings and summarizes its history. It was a somewhat biased text because he wanted to demonstrate that photography was a purely French invention, due almost exclusively to Nicéphore Niepce and to Louis Jacques Mandé Daguerre (1787–1851).[31]

---

[29] **Comptes rendus de l'Académie des sciences* 11 (1840), p. 101–115.

[30] *Arago, F. *Œuvres complètes* t. 7 p. 455–517. See also Tobin (2003) Chaps. 3–5.

[31] See on this topic and for political aspects McCauley (1991), p. 1–21.

While I will note that the English contribution to photography was far from negligible, other Frenchmen, like Hippolyte Bayard (1807–1887), were the first to expose photographs in public, playing an important role in parallel with Daguerre.

The Beginnings

Scientists knew for a long time that several chemical components of silver were sensitive to light, and attempts have been made at the beginning of the nineteenth century to fix images using these components. For example, Laplace, Malus and Arago himself tried to obtain an image of the Moon around 1811, but without success. In England, Thomas Wedgwood (1771–1805) and Humphry Davy succeeded in obtaining images, but these quickly disappeared for lack of fixation. Niepce was first to obtain a permanent image; he made the first real photograph in 1816, and the first engraving from a photograph in 1829. According to Arago, he even succeeded in 1829 to make a negative from which positive images could be obtained by contact.

Daguerre was a painter of "panoramas," very appreciated at the time. He was familiar with the *camera obscura,* which he used in his work, and he began an association with Niepce that lasted until the death of the latter in 1833. Then Daguerre continued by himself, using as a sensitive layer silver iodide obtained by placing a plate of silver in iodine vapor. The exposure times were very long, until Daguerre discovered by chance that the latent image of an exposed plate, on which nothing could be seen, could be made visible (we now say "developed") by mercury vapor. It took some time before Daguerre discovered that the developed image could further be fixed by washing in concentrated brine, which removed the excess silver iodide. The daguerreotype was born.

Arago's Presentations at the Academy of Sciences

On 7 January 1839, Arago presented Daguerre's invention at the Academy. Extracts of the minutes[32] read as follows:

The principal products of his new processes that M. Daguerre has shown to three members of the Academy, MM. de Humboldt, Biot and Arago, were a view of the large gallery that connects the Louvre with the Tuileries; a view of Île de la Cité and of the towers of the Notre-Dame cathedral; views of the Seine river and several of its bridges, views of several barriers of the capital.[…] In summer at noon, eight to ten minute exposures are sufficient.[…]

M. Arago tried to emphasize all that the invention of M. Daguerre would offer to travelers, and above all to the scientific bodies and to the people who work so zealously

[32] **Comptes rendus de l'Académie des sciences* 8 (1839), p. 4–7.

at the graphical representation of the monuments of our country. The easiness and exactness that will result from these new processes, far from being prejudicial to the so interesting corporation of drawers, will give them more work. They will certainly work less outside, but much more in their studios.

The new reactive seems also to offer very precious means of investigation to physicists and astronomers. At the request of the Academicians we have cited above, M. Daguerre has projected the image of the Moon produced by a simple lens on one of his screens [sic], and it left on it a clear white trace.[…]

The invention of M. Daguerre results from an assiduous work of several years, during which he collaborated with his friend, the late M. Niepce from Châlons-sur-Saône. Searching how he could be paid out for his efforts and his expenses, this eminent painter soon recognized that a patent would not be appropriate: it thus seems necessary that the government compensates directly M. Daguerre, and then offers to the whole world a discovery that would contribute so much to the progress of arts and sciences. M. Arago announces that he will send a request in this direction to the Ministry and to the Chambers.[…]

However, on 4 February, Arago had to read a claim for priority from the Englishman William Henry Fox Talbot (1800–1877). Of course, he stepped up to defend Daguerre. It appears that Fox Talbot discovered the latent image and its development only in 1840 while Daguerre did it in 1835.

As promised to the Academy, on 3 July 1839 Arago gave a report on the invention to the Chamber of Deputies, which voted on 9 July the pension promised to Daguerre, of 6000 francs (about $25,000) per year, while 4000 francs were given annually to the heirs of Niepce. All the details of Daguerre's process were finally disclosed at a joint meeting of the Academy of Sciences and of the *Académie des Beaux-Arts* (Academy of Fine Arts). Daguerre was so emotional that his text had to be read by Arago (see the figure at the beginning of Chap. 1). Arago took the opportunity to cite in his report[33] a note written "at his request" by the then famous painter Paul Delaroche (1797–1856):

The painter will find in this process a fast means to obtain collections of studies that would be impossible to get without much time and much effort, and in a less perfect way whatever his skill.[…] In summary, the admirable discovery of M. Daguerre is an immense service provided to the arts.

Somewhat surprisingly, the reactions of the painters and engravers were rather favorable to the invention, and the words attributed later to Delaroche, "Painting is dead this day," were never actually said.

After 1850, the daguerreotype disappeared gradually, to be replaced by the more familiar process of the negative photograph, first on paper then on a glass plate and much later on film, followed by photo printing on paper or film.

---

[33] *Arago (1839).

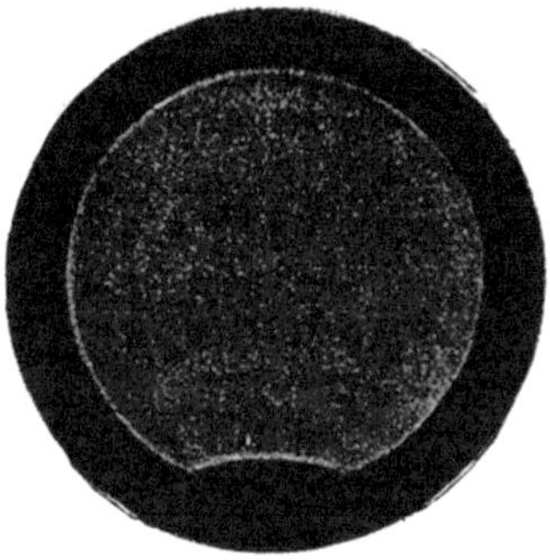

Fig. 11.12 Reproduction of the first photograph of the Sun obtained by Daguerre and Arago, very probably during the partial eclipse of 15 March 1839. Author's collection

The First Astronomical Photographs

In his presentations of 1839, Arago foresaw some applications of photography to astronomy. He thought of mapping the Moon and to photograph the spectrum of the Sun in the ultraviolet inaccessible to human vision; such a spectrum was first obtained in 1842 by Edmond Becquerel (1820–1891). Arago also considered photometry, which would be more objective if using the photographic plate rather than the human eye. And he also dreamed of fortuitous discoveries made possible by the new means.

However, things proceeded rather slowly. The photograph of the Moon obtained by Daguerre in 1838 was very poor. The first successful astronomical photograph was probably that obtained by Daguerre and Arago together during the partial solar eclipse of 15 March 1839. This daguerreotype was still mentioned by Admiral Ernest Mouchez (1821–1892), then Director of the Paris Observatory,[34] and there exists a reproduction dating from about 1893 (Fig. 11.12).[35] Unfortunately the plate is lost, but Mouchez said that it was already almost erased when he saw it. Arago also mentioned that "very soon after the publication of the bright discovery of M. Daguerre, [he had] seen, by forming the image of the Sun on a silver plate, that the rays coming from the central part of the solar disk had a stronger photographic action than those from the edges."[36] Was this photograph that of the eclipse, or another one?

An Italian physicist, Alessandro Majocchi (1795–1854), obtained a daguerreotype of the Sun during the partial phase of the total eclipse of 8 July 1842. Lerebours obtained another one, but probably overexposed. This was not the case of the remarkable photographs taken from 1843 to 1855 by Fizeau and Foucault (Fig. 8.15). They used a rudimentary shutter curtain, and their daguerreotypes showed clearly

[34] Mouchez (1893); see also *Mouchez (1887) *La photographie astronomique à l'Observatoire de Paris et la Carte du Ciel,* Gautier-Villars, Paris, p. 11.

[35] In Fonveille (s.d.), p. 289.

[36] See Chap. 8.

some sunspots and the limb darkening of the Sun. It might seem strange that they did not show their photographs at the Academy, but the subject was in a way the property of Arago.

The first photograph of the totally eclipsed Sun was obtained on 18 July 1851 by August Busch (1804–1855), director of the Königsberg observatory, assisted by the photographer Wolfgang Berkowski. One could see on this daguerreotype protuberances and the lower corona. Other photographs were obtained, now on wet collodion on glass, by Warren de la Rue (1815–1889), Father Antonio Secchi and Foucault during the total solar eclipse of 18 July 1860 in Spain.

Photographing the Moon was easier because there was no need for a fast shutter; but one needed a telescope with an equatorial mount to track the target. As early as 1840, the American chemist John William Draper (1810–1882), the father of the famous astronomer Henry Draper (1837–1882), obtained in 20 min a reasonably good daguerreotype of the Moon. Comets were much more difficult, and the first photographs, of rather poor quality, were only obtained in 1858 by an English photographer, William Usherwood (1821–1915), and in the United States by the astronomer George Phillips Bond (1825–1865). The latter had obtained the first photographs of stars a year earlier. As for nebulae, the first one, the Orion nebula, was photographed in 1880 by Henry Draper.

The Water Supply of Paris and the Artesian Wells

As for all the great cities, the water supply of Paris always posed big problems and, not surprisingly, Arago showed an interest in the subject as a member of the city council. He proposed improvements in two directions: using the water of the Seine river, and the drilling of artesian wells.

In a speech at the Chamber of Deputies on 2 March 1846,[37] Arago proposed replacing an old pump, which elevated the water of the Seine to some reservoirs, with a new one of 4000 hp, which would use a turbine invented by his friend Benoît Fourneyron (1802–1867). This turbine (Figs. 11.13 and 11.14) was supposed to have an efficiency of the order of 90 %, but the actual value was lower. Arago also proposed, in order to defend Paris against possible enemies, that the trench around the fortifications of the city be flooded rapidly in case of attack. Six Fourneyron turbines would have raised the water to large reservoirs in the Belleville quarter, from which it would have descended in case of danger. In time of peace, the system would have been used for water supply, mills, etc. These grandiose projects were not realized, but fortunately the turbines could find other applications.

More reasonable was the idea of supplying water by artesian wells. A long, late text by Arago is a real encyclopedia on the subject.[38] The first such well is supposed to have been drilled in 1126 at Lillers, a village located to the north of France, in the

[37] *Arago, F. *Œuvres complètes* t. 5, p. 527–560.

[38] *Arago, F. *Œuvres complètes* t. 6, p. 263–480.

Fig. 11.13 The Fourneyron turbine. This turbine with a vertical axis is particularly adapted to low level differences. Water enters in the cylinder BB and exits on the periphery DD of the rotating wheel EE that powers the shaft F. A front view of this wheel can be found in Fig. 11.14. A valve driven by the rods bb regulates the flow, and the lever driven by the rod L insures that the wheel is well placed with respect to the flow in CC. From Delaunay (1850), author's collection

Fig. 11.14 The wheel of the Fourneyron turbine. The central part B is fixed, and its blades send the water to the mobile external part D. The measured efficiency varied from 60 % for a level difference of 60 cm to 80 % for a large difference. From Delaunay (1850), author's collection

province of Artois, hence the name. The Chinese also knew the principle. In the Parisian basin, the geological layers form gigantic concentric bowls. Rainwater falling in the external parts of the basin can be trapped between two impermeable layers. To obtain water in the central parts of the basin, it suffices to drill a well, pressurized because of the level difference between the place where it infiltrates and that where

Fig. 11.15 The drill of the Grenelle artesian well. Private collection

it is collected. This water is pure and drinkable because it has been filtrated along its long path; this is a considerable advantage with respect to river water. It is necessary to install a tube along the well in order to avoid the collapse of the sides. For the first wells like that of Lillers, these tubes were in oak wood, but they were replaced later by metallic ones.

Arago was the successful promoter of artesian wells to feed Paris, and also to feed his department of Pyrénées-Orientales, where several artesian wells were drilled. In Paris, the town council voted on 28 September 1832 a subvention of 18,000 francs (about $75,000) to drill three wells. After some study, it seemed more reasonable to drill only one near the slaughterhouse of Grenelle, which needed much clean water. The drilling started on 30 December. But difficulties were largely underestimated, and the operation was to extend over many years and to cost much more than foreseen.

The contractor Georges Mulot (1792–1872) was supposed to drill to a depth of 400 m, for an amount raised to 80,000 francs. Initially, the rotary drill was driven by five or six men. But, when a depth of 150 m was reached, seven horses replaced the men (Fig. 11.15). On 21 March 1837, the contract depth of 400 m was reached, but there was no water. Arago obtained from the council a supplement of 52,000 francs which should have allowed to reach 500 m, but without the provisional tubing that would cost an additional 53,000 francs. A new contract was signed on 1 September, but many incidents occurred so that on 3 August 1838 the depth reached was only 408 m, and 450 m on 29 April 1839. Still no water at 500 m! Another contract of

Fig. 11.16 The medallion of George Mulot on the fountain commemorating his artesian well. There are medallions of three other celebrities around the fountain. Wikimedia Commons, Siren-Com

84,000 francs was set to drill to 600 m. Finally, water shot out on 26 February 1841 from a depth of 548 m. This was not the end: the provisional tubing had to be replaced by the permanent one. Copper tubes were first used, but they crushed and had to be taken out and replaced by stronger iron tubes. Everything was finished on 30 November 1842, after 10 years and 362,432.65 francs (about $1.5 million), 20 times the initial subvention. The well supplied 2.2 m^3 of water per minute, at a temperature of 27.5 °C.

As expected, this operation, according to Arago, raised "truly unbelievable" commentaries, which were inserted in several newspapers in 1841 and 1842. Some journalists even thought that the well could collapse and that the water of the Seine would then disappear in the chasm. However, Arago had always been confident in the success of the enterprise. He added that the relatively high temperature of water coming from large depths would serve to heat greenhouses, hospitals and jails. Finally, the remarkable technical feat of this drilling was the subject of many eulogies. A cast iron column, 43 m high, was raised on the Place de Breteuil near the well, so that water could be sent by gravity to tanks near the Panthéon, from which it was distributed, and also to celebrate the feat. It was rather ugly, and was replaced in 1905 by a fountain installed Place Georges Mulot, very close to the Place de Breteuil (Fig. 11.16).

Arago made use of the drilling for thermometric measurements with his friend François-Hippolyte Walferdin (1795–1880). On 2 November 1840, when the reached depth was 505 m, he announced to the Academy that the temperature increase was one degree per 26.4 m. This was one of the few reliable measurements of this phenomenon. The drilling also provided the first geological cross-section of the Parisian basin.

The Grenelle well was the first drilled in Paris, but not the last one. A dozen exist at present.[39] The well finished in 1861 in Passy to feed the lakes in the *Bois de Boulogne* caused a decrease in the flow rate of the Grenelle well. Other wells still feed three swimming pools,[40] at least in part; the well near the Radio house has insured its heating and air conditioning since 1963, thanks to heat pumps. The geothermal use of the wells foreseen by Arago has become reality. However, the Parisian wells have lost their artesian character; the pressure has decreased due to extensive usage, and water has now to be pumped to the surface.

[39] For a history of the artesian wells in Paris, see Clément and Thomas (2001), p. 150–151.

[40] That of the Butte-aux-Cailles, and the Hébert and Blomet swimming pools.

Chapter 12
Arago's Legacy

The time has come to see how Arago's image and that of his work has evolved after his death. The texts that have been devoted to this are rather numerous, and it suffices to follow them. Let us discuss what occurred immediately after his death.

Funeral and Discourses

Arago's death caused much emotion at the Academy of Sciences, which suspended its meeting of 3 October 1853 when his death was first known. A short speech by the botanist Charles Combes (1801–1872) expressed a sincere grief. Arago's fame was such that it was difficult for Emperor Napoléon III to neglect the scientist's death nor avoid ordering a solemn funeral service, delegating Marshall Vaillant to represent him. The Minister of the Navy and Colonies was present as a reminder of Arago's brief appointment to that office. There were also many representatives of the Institute, of the École Polytechnique and of the Council of the Département de la Seine. An enormous crowd, which learned of the event in an announcement by the journal *Le Globe*, accompanied the funeral procession from the Observatory to the Père Lachaise cemetery, passing in front of the church of Saint-Jacques-du-Haut-Pas, where the funeral mass was said. The army was there, not so much to pay homage to the deceased man than to avoid potential excesses of the crowd, which had come more for Arago the politician than for Arago the scientist.

However, it is the scientist who was celebrated in the official speeches, in order to avoid speaking of the politician. His friends had expected that, as the famous writer Alexandre Dumas (1802–1870) wrote in his diary. The only political allusion was in the speech of Admiral Charles Baudin, the president of the Board of Longitudes, who alluded to Arago's activity as Minister of the Navy and Colonies. A long necrology by Barral,[1] in which one could infer political convictions close to

[1] *Barral (1853).

J. Lequeux, *François Arago*, Astrophysics and Space Science Library 421,
DOI 10.1007/978-3-319-20723-0_12

The model of Arago's statue, erected in Paris in 1893. The sculptor, Alexandre Joseph Oliva (1823–1880) is visible on the left. The statue was placed close to the Paris Observatory on a pedestal by Joannis. Only this pedestal survives, because the bronze statue was melted during the Second World War. Bibliothèque de l'Observatoire de Paris

those of the Arago's, only mentions the abolition of slavery and contains the following sentences:

We should not give here an appreciation on the statesman, and recall the acts of one of the leaders of the revolutionary government. He wrote: "As far as politics is concerned, one cannot without danger follow the inspirations of one's heart." We will not add anything to this, and we will only emphasize that, in all the positions he took up, he was always a great citizen. After the [coup d'état of] 2 December [1852], he refused the oath, and the political rigors stopped at his name: politics should never interfere with science.

There was nothing more in any of the numerous articles published in the daily newspapers. In all these papers, the scientist was glorified, in particular as a promoter of science, while his political activities remained in the shadow. The famous critic Sainte-Beuve wrote 5 months after Arago's death[2]:

I would readily imagine a double statue of Arago: one of him young, in the beauty of his ardor and in his most virile ascent, devoted to pure science, to the measurement of the globe, to the discovery of the celestial spaces and of the laws of light, as he could be, aged 21, in his quiet watches on the plateau of the Desierto de las Palmas. The other would represent him after the end of his double career, a saddened figure, but still imposing; a tall man; his head bowed, as half-struck by lightning; seemingly warning by some gesture the scientists to avoid the reef of the crowd: but even so, and on all sides, let us engrave and read again and again on the pedestal the memorable dates of the rendered services.[3]

Eugène de Mirecourt (1812–1880), a sulfurous polemist who wrote a small book on Arago,[4] was probably the originator of a tenacious legend that pretends that Arago's illness was due to his discouragement after the failure of the Second Republic, although diabetes struck him well before 1848.

In spite of the recommendation of Sainte-Beuve, one did not engrave on Arago's tombstone at the Père Lachaise cemetery "the memorable dates of the rendered services," but a list of what was considered as his main scientific works, i.e.:

Extension of the meridian
Colored polarization
Magnetism of rotation
Photometric method and observation

The project for the grave, planned by David d'Angers, a close friend of Arago's, was not entirely realized. The grave (Fig. 12.1) consists only of a bust, a copy of that sculpted by d'Angers in 1843 for Estagel town hall, and placed on a granite pedestal created by Félix Duban (1797–1870). In 1858, the painter and sculptor Paul-Joseph Chenavard (1808–1895) proposed adding to the grave a small building with a recumbent statue (Fig. 12.2). This was not realized either, although it would have had "a great impact," according to its creator.

---

[2] *Sainte-Beuve (1854), see p. 18. First a socialist, Sainte-Beuve joined the service of the Empire.

[3] Probably only the services rendered to science!

[4] *Mirecourt (1859).

Fig. 12.1 Arago's grave at the Père Lachaise cemetery. Wikimedia Commons, Pierre-Yves Beaudouin

Other Times, New Dispositions

Some reservations soon appeared as to Arago's actions at the Chamber of Deputies. They were not relative to his discourse on universal suffrage, which had facilitated the accession of Napoléon III to absolute power, but rather to railways. Here is an example from 1857[5]:

[5] *Audiganne (1857), see p. 90.

Fig. 12.2 A project by Chenavard for Arago's grave. Bibliothèque de l'École Polytechnique

From 1830 to 1848, Arago used with success in many occasions his influence as a scientist on the promotion of useful measures. However, we should except the question of railways in which, in spite of many luminous ideas that fill his writings, he obviously let the political elements dominate. It would have been better for science if he had served it exclusively until the end.

Péreire at Estagel

In 1863, the wealthy banker Isaac Péreire, who knew Arago well and even attended several of his astronomy lessons, entered politics and ran for election as deputy of the Département des Pyrénées-Orientales, which had been Arago's stronghold.

Fig. 12.3 Arago's statue in his birthplace, Estagel. This is not the original bronze one, which was melted during the Second World War, but its replacement of 1955 by the sculptor Marcel Homs (1910–1995). Wikimedia Commons, BenS67

Of course, he spoke very highly of Arago in his speeches and this helped enormously in his election. The only promise he fulfilled after being elected (there were many he made) was to have a statue of the great man erected in his birthplace, Estagel. This statue (Fig. 12.3) cost him very little because the money came from a public subscription: of a total of 2500 inhabitants of Estagel, as many as 2000 subscribed, flattered by this "novel initiative in the Pyrénées-Orientales, where no other mortal, king or very important person, had acceded at such an honor, with the exception of François Arago whose bust was installed long ago at Estagel."[6]

The inauguration of the statue took place on 30 August 1865 in the absence of Arago's relatives, who did not want to endorse Péreire's maneuvering. But Péreire succeeded in attracting eminent scientists like the chemist Henri Saint-Claire

[6] Frénay (1986).

Deville and the mathematician Joseph Bertrand (1822–1900), both sent by the Academy of Sciences. Pereire's speech[7] was well documented and quite skillful. Hiding nothing of Arago's political activities, he rather exploited them, even claiming that the social ideas of the Emperor were close to those of Arago! His only reservations were once again about the railways:

[Arago] would have wished that one waited for new progress in the construction of railways before starting work. Yet it is not in human nature to reach perfection[...] The inconveniences of an imperfect application are less than that resulting from an infinite delay.[...] Arago would have been better inspired to try to define to whom the construction of new railway lines would have been given. [Of course, because Pereire was one of the great beneficiaries!]

The recuperation of Arago did not stop there. Now that he could no longer oppose the Emperor and was still famous, there were only advantages to give his name to the large boulevard that Baron Georges Haussmann (1822–1891) had established close to the Observatory.

The Third Republic

Things were to change after the fall of Emperor Napoléon III and the advent of the Third Republic in 1870. Another recuperation of Arago as a politician took place, but in the other direction. The remarks made at the inauguration of a statue of Arago in Perpignan (Pyrénées-Orientales) are telling[8]: Arago's reputation and his work were resurrected as the third Republic was founded. Here is an extract of the speech[9] of the Minister of Public Instruction, Jules Ferry (1832–1893):

The immortal works and the scientific glory of François Arago are parts of Humanity's patrimony. Posterity began for him during his lifetime; the years could only confirm the opinion of his contemporaries.[...] There was in this scientist a citizen, a patriot, a politician with high aims, an ardent democrat, a republican. He belonged to the heroic and tormented age of the militant republic; in the first half of this century, no other man has better represented the grandeur and the vicissitudes of our country. He was amongst the best, the purest, the most farsighted; he was one of the most determined, and at the same time one of the wisest.

On 11 June 1893, another statue of Arago was inaugurated in Paris thanks to a subscription open in 1886, the year of the centenary of Arago's birth (see the photograph

[7] Péreire (1865).

[8] This statue by Antoine Mercié (1845–1916) rests on a pedestal with three bronze bas-reliefs representing episodes of Arago's life. On one of them, he is figured appeasing the anger of the mob on the barricades in 1848, and on another one he is supposed to dictate his *Astronomie populaire* to his niece Lucie Laugier.

[9] Cited by Sarda (2000).

at the beginning of this chapter). Political disputes had delayed the ceremony.[10] By then the sculptor Oliva and Admiral Mouchez, the director of the Observatory, were dead. Seven speeches were given, emphasizing the scientific work of Arago, and to a lesser extent his political activities. Particularly interesting was that by François de Mahy (1830–1906), deputy of Réunion, an island department of France, who recalled Arago's role in the abolition of slavery.[11] Schœlcher was still alive but could not come because of illness, and the two men were put on the same footing in the speech.

The speech by the astronomer Félix Tisserand[12] was the first to emphasize Arago's contribution to astronomy, although limited to the Solar system; he mentioned the gaseous nature of the surface of the Sun, the measurements of the diameters of planets, the scintillation. He added:

Arago must be considered as the real founder of physical astronomy, that part of science so marvelously developed since. He foresaw the future importance of photography for the study of celestial bodies.

However, he could not imagine that, beside physics, chemistry would penetrate into the domain of astronomy and would unveil the intimate constitution of celestial bodies: indeed, spectral analysis has been only discovered after the death of Arago.

In his posthumous address,[13] Admiral Ernest Mouchez was more vague and expressed some reservations about the scientist:

Arago has no doubt contributed by his own researches to the progress of science; but what characterizes best his work and his genius is his remarkable intuition that allowed him to foresee rapidly the most remote consequences of the new ideas.[…] His many occupations, his feverish activity in searching for new experiments, did not often allow him to deepen the facts he was discovering: but he mentioned them with a rare selflessness to colleagues or to young scientists, that he helped with his advices and his fecund collaboration in order to derive the foreseen consequences of these facts.

However, speaking in the name of the Academy and of the Board of Longitudes, Alfred Cornu defended the memory of Arago against the doubts about the quality of his personal research, and emphasized rightly his skills as a popularizer of science.[14] To tell the truth, Jules-Célestin Jamin had already made the same reservations as Mouchez in his eulogy of Arago at the Academy of Sciences,[15] a eulogy delayed to

[10] In February 1886, the town council of Paris did not authorize the organizing committee to give a public banquet, and refused a subvention for the statue. The opposition was essentially due to the fear that the much hatred Étienne Arago would benefit from the glory of his brother. Things turned better after his death in 1892, but the right-wing opposition succeeded in delaying again the ceremony: see *Fonveille (s.d.), see p. 280–289, in which large extracts of the speeches are reproduced.

[11] Mahy (1893).

[12] Tisserand (1893).

[13] Mouchez (1893).

[14] Cornu (1893).

[15] Jamin (1885).

1885 for various reasons, political in particular. These reservations are rather harsh, but fully justified:

It is rare that an inventor has reached the limits of his discovery: he looked for consequences where they were not, he was lost in a labyrinth where no thread could guide him, he came close to the truth without seeing it and finally he left to his successors the crop he had sowed. As many others before him, Arago left without completion the great discovery he had just made [this was the rotational polarization]. He was gifted with an unparalleled insight, he guessed the discoveries before making them, but he lacked patience for the details; he opened mines without exploiting them, he began works without achieving them. Once his curiosity satisfied, he looked for other topics. He was like a busy traveller riding through a virgin country, just giving a name to it and then hurrying towards more remote places. Experimenter by inspiration, discoverer by instinct, he had too much passion and not enough leisure; a too fertile mind, and not enough of this stubborn persistence that allows finishing what was begun. How many others would be content with such defects? As to the theoretical ideas that can confine an entire science into a few general hypotheses and leave a indelible trace, he produced none, and sometimes he even pushed them away while his own experiments inspired others.[...] What we should admire above anything in Arago's scientific work, it is the amazing fecundity of this unrivalled inventor. He was fit and ready for any kind of research.

At the beginning of the twentieth century, an interesting testimony of the feelings of the time was given by Wilfrid de Fonveille (1824–1914), a popularizer of science who had known Arago and whose father, a "distinguished industrial," had been "strongly encouraged" by him.[16] As a convinced republican, Fonveille had much sympathy for Arago. He emphasized very clearly the fact that the political activities of Arago had eclipsed to some extent his scientific side, and he seemed to suggest that his discoveries would perhaps not have the longevity foreseen by most preceding commenters. He hid nothing of some errors of Arago, in particular (again!) about railways.

Modern Times

Let us jump another half century to come to a text by André Danjon (1890–1967),[17] one of Arago's brightest successors as the head of the Paris Observatory. Of all commenters, he is probably, with Jamin, the one who best appreciated Arago's qualities and defects:

The great popularity, he found it, he enjoyed it with delight. His contemporaries have seen in him one of the greatest men of his century, the Napoléon of science. The point of view has changed. More than talent, as he had much of it, perseverance and a taste for diligent work were lacking to Arago who could not equal Laplace, Fresnel or

---

[16] Fonveille (s.d.). The citation is from Péreire (1865) p. 12.

[17] Danjon (1853).

Fig. 12.4 One of the 135 medallions placed by the Dutch conceptual artist Jan Dibbets along the Paris meridian, from Cité Universitaire to Porte Montmartre. These enigmatic medallions trace in the city a line, the symbolic center of which is the Observatory. They replace from 1994 the destroyed statue of 1893 (see the figure at the beginning of this chapter). Wikimedia Commons, Poulpy

Ampère. One cannot place Arago at the level of these great geniuses; his work does not have the radiance one would like. However, it contributed to the progress of science by opening new chapters... that he left to others to develop. Arago was much more attracted by the discovery of new phenomena than by the deepening and the practical application of his own discoveries.[...] More than any other, he contributed to the prosperity of the great institutions and of the great scientific establishments. He recognized Fresnel, Foucault or Le Verrier, amongst others. He extended a hand to them and helped them in their ascent.

One can only share this opinion. However, one should not underestimate Arago's personal contribution to physics and astronomy, which is far from negligible; to discover the gaseous nature of the Sun and stars is not a small feat! Today, the politician has been somewhat forgotten in the history books, but astronomers and physicists keep alive the memory of the scientist, whose name can be read on medallions that the pedestrian encounters on the Parisian ground (Fig. 12.4).

Magnificence and Decadence of French Physics and Astronomy in the Nineteenth Century

The first half of the nineteenth century saw in France an incredible blossoming of first-rate scientists and engineers. Of course, some of the great "geometers" of the end of the eighteenth century, like Delambre, Lagrange, Laplace or Legendre, survived the Revolution, as well as scientists like Coulomb, Cuvier, Lamarck or Monge, who did not experience the sad fate of Condorcet or Lavoisier. But new stars were born, thanks to the social promotion allowed by the free-tuition *grandes écoles* created during the Revolution. Moreover, as shown brightly by Jacques Debyser,[18] time was ready for a fast development of science, because the Age of Enlightenment had removed from scientific thinking metaphysical aspects inherited from medieval times. Limiting ourselves to physics, one could see the birth of all the science that is presently useful in daily life (with of course the exception of electronics and data processing): optics, thermodynamics, electromagnetism, remote transmission. Technology was not out of step: photography, lighthouse optics, turbines, combustion and electric motors were born before 1850. The first automobile was to run soon after. Only the telephone, the dynamo and the alternator had to wait for a few decades, not to forget the airplane.

However, this magnificent blossoming turned anemic in the second half of the nineteenth century, where one would search in vain in France for geniuses like Laplace, Fresnel, Ampère, Sadi Carnot or Joseph Fourier. The fault is in the sclerosis of men and institutions, after the enthusiasm of the romantic period, of which Arago was one of the heralds. The possibilities for social promotion had shrunk considerably; the Academy of Sciences and the universities sought refuge in their ivory towers; the bourgeois civilization showed much more interest in technology, which experienced strong progress, than in fundamental research. One could not find in France at that time the creative daring of foreign scientists like Maxwell, Helmholtz, Mach, Lorentz, Planck or Einstein, to cite only a few examples. The only physicists of similar level were Henri Poincaré, Paul Langevin, and to some extent the Curies. France had lost its supremacy, which it would not recover despite the work of scientists like Louis de Broglie, Jean Perrin, Louis Néel and the Joliot-Curies. In the 1950s—and I can testify to this myself—the teaching of physics at the university level was of a distressing mediocrity.[19] The situation is now much better but has required much effort.

Astronomy, which was an older and more mature science than physics, did not experience changes of the same importance at the beginning of the nineteenth century, at least in France. After Laplace and Delambre, the list of French astronomers is not impressive. Indeed, Danjon commented[20]:

Arago had close to him devoted collaborators, who admired him up to a degree close to adoration. The oldest were Mathieu, his brother-in-law, and Laugier who was the son-in-law of Mathieu.[...] Eugène Bouvard, Mauvais, Goujon, Barral were amongst the devotees. If one excepts Mathieu, who was a good mathematician, this environment was mediocre. It was lacking, to serve well the Master, qualities complementary of his defects.

[18] Debyser (2007).

[19] See on this topic Abragam (1989).

[20] Danjon (1953).

Of course, there was Le Verrier, who did not belong to the Arago clan.[21] Famous for his discovery of Neptune in 1846, he spent the rest of his life building a complete theory of the motions in the Solar system, an enormous work that served for a century as a basis for the ephemerides of the Board of Longitudes. While doing so, he discovered an anomaly in the motion of Mercury that gave the first observational proof of General Relativity. In parallel, or following Le Verrier, we find other talented specialists of celestial mechanics like Delaunay or Tisserand, but the old French supremacy in this domain vanished: England, Germany, and especially the United States, with Simon Newcomb, provided formidable competitors. Only Henri Poincaré, a visionary mathematician, precursor of the chaos theory, emerged at the summit of international astronomy.

In the domain of observation, the situation was not bright in the nineteenth century France. I have largely discussed this point. Although Foucault created the modern reflecting telescope, the best realizations after him were in foreign countries. France can only take pride in two important discoveries: that of emission-line stars by Wolf and Rayet, made in 1867 with one of Foucault's telescopes, and, the following year, that of a new solar line identified later as due to a new element, helium: this discovery was made simultaneously by Janssen in France and by Norman Lockyer (1836–1920) in England. The talents of French astronomers as photographers were soon exhausted in the realization of the *Carte du ciel*, a gigantic project that was a half fiasco. As to astrophysics, it disappeared, stillborn, with Arago. A shy revival came with Janssen 10 years after, then later with Henri Deslandres (1853–1948), but astrophysics did not raise much interest with the "classical" astronomers, and even less with the physicists. France missed most of the great discoveries that were made in England, in Germany and in the United States. It is only after 1936 that France regained very progressively its status amongst the countries where astronomy flourishes.[22]

[21] See Lequeux (2013).

[22] See Pannekoek (1989) for the best history of astronomy in existence for the period before 1930, and Lequeux (2005) for a history after 1910.

Bibliography[1]

Writings by Arago

*Arago, F. (1854-1862) *Œuvres complètes de François Arago*, ed. par J.-A. Barral, 13 vol. Gide, Paris et T. O. Weigel, Leipzig (*OC*).

Arago, F. (1855) *History of my youth*, transl. Baden-Powell, Longman; Brown, Green & Longmans, London. Accessible via google books.

*Arago, F. (1854-1857) *Astronomie populaire*, ed. par J.-A. Barral, Gide et J. Baudry, Paris (*AP*).

*Arago, F. (1839) *Rapport de M. Arago sur le daguerréotype...* Bachelier, Paris.

Biot, J.-B., Arago, F. (1821) *Recueil d'observations géodésiques, astronomiques et physiques exécutées par ordre du Bureau des longitudes de France, en Espagne, en France, en Angleterre et en Écosse, pour déterminer la variation de la pesanteur et des degrés terrestres sur le prolongement de la Méridienne de Paris, faisant suite au troisième volume de la Base du système métrique, rédigé par MM. Biot et Arago*, Courcier, Paris. Accessible via google books.

*Hamy, E.-Th., ed. (1907) *Correspondance d'Alexandre de Humboldt avec François Arago (1809-1853)* R. Guilmoto, Paris.

Books and Documents on Arago[2]

Archives départementales des Pyrénées-Orientales (2009) *Les Arago, acteurs de leur temps*, Archives départementales, Perpignan.

*Audiganne, A. (1857) *François Arago, son génie et son influence*, Garnier, Paris.

---

[1] <Footnote ID="Fn1"><Para ID="Par2">Many documents in French are accessible freely on Gallica, the site of numerised documents of Bibliothèque Nationale de France (<ExternalRef><RefSource>http://gallica.bnf.fr</RefSource><RefTarget TargetType="URL" Address="http://gallica.bnf.fr/"/></ExternalRef>). They are marked with an asterisk.</Para></Footnote>

[2] <Footnote ID="Fn2"><Para ID="Par3">The vast majority is in French. This bibliography is almost complete. A few other references of lesser interest are cited in the biography by Sarda, 2002.</Para></Footnote>

J. Lequeux, *François Arago*, Astrophysics and Space Science Library 421,
DOI 10.1007/978-3-319-20723-0

Audubert, R. (1953) *Arago et son temps*, Imprimerie Alençonnaise, Alençon.
*Barral, J. A. (1853) François Arago, *Journal d'Agriculture pratique*, Ser. 3, t. 7, part 2, p. 309-317.
Barthalot, R. (1987) Arago et l'Observatoire de Paris, in *François Arago, Actes du colloque national des 20, 21 et 22 octobre 1986*, Cahiers de l'Université de Perpignan, N° 2, p. 11-22.
Bijaoui, A. (2003) François Arago, un astronome à l'esprit universel, *L'Astronomie* 117, p. 402-407 (September 2003).
Bijaoui, A. (2006) François Arago, l'optique et les théories de la lumière, in *Arago, Journée scientifique du Bureau des longitudes*, Académie des Sciences, Paris, p. 35-46.
Bobis, L., Lequeux, J. (2003) *Catalogue de l'exposition « François Arago et l'Observatoire de Paris »,* Observatoire de Paris. See also the virtual exhibition on http://expositions.obspm.fr/F.Arago/intro.html.
Cawood, J.A. (1974) *The Scientific work of D.F.J. Arago*, Ph.D. Thesis, University of Leeds.
Cawood, J. (1985) François Arago, homme de science et homme politique, *La Recherche*, 172, p. 1464-1471 (December 1985).
Cayrel, R. (2006) Arago et la diffusion de la connaissance, in *Arago, Journée scientifique du Bureau des longitudes*, Académie des Sciences, Paris, p. 59-68.
Chauvet, H. (1954) *François Arago et son temps*, Ed. des amis de François Arago, Paris.
Cornu, A. (1893) Discours au nom de l'Académie des sciences et du Bureau des longitudes, *Inauguration de la statue de François Arago à Paris le Dimanche 11 juin 1893,* Firmin-Didot, Paris, p. 31-40.
Danjon, A. (1953) François Arago, *L'Astronomie* (December 1953), p. 445–464. Accessible via http://cdsads.u-strasbg.fr/abs/1953LAstr..67..445D.
Daumas, M. (1987) *Arago, La jeunesse de la science,* 2e édition, Paris, Belin, Collection *Un savant, une époque.*
Débarbat, S. (2003) Arago et les hommes de science, *L'Astronomie* 117, p. 386-392 (September 2003).
Débarbat, S. (2006) Arago et l'*Astronomie populaire*, in *Arago, Journée scientifique du Bureau des longitudes,* Académie des Sciences, Paris, p. 49-57.
Dumont, S. (2003) François Arago et l'Astronomie populaire, *L'Astronomie* 117, p. 394-400 (September 2003).
*Fonvielle, W. de (s.d.) *La jeunesse d'un grand savant républicain,* E. Gaillard ou Société française d'éditions d'art, Paris.
Frachon, B., Lefebvre, M. (2006) Arago et la physique de la mer, in *Arago, Journée scientifique du Bureau des longitudes*, Académie des Sciences, Paris, p. 7-21.
Frénay, E. (1986) *Arago et Estagel, son village natal*, Mairie d'Estagel.
Gerest, R. (1988) *Arago, le plus grand cœur et la plus forte tête de son époque*, s.n.
Grison, E. (1987) Arago et l'École polytechnique, in *François Arago, Actes du colloque national des 20,21 et 22 octobre 1986*, Cahiers de l'Université de Perpignan, N° 2, p. 47-66.
Grison, E. (1989) François Arago et l'École polytechnique, in *Bulletin de la Société des amis de la Bibliothèque de l'École polytechnique,* N° 4.
Hahn, R. (1970) Arago, in *Dictionary of scientific biography*, ed. Charles Coulson Gillespie, vol. 1 p. 200-203, Charles Scibner's sons, New York.
Jamin, J.-C. (1885) *Éloge historique de M. François Arago*, Firmin-Didot, Paris.
Le Mouël, J.-L. (2006) Arago et le magnétisme terrestre, in *Arago, Journée scientifique du Bureau des longitudes*, Académie des Sciences, Paris, p. 23-32.
Lequeux, J. (2003) *François Arago, un savant généreux; physique et astronomie au XIXe siècle,* EDP Sciences/Observatoire de Paris, Paris.
Mahy, F. de (1893) Discours au nom des colonies françaises, *Inauguration de la statue de François Arago à Paris le Dimanche 11 juin 1893,* Firmin-Didot, Paris, p. 41-55.
McCauley, A. (1991) François Arago and the Politics of the French Invention of Photography, in Younger, D.P., ed. *Multiple views, Logan Grant Essays on Photography*, 1983-1989, University of New Mexico Press, Albuquerque.
*Mirecourt, E. de (1859) *François Arago*, G. Havard, Paris.

Mouchez, amiral (1893) Discours [posthumous], *Inauguration de la statue de François Arago à Paris le Dimanche 11 juin 1893,* Firmin-Didot, Paris, p. 11-21.
*Pereire, I. (1865) *Discours prononcé le 31 août 1865 à l'occasion de l'inauguration de la statue de François Arago à Estagel*, Paul Dupont, Paris.
Pinet, G. (s.d., 1907) *Arago, notice biographique*, coll. Panthéon polytechnicien, impr. De Lafolye, Paris.
Poincaré, R. (1893) Discours, *Inauguration de la statue de François Arago à Paris le Dimanche 11 juin 1893,* Firmin-Didot, Paris, p. 31-40.
*Poirier, J.-P. (2003) *Arago, un savant et un homme généreux*, discours prononcé le 10 juin 2003 devant l'Académie des sciences; accessible via http://www.academie-sciences.fr/activite/conf/exposePoirier_100603.pdf.
Poirier, J.-P. (2006) Arago et les notices de l'*Annuaire du Bureau des Longitudes,* in *Arago, Journée scientifique du Bureau des longitudes*, Académie des Sciences, Paris, p. 69-75.
*Sainte-Beuve (1854) *Causeries du Lundi*, Garnier, Paris: t. 10, p. 1-18.
Sarda, F. (2000) *Les Arago: François et les autres,* Paris, Taillandier.
Taton, R. (1987) Arago et l'Académie des sciences, in *François Arago, Actes du colloque national des 20, 21 et 22 octobre 1986*, Cahiers de l'Université de Perpignan, N° 2, p. 23-46.
Ten, A.E. (1987) Arago géodésien, in *François Arago, Actes du colloque national des 20, 21 et 22 octobre 1986*, Cahiers de l'Université de Perpignan, N° 2, p. 67-88.
Tisserand, F. (1893) La statue d'Arago, L'Astronomie 12, p. 241–246, accessible via http://adsabs.harvard.edu/full/1893LAstr..12..241T.
Tisserand, F. (1893-2) Discours au nom du comité de souscription de l'Observatoire, *Inauguration de la statue de François Arago à Paris le Dimanche 11 juin 1893,* Firmin-Didot, Paris, p. 3-9.

Other Ancient Books

*(Ampère, A.-M.) (s.d.) *Exposé méthodique des phénomènes électro-dynamiques et des lois de ces phénomènes,* Plassan, Paris.
Ampère, A.-M., Babinet, J. (1822) *Exposé des nouvelles découvertes sur l'électricité et le magnétisme de MM. Oersted, Arago, Ampère, H. Davy, Biot, Erman, Schweiger, de la Rive, etc.* Méquignon-Marvis, Paris. Accessible via Google books.
*Anonyme (Ampère) (1824) *Description d'un appareil électrodynamique, construit par M. Ampère,* Crochard et Bachelier, Paris.
Anonyme (1823) *L'astronomie enseignée en vingt-deux leçons*, traduit de l'anglais par Ph. Cr., Audin, Canel, Paris.
Arago, F. & Lardner, D. (1845) *Popular lectures on Astronomy: by M. Arago, with additions and corrections by Dionysius Lardner*. Greeley & McElrath, New York.
*Bailly, E.-M. (1830) *Manuel d'astronomie*, Roret, Paris.
Bartholin, E. (1669) *Experimenta crystalli islandici disdiaclastici Quibus mira and insolita refractio detegitur,* Hafniae [Copenhagen], accessible via Google books. Abridged English translation: *Philosophical transactions* (1670) 5, p. 2039-48, accessible as http://www.jstor.org/stable/101617.
*Becquerel, [Antoine-César] & Becquerel, E. (1847a) *Traité d'électricité et de magnétisme,* Firmin-Didot, Paris, 3 vol.
*Becquerel, [Antoine-César] & Becquerel, E. (1847b) *Éléments de physique terrestre et de météorologie,* Firmin-Didot, Paris.
Bernoulli, D. (1738) *Hydrodynamica*, Dulsecker, Basel; English translation (1968), Dover; latine original accessible via http://num-scd-ulp.u-strasbg.fr:8080/179/ English translation (1968): Dover, New York.
Berthaut, Colonel (1898) *La Carte de France 1750-1898, étude historique*, 2 Vol., Imprimerie du Service Géographique de l'Armée.

*Bigourdan, G. (1928-1932) Le Bureau des longitudes: son histoire de l'origine à ce jour, in *Annuaire du Bureau des longitudes* pour 1928, p. A.1-A.72; 1929, p. C.1-C.92; 1930, p. A.1-A.110; 1931, p. A.1-A.151; 1932, p. A.1-A.91.

*Bouguer, P. (1760) *Traité d'Optique sur la Gradation de la Lumière,* Paris, Guérin et Delatour. English translation by Middleton, W.E.K. (1961) University of Toronto Press.

*Buffon, G.-L. de (1778) *Des Époques de la Nature*, ed. critique par J. Roger, in *Mémoires du Muséum d'Histoire Naturelle*, t. 10 (1962); ed. orig. accessible par http://www.buffon.cnrs.fr Anonymous English translation available.

*Carnot, S. (1824) *Réflexions sur la puissance motrice du feu et sur les machines propres à développer cette puissance*, Bachelier, Paris. Translation of the 1872 edition: http://en.wikisource.org/wiki/Reflections_on_the_Motive_Power_of_Heat/Chapter_3.

*Cassini, J.D. (1810) *Mémoires pour servir à l'histoire des Sciences et à celle de l'Observatoire Royal de Paris, suivis de la vie de J. D. Cassini, écrite par lui-même, et des éloges de plusieurs académiciens morts pendant la Révolution,* Bleuet, Paris.

*Cassini, J.D., Méchain, P.A. & Legendre, A.M. (s.d., 1791?) *Exposé des opérations faites en France en 1787, pour la jonction des Observatoires de Paris et de Greenwich,* Institution des Sourds-Muets, Paris.

*Clairaut, A. (1743) *Théorie de la figure de la Terre, tirée des principes de l'hydrostatique,* David Fils, Paris.

*Delaunay, Ch. (1850) *Cours élémentaire de mécanique théorique et appliquée*, Victor Masson, Langlois et Leclercq, Paris (many editions).

*Delambre J.B.J., Méchain P.A. (1806-1810) *Base du Système métrique décimal:* voir Méchain.

*Delambre, J.B.J. (1912) *Grandeur et Figure de la Terre,* ouvrage [posthume] augmenté de notes, de cartes et publié par les soins de G. Bigourdan, Gauthier-Villars, Paris.

*Euler, L. (1746) *Nova theoria lucis et colorum,* in *Opera Omnia,* Basel (1911-1956).

*Euler, L. (1761) *Lettres à une princesse d'Allemagne sur divers sujets de physique et de philosophie,* rééd. Presses polytechniques et universitaires romandes, Lausanne, 2003: accessible sur http://num-scd-ulp.u-strasbg.fr:8080/273/ English translation by Hunter, H. (1902) accessible via https://archive.org/details/letterseulerond00eulegoog.

Faraday, M. (1855) *Experimental researches in electricity*, London, Taylor and Francis, vol. 3, accessible via Google books.

*Figuier, L. (s.d., ca.1870) *Les Merveilles de la Science*, 4 Vol., Furne, Jouvet et Cie, Paris.

*Flammarion, C. (1892) *Astronomie élémentaire*, Bibliothèque scolaire de la ligue franco-américaine de l'enseignement, Paris.

*Flammarion, C. (1880) *Astronomie populaire*, Marpon et Flammarion, Paris.

Foucault, L. (1878) *Recueil des travaux scientifiques de Léon Foucault,* ed. par Gariel C.-M., Gauthier-Villars, Paris, accessible via http://jubilotheque.upmc.fr/ead.html?id=PC_000326_001.

*Fresnel, A. (1822) *Mémoire sur un nouveau système d'éclairage des phares*, Paris, Imprimerie royale.

Fresnel, A. (1866, 1868, 1870) *Œuvres complètes d'Augustin Fresnel,* publiées par MM. Henri de Senarmont, Verdet et Léonor Fresnel, Paris, Imprimerie impériale, 3 tomes. T. 1 accessible via Google books, t. 2 via Gallica.

Gauss, C. F. & Weber, W. (1840) *Atlas des Erdmagnetismus nach den Elementen des Theorie entworfen*, Weidmann, Leipzich, accessible via http://www.maproom.org/00/16/index.php.

Gay-Lussac J. & Thenard, L.-J. (1811) *Recherches physico-chimiques, faites à l'occasion de la grande batterie voltaïque donnée par S.M.I. et R. à l'École polytechnique*, 2 vol., Deterville, Paris, accessible via http://cnum.cnam.fr.

Herschel, J.F.W. (s.d.) *Treatises on physical astronomy, light and sound contributed to the Encyclopaedia metropolitana*, London and Glasgow, Griffin, available as https://archive.org/details/treatisesonphysi00hersrich.

Herschel, J.F.W. (1847) *Results of astronomical observations made during the years 1834, 5, 6, 7, 8 at the Cape of Good Hope*, Smith Elder and Co, London, accessible via Google books.

*Herschel, J.F.W. (1853) *Nouveau manuel complet d'astronomie*, trad. A.-D. Vergnaud, Roret, Paris; this is the translation into French of his *Outlines of Astronomy* (1849), 2d. edition available via Google books.

Highton, E. (1852) *The electric telegraph, its history and progress*, London, Weale, accessible via Google books.

*Hooke, R. (1665) *Micrographia*..., John Martyn, James Allestry, London, p. 55.

Hugo, V. (1848) *Choses vues*, rééd. Gallimard, Paris (2002).

Hugo, V. (1863) *Promontorium Somnii*, in *Proses philosophiques de 1860-65*, accessible via http://fr.wikisource.org/wiki/Proses_philosophiques/Promontorium_somnii.

*Hugo, V. (1863a) *Napoléon le petit*, W. Jeffs, Londres.

*Humboldt, A. de (1816) *Voyage aux régions équinoxiales du nouveau continent*, 4 t., Schoell, Paris, et Librairie grecque – latine – allemande, Paris.

*Humboldt, A. de (1846-1851) *Cosmos, Essai d'une description physique du monde*, traduit par H. Faye et Ch. Galusky, 4 vol., Gide et Baudry, Paris.

*Huygens, C. (1690) *Traité de la lumière*, vol. XIX of *Œuvres complètes*, p. 451-550, Nijhoff, La Haye (1888-1950); English translation by Thomson, S.P. (s. d.) University of Chicago Press, kindle edition.

*Huygens, C. (1698) *Kosmotheoros*, vol. XXI of *Œuvres complètes*, Nijhoff, La Haye (1888-1950); Anonymous English translation (2010) as *Cosmotheoros*, Gale ECCO Print edition.

*Jamin, J.-C. (1858-1866) *Cours de Physique de l'École polytechnique*, 3 Vol., Mallet-Bachelier, Paris.

Lambert, J. H. (1760) *Photometria sive de mensura et gradibus luminis, colorum et umbrae*, Eberhardi Klett, Augsburg, accessible via Google books; *Photometry, or on the measure and gradations of light, colors and shade*, English translation by DiLaura D. L., The illuminating Engineering Society of North America (2001).

*Lamé, G. (1840) *Cours de physique de l'École polytechnique*, 3 t., Bachelier, Paris.

*Laplace, P. S. (1796) *Exposition du système du Monde*, de l'Imprimerie du Cercle social; éd. rév. en 1799, 1808 (chez Courier), 1813 (Mme Vve Courier), 1824 et 1836 (Bachelier). Cette dernière est accessible par Gallica en mode texte.

*Larousse, P.: *Grand dictionnaire universel du XIXe siècle*, 24 vol. and suppléments.

*Lavoisier, A.-L. de: *Œuvres de Lavoisier*, 6 vol., Imprimerie impériale (1862-1893).

*Lavoisier, A.-L. de (1790) *Traité élémentaire de chimie*, reproduit dans *Œuvres de Lavoisier*, 2 t., Imprimerie impériale, Paris, 1864; aussi Ed. Culture et civilisation, Bruxelles, 1965, accessible par http://num-scd-ulp.u-strasbg.fr:8080/199/et http://num-scd-ulp.u-strasbg.fr:8080/200/.

*Mascart, E. (1889-1893) *Traité d'Optique*, 3 t., Gauthier-Villars, Paris.

*Maxwell, J.-C. (1873) *A treatise on electricity and magnetism*, 2 vol., Clarendon press, Oxford.

*Méchain, P.F.A. & Delambre, J.B.J. (1806-1810) *Base du Système métrique décimal, ou mesure de l'arc du méridien compris entre les parallèles de Dunkerque et Barcelone, exécutée en 1792 et années suivantes*, 3 vol., Baudouin [Paris].

*Mouchez, F. (1887) *La photographie astronomique à l'Observatoire de Paris et la Carte du Ciel*, Gautier-Villars, Paris.

*Newton, I. (1687) *Philosophiae naturalis principia mathematica*, Jussu Societatis Regia ac Typis Josephi Streater. Original accessible par Gallica. 1729 translation into English accessible via Google books.

*Newton, I. (1704) *Opticks*, many editions. The fourth one is available via http://www.gutenberg.org/ebooks/33504?msg=welcome_stranger.

*Picard, J. (1684) *Traité du Nivellement*, Michallet, Paris.

Prony, F.M. Riche de (1791) *Description des opérations faites en Angleterre pour déterminer les positions respectives des Observatoires de Greenwich et de Paris*, Firmin-Didot, Paris.

*Proudhon, P.J. (1846) *Système des contradictions économiques ou Philosophie de la misère*, Guillaumin, Paris. English translation available via https://www.marxists.org/reference/subject/economics/proudhon/philosophy/index.htm.

Repsold J. A. (1908, 1914) *Zur Geschichte der astronomischen Messverkzeuge;* Vol. 1: *von Purbach bis Reichenbach, 1450 bis 1830*; Vol. 2: *von 1830 bis um 1900,* Reinicke, Leipzig.

*Secchi, A. (1875 et 1877) *Le Soleil*, 2 t., Gauthier-Villars, Paris.

Tyndall, J. (1868) *Faraday as a discoverer*, many re-editions. Accessible via http://gutenberg.org/ebooks/1225.

Wells W.C. (1814). *An essay on dew*. Taylor & Hessay, London; many editions; the second edition is accessible via https://archive.org/details/essayondewsevera00welliala.

Whewell, W. (1866) *History of inductive sciences*, 3d ed., Appleton, New York, accessible via Google books.

Wolf C. (1902) *Histoire de l'Observatoire de Paris de sa fondation à 1793,* Gauthier-Villars, Paris, accessible via https://archive.org/details/histoiredelobse00wolfgoog.

Young, T. (1807) *A course of lectures on natural and experimental philosophy,* Royal Institution, Londres; accessible via http://www.biodiversitylibrary.org/item/63005.

Some Selected Other Books

Abragam, A. (1989). *Time Reversal: an Autobiography*, Oxford University Press.

Adler, K. (2002) *The Measure of all Things. The seven-year odyssey and hidden error that transformed the world,* Free Press, New York.

Blondel, C. (1982) *A.-M. Ampère et la création de l'électrodynamique (1820-1827).* Éditions du CTHS, Paris.

Bobis, L. & Lequeux, J. ed. (2012) *L'Observatoire de Paris, 350 ans d'histoire*, Gallimard/Observatoire de Paris, Paris.

Caron, F. (1997) *Histoire des chemins de fer en France*, t. 1 (1740-1883), Fayard, Paris.

Chappert, A. (1977) *Étienne Louis Malus,* Vrin, Paris.

Clément, A. & Thomas, G., dir. (2001) *Atlas du Paris souterrain*, Parigramme, Paris.

Darrigol, O. (2003) *Electrodynamics from Ampère to Einstein*, Oxford University Press, Oxford.

Debyser, J. (2007) *Un nouveau regard sur la Nature; temps, espace et matière au Siècle des Lumières*, EDP Sciences/Observatoire de Paris, Paris.

Dupont, J.-Y. (2000) Le cours de machines de l'École polytechnique, de sa création jusqu'en 1850, *SABIX (Bulletin de la Société des Amis de la Bibliothèque de l'École polytechnique)* N° 25, p. 1-101.

Eisenstaedt; J. (2005) *Avant Einstein: Relativité, lumière, gravitation,* Ed. du Seuil, Paris.

Feynman, Leighton & Sands (1964) *Lectures on Physics*, available via http://www.feynmanlectures.info/.

Fierro, A. (1991) *Histoire de la météorologie*, Denoël, Paris.

Frankel, E. (1976) Corpuscular Optics and the Wave Theory of Light: The Science and Politics of a Revolution in Physics, *Social Studies of Science,* 6, p. 141-184.

Gillispie, C.C., ed. (1970-1990) *Dictionary of scientific biography,* 18 vol., American Council of Learned Societies, C. Scribner's sons, New York.

Guedj, D. (2001) *The Measure of the World: a novel,* University of Chicago Press.

Hakfoort, C. (2006) *Optics in the age of Euler, Conceptions on the Nature of Light, 1700-1795,* Cambridge University Press.

Fabry, Ch. (1938) La vie et l'œuvre scientifique d'Augustin Fresnel, in *La vie et l'œuvre de Charles Fabry, Œuvres choisies publiées à l'occasion du jubilé scientifique de M. Charles Fabry,* Gauthier-Villars, 1938.

Lamy, J. (2007) Le Bureau des Longitudes: la gestion des instruments et les rimes de savoir au XIXe siècle, *Revue d'antrhropologie des connaissances, 1 p. 167-188,* accessible via http://www.cairn.info/revue-anthropologie-des-connaissances-2007-2-page-167.htm

Lequeux, J., (2005) *L'Univers dévoilé*, EDP Sciences, Les Ulis.

Lequeux, J. (2013) *Le Verrier – Magnificent and Detestable Astronomer*, Springer, New York.
Lequeux, J., (2014) *Hippolyte Fizeau, physicien de la lumière*, EDP Sciences, Les Ulis.
Levallois, J.-J. (1988) *Mesurer la Terre – 300 ans de géodésie française*, École des Ponts et Chaussées/Association Française de Topographie, Paris.
Mostert, M. (1995) *The life and times of Guglielmo Libri*, 1802-1869, Verloren Pub.
Murdin, P. (2009) *Full Meridian of Glory: Perilous Adventures in the Competition to Measure the Earth,* Springer, New York.
Pannekoek, A. (1989) *A History of Astronomy*, Dover, New York.
Ronchi, V. (1970) *The Nature of Light: An Historical Survey,* Harvard University Press.
Tobin, W. (2003) *The Life and Science of Léon Foucault*, Cambridge University Press.
Todhunter, I. (2010) *The Problem of the Earth's Shape from Newton to Clairaut*, Cambridge University Press.
Vaucouleurs, G. de (1961) *Astronomical photography: from the daguerreotype to the electronic camera*, Macmillan, New York.

Some Useful Internet Sites

http://cdsads.u-strasbg.fr
http://cnum.cnam.fr
http://gallica.bnf.fr
http://expositions.obspm.fr/
http://www.academie-sciences.fr
http://www.ampere.cnrs.fr (contains an interesting history of electricity in the 19th century)
http://www.bibliotheque.polytechnique.fr/

Index

J. Lequeux, *François Arago*, Astrophysics and Space Science Library 421,
DOI 10.1007/978-3-319-20723-0

C

M

MIX
Papier aus verantwortungsvollen Quellen
Paper from responsible sources
FSC® C105338
FSC
www.fsc.org

If you have any concerns about our products,
you can contact us on
ProductSafety@springernature.com

In case Publisher is established outside the EU,
the EU authorized representative is:
Springer Nature Customer Service Center GmbH
Europaplatz 3, 69115 Heidelberg, Germany

Printed by Libri Plureos GmbH
in Hamburg, Germany